INTRODUCTION TO OPTICS AND OPTICAL IMAGING

Books of Related Interest from IEEE PRESS . . .

THE UNDERSTANDING LASERS: An Entry-Level Guide, Second Edition
Jeff Hecht
1994   Softcover   448 pp   IEEE Order No. PP2931   ISBN 0-87942-298-X

LASER AND EYE SAFETY IN THE LABORATORY
Copublished with SPIE Optical Engineering Press
Larry Matthews and Gabe Garcia
1995   Softcover   168 pp   IEEE Order No. PP3863   ISBN 0-7803-1037-3

AMAZING LIGHT: A Volume Dedicated to Charles Hard Townes on His 80th Birthday
Copublished with Springer-Verlag and SPIE Press
Edited by Raymond Y. Chiao
1996   Hardcover   712 pp   IEEE Order No. PC5658   ISBN 0-7803-1181-7

INTRODUCTION TO OPTICS AND OPTICAL IMAGING

Craig Scott

IEEE
PRESS

The Institute of Electrical and Electronics Engineers, Inc., New York

This book and other books may be purchased at a discount from the publisher when ordered in bulk quantities. Contact:

IEEE Press Marketing
Attn: Special Sales
Piscataway, NJ 08855-1331
Fax: (732) 981-9334

For more information about IEEE PRESS products,
visit the IEEE Home Page: http://www.ieee.org/

Printed in the United States of America

10   9   8   7   6   5   4   3   2   1

ISBN 0-7803-3440-X
IEEE Order Number: PC4309

Library of Congress Cataloging-in-Publication Data
Scott, Craig
    Introduction to optics and optical imaging/Craig Scott.
      p.   cm.
    Includes bibliographical references and index.
    ISBN 0-7803-3440-X
    1. Physical optics. 2. Imaging systems. II. Title.
QC395.2.S36 1997
535′.2—DC21   97-23257
                                    CIP

To
Tracy and Lee, Bob and Karyl, and Barbara and Jim

Contents

Preface

Introduction to Optics and Optical Imaging is designed for students, engineers, and optical scientists seeking a comprehensive background in physical optics in the minimum time with the minimum number of mathematical and physical concepts. Maxwell's equations are taken as the basic foundation of the book, and this set of four vector differential equations is shown to spawn the three major trends in understanding and describing optical phenomena today: ray theory, diffraction integral theory, and plane wave spectrum theory. All of the material contained in this book is explained by one (and frequently, more than one) of these three approaches. When possible, the interrelationships between these three methods are detailed. In this way, many diverse physical optics phenomena are explained in terms of only three different fundamental viewpoints, which themselves stem from the four Maxwell equations.

Since the optical theory is developed from the broader viewpoint of electromagnetic theory (i.e., from Maxwell's equations) rather than from the traditional approach of classical optics, certain seemingly disjointed historical concepts such as "Huygens sources," "Kirchhoff Integrals," "light rays," and "the Abbe plane wave viewpoint" are replaced by the more modern concepts of "Greens functions," "the Stratton-Chu (and vector potential) formulation," "the high frequency (zero wavelength) limit of the diffraction integral approach," and "the plane wave spectrum viewpoint," respectively. In this way, the ideas of classical and modern optics are integrated within a unified framework that uses general electromagnetic theory as its foundation. This provides a broad-based understanding of the subject along with a solid foundation for future study and research.

The presentation in this book is based on a combination of intuitive explanations and vector mathematics. The purpose of the text is to explain, via a minimum number of fundamental principles, the concepts being presented. The purpose of the equations is to provide a basis for definitizing the ideas in the text, as well as for making practical calculations (and, of course, for developing software, if the reader is so inclined). In this way, readers primarily interested in acquiring the concepts of optics and optical imaging can read the text only and skim through the mathematics. Readers interested in a more in-depth analytical knowledge of the subject can get that from working through the equations.

The dual presentation, math plus exposition, is at the heart of what this book is all about—namely, *understanding*. The purpose of this book is to open up the field of optics and optical imaging so that, having read the book, the reader may understand not only modern imaging systems, but also the underlying principles of optics enough to become a developer of future optical technology or software.

The mathematical prerequisites for this book include a sophomore-level background in vector differential and integral calculus, as well as an introductory background in Fourier series techniques. The recommended electromagnetic background would include a junior-level course in general electromagnetics. Beyond this, the book is essentially self-contained, having been specifically designed for self-study as well as for classroom use.

I would like to thank Linda Matarazzo and Sovoula Amanatidis at IEEE Press for their efforts in making this book become a reality.

Craig Scott

PART I
ELECTROMAGNETICS
FOR OPTICS:
THREE VIEWPOINTS

1

Foundations of the Diffraction Integral Method

This chapter introduces some fundamental electromagnetic concepts that can be obtained reasonably quickly from Maxwell's equations. These concepts will be useful in understanding much of the material on optics presented later in the text. Since Maxwell's equations form the basis of all classical optical phenomena, it is worthwhile spending some time with these relations to see how they can be applied in understanding optics.

Maxwell's equations define the field of study known as classical electromagnetics. This field unfortunately excludes many types of active optical phenomena and devices, including lasers and doped fiberoptic amplifiers. On the other hand, it still includes a vast array of theory and technology in optics. In fact, classical optics is even used in at least one aspect of laser design, that is, in the design of the resonator mirrors and external diffraction gratings used to confine the optical fields within the active laser medium.

Historically, the study of optics has not been approached from the point of view of Maxwell's equations and general electromagnetic theory. Many optical instruments were developed long before the consolidation of Maxwell's equations into their current form, so traditional means of understanding optical phenomena have developed in rather ad hoc ways. For example, the concept of *light rays* is quite ancient and certainly predates the modern understanding of ray theory as an asymptotic, high-frequency limit of Maxwell's equations. The concept of *Huygens sources* is also a historical notion, having since been replaced by the much more precisely defined concept of *Green's functions* (which is presented later in this chapter).

In this book, optical theory and technology will not be presented in the traditional, historical fashion, not because this approach is ineffective, but because it is simply not the best way to examine the subject. In addition, the historical concepts are rarely used today in the technical literature—in either optics or general electromagnetics. They have been replaced by something better, and that is a vector formulation of classical optics based on Maxwell's equations. This formulation is the subject of the present chapter.

3

1.1 MAXWELL'S EQUATIONS

For the purposes of this book then, Maxwell's equations will be regarded as being *given*. Every electromagnetic analyst knows Maxwell's equations by heart. If the reader hasn't yet committed them to memory, they're repeated below for reference. Faraday's law is

$$\nabla \times E = -\frac{\partial B}{\partial t} \tag{1.1}$$

Ampère's law is

$$\nabla \times H = \frac{\partial D}{\partial t} + J \tag{1.2}$$

Gauss' law for electricity is

$$\nabla \cdot D = \rho \tag{1.3}$$

Gauss' law for magnetism is

$$\nabla \cdot H = 0 \tag{1.4}$$

The continuity equation is

$$\nabla \cdot J = -\frac{\partial \rho}{\partial t} \tag{1.5}$$

and the constitutive (material) equations are

$$D = \epsilon E, \qquad B = \mu H \tag{1.6}$$

In the equations above,

E = electric field intensity (voltage/distance)

D = electric displacement (charge/area)

H = magnetic field intensity (current/distance)

B = magnetic induction (voltage/velocity/distance)

J = volume current density (current/area)

ρ = volume charge density (charge/volume)

ϵ, μ = permittivity, permeability tensors
(capacitance/distance, inductance/distance, respectively)

Some readers may not be familiar with the continuity equation (1.5). This equation is shown in its common differential form, but Gauss' law allows it to be rewritten in integral form as

$$\iint J \cdot dS = -\frac{dQ}{dt}$$

This version of the equation is obtained almost immediately from the definition of electrical current, which states that current is equal to the quantity of electrical charge passing across (i.e., passing *normal to*) a given plane in a given amount of time. If we imagine a cubic volume in three dimensions, with a normal component of current passing through each of its six faces, we may apply the definition of electrical current to all three sets of opposing faces of the cube. Doing so yields the integral form of the continuity equation above, namely, that the rate of charge accumulation between all three sets of opposing faces is equal to the integral of the normal component of current over all six faces.

The two field quantities that have definite physical significance are the electric field strength E (measured in voltage/distance) and the magnetic induction B (measured in terms of voltage/velocity/distance). These two fields exert a definite *mechanical* force on a charged particle (of charge q) that is given by the relation

$$F = q(E + v \times B) \tag{1.7}$$

where

$$v = \text{particle velocity vector}$$

The other two field quantities (D, H) are related to E, B via the material parameters (permittivity, ϵ and permeability, μ). D is a quantity that is related to electrical charge, and H is related to electrical current, both independent of the particular medium involved. Thus, both quantities are related to electrical charge, D to *static* charge and H to charge in *motion*.

In this book we'll be dealing with fields that are harmonic in time (single-frequency sinusoids having infinite temporal duration). This assumption has been implicit in the study of optical phenomena literally for centuries, yet only within the last three decades have laser light sources been available which can produce this type of *coherent radiation* in the optical regime. This has enabled longstanding mathematical models of optical fields to finally be in step with the real phenomena.

The assumption of sinusoidal time variation is generally not considered that restrictive since an arbitrary time-varying field may be expressed in terms of a Fourier transform superposition of single-frequency sinusoids. Thus, field solutions at many frequencies may (in principle) be combined to yield the solution for a complex time-dependent field. (In practice, such a superposition is rarely used to solve time-domain problems.)

Under the assumption of time-harmonic fields, we may write (for the electric field)

$$E(x, y, z, t) = E(x, y, z)\, e^{j\omega t}$$

In addition to specifying time-harmonic fields, we'll also limit ourselves to isotropic (scalar) media, for which the permittivity and permittivity tensors are simple scalar quantities. Under the joint assumptions of time-harmonic fields and scalar media, Maxwell's equations become (see Appendix B):

$$\nabla \times E = -j\omega\mu H \tag{1.8}$$

$$\nabla \times H = j\omega\epsilon E + J \tag{1.9}$$

$$\nabla \cdot E = \rho/\epsilon \tag{1.10}$$

$$\nabla \cdot H = 0 \tag{1.11}$$

$$\nabla \cdot J = -j\omega\rho \tag{1.12}$$

$$D = \epsilon E, \qquad B = \mu H \tag{1.13}$$

This is the form of Maxwell's equations that we'll use in this book. It is the form most often used in optics and electromagnetics. The fields in this format are no longer purely real quantities, and they are certainly no longer temporal quantities. They may now take on complex number values. These complex field values are known as *phasor* quantities, inasmuch as it is the *phase* of these quantities that distinguishes them from ordinary real-valued oscillatory temporal field quantities.

The magnitude part of the phasor is related to the energy density of the field (as in the non-time-harmonic case). The phase part of the phasor carries the temporal information (though in a somewhat coded form). *Time shifts* are translated into *phase shifts* (phase retardation for time delay and phase advance for forward time shifts). Time advances and delays may take on any numerical values, whereas phase shifts may only occupy the range from 0 to 2π.

For readers who are not already familiar with phasor quantities, we may readily relate time functions with their phasor counterparts. Let

$$E(t) = E(r, t), \text{ for some fixed point, } r$$

We'll use the Fourier transform pairs

$$E(t) = \int_{-\infty}^{\infty} E(\omega) \, e^{j\omega t} \, d\omega \tag{1.14a}$$

$$E(\omega) = \frac{1}{2\pi} \int_{-\infty}^{\infty} E(t) \, e^{-j\omega t} \, dt \tag{1.14b}$$

to relate the temporal and spectral representations of the field. According to these representations, the temporal dependence

$$E(t) = e^{j\omega_0 t} \tag{1.15a}$$

is given in the spectral domain as

$$E(\omega) = \delta(\omega - \omega_0) \tag{1.15b}$$

where we've used the identity below, which is obtained in Appendix B on Fourier analysis.

$$\delta(\omega - \omega_0) = \frac{1}{2\pi} \int_{-\infty}^{\infty} e^{j(\omega - \omega_0)t} \, dt$$

A time-delayed field of the form

$$E(t - t_0)$$

will have a phasor (spectral) form obtained via Eq. (1.14b) as

$$E'(\omega) = e^{-j\omega t_0} E(\omega)$$

where $E(\omega)$ is given by (1.15). Thus, a time-delayed field produces a linearly phase-retarded spectrum. This is one way in which temporal information is *phase-coded* in the phasor domain. That is, a linear *phase* in the spectral domain relates to either a time delay or an advance. (In causal systems, only time *delays*—or phase lags—are permitted.) It's interesting to note that in electrical circuit theory, "linear-phase," or "constant group delay'' filters are often designed for the purpose of producing a desired time delay for a signal consisting of many frequency components.

The reader may also verify directly from (1.14a) (using the product rule of integration, along with an assumption that $E(t) \to 0$ as $t \to \pm\infty$) that derivatives in the time domain correspond to multiplication by $j\omega$ in the spectral domain. This is one more way in which time-domain information is coded in the spectral domain via the phase of the phasor spectrum. Other relationships between time functions and their spectra are given in Appendix B.

1.2 THE WAVE EQUATIONS

The first step in mathematical refinement up from Maxwell's equations is given by the wave equations. Whereas Maxwell's equations contain the electric and magnetic field vectors all coupled together in an inconvenient fashion, the wave equations are a set of equations in which the electric and magnetic fields have been separated, or *decoupled*. This decoupling feature allows the electric and magnetic fields to be solved quickly in terms of the currents (a process we'll carry out in the next session).

The electric field wave equation is derived in about three lines. Take the curl of Faraday's law (1.8) and substitute Ampère's law (1.9) into it to obtain

$$\nabla \times \nabla \times E - \omega^2 \mu\epsilon E = -j\omega\mu J \qquad (1.16)$$

It will be convenient for our purposes to remove the double curl operator using the following vector identity. (Later, we'll talk more about the meaning of the second term on the right-hand side.)

$$\nabla \times \nabla \times E = \nabla(\nabla \cdot E) - \nabla^2 E \qquad (1.17)$$

to obtain

$$\nabla^2 E + k^2 E = j\omega\mu J + \frac{1}{\epsilon}\nabla\rho \qquad (1.18)$$

where

$$k^2 = \omega^2 \mu\epsilon$$

and we've used Gauss' law (1.10) to obtain (1.18). Equation (1.18) is the electric field wave equation we sought. Note that only the electric field is present in this equation on the left-hand side. A similar process may be used to obtain a wave equation for the magnetic field. This is given as

$$\nabla^2 H + k^2 H = -\nabla \times J \qquad (1.19)$$

The wave equations are linear partial differential equations for the fields. The forcing functions for the equations are the currents and charges. [*Note:* We could have expressed (1.18) entirely in terms of the electric current, by invoking the continuity condition (1.12).]

Equation (1.17) may be regarded as the defining equation for the *vector Laplacian* [it's the second term on the RHS of (1.17)]. According to (1.17), the vector Laplacian is related to the "curl curl" operator and the "gradient divergence" operator. The vector

Laplacian defined by (1.17) is of little use in most coordinate systems. It is really useful only in rectangular coordinates, where it takes on an exceedingly simple form (see Appendix A). In the next section, we'll make use of the simple form of the vector Laplacian in rectangular coordinates to show how the wave equations above may be solved quickly.

1.3 SCALAR AND VECTOR POTENTIALS

In this section, we'll solve the two wave equations and in so doing introduce the concept of the scalar and vector potentials. Most readers are undoubtedly familiar with the ordinary scalar potential from electrostatic theory. In electrodynamics, however, there is a second potential—the vector potential—which takes on far greater importance than the scalar potential. This is because fields generated via the vector potential may radiate in space, whereas the fields generated via the scalar potential are "bound" to electrical charges and cannot radiate.

The reader is probably familiar with Huygens' principle from elementary optics. This principle provides a graphical means for determining the field produced by an optical wavefront. According to Huygens' principle, a wavefront may be divided up into an infinite number of point source radiators. Each of these point radiators may be regarded as a source of spherical waves; therefore, the radiated field may be determined graphically by moving a compass along the wavefront and drawing circles about a series of closely spaced points on the front. The intersections of the various circles so drawn are tangent to the new wavefront and therefore indicate its contour.

In this section, we're going to do the same sort of thing mathematically—it's really not that hard to do—in order to obtain a mathematical version of Huygens' graphical construction. The mathematical version we'll be looking at is based on the same principle as Huygens' construction. That is, we'll regard the source distribution as being composed of an infinite number of point source radiators and then sum up (actually, *integrate*) the contributions due to this distribution of point sources.

In retrospect, Huygens' discovery of this principle (in 1678!) is quite remarkable. Today, this same idea is used in numerous branches of engineering and physics (from electromagnetic scattering to digital signal processing), though now it generally goes by the name, *the superposition principle.* The concept simply states that a time-varying signal (or spatial distribution of charge, mass, etc.) may be regarded as the superposition integral of an infinite number in infinitesimally short impulses (or, for spatial distributions, point sources). And if the response of an electronic circuit to one impulse (or the response of an electromagnetic system to a single point charge, or a mechanical system to a point mass) is known, then it is possible (in principle) to integrate over the entire source distribution to get the total response of the system (due to the entire distribution). This is the principle we'll explore in this section.

The principle of superposition applies only in connection with *linear* differential equations. (Both the wave equations above are linear, since the fields only appear to the first power, e.g., there are no terms of the form E^2 present.) We'll describe the superposition principle briefly here (in connection with time signals) before applying it to the electromagnetic wave equations.

Say we want to find the solution to the linear differential equation

$$\frac{d^2}{dt^2} S(t) + k^2 S(t) = f(t) \tag{1.20}$$

where $S(t)$ is some unknown signal response function and $f(t)$ is a specified forcing function. By the sifting property of the impulse function, we may just as well write $f(t)$ in the form

$$f(t) = \int_{-\infty}^{\infty} f(t')\delta(t - t')\, dt' \tag{1.21}$$

that is, as a superposition of impulse functions, weighted by the value of the forcing function. With (1.21) in hand, it's now evident that all we have to do in order to solve (1.20) is to solve the equation

$$\frac{d^2}{dt^2} S(t) + k^2 S(t) = \delta(t - t') \tag{1.22}$$

and then add up all of the individual responses due to each of the various weighted impulse function inputs. So, say the solution to (1.22) is given by $h(t - t')$. This is called the *impulse response* of the differential operator on the LHS of (1.20). The solution to (1.20) is now just a matter of adding up all the impulse *responses* due to each of the impulse *inputs* in (1.21), that is,

$$S(t) = \int_{-\infty}^{t} f(t)h(t - t')\, dt' \tag{1.23}$$

So, all we have to do to solve (1.20) is to find the impulse response solution to (1.22). Once that is known, the response to an arbitrary forcing function may be obtained by superposition of impulse responses, using the same technique Huygens used. [*Note:* Huygens assumed that the field from a point source radiator—his impulse response—was a perfectly spherical wave. In this section, we'll verify that assumption.]

Note that the upper limit to the integral in (1.23) is set at t, since inputs after time t do not contribute to the output of the system at t. When the independent variables are spatial rather than time coordinates, a two-sided infinite range of integration may be employed.

With this brief introduction to the superposition principle, let's now apply it to the solution of the magnetic field wave equation (1.19). We begin by seeking a solution to the equation

$$\nabla^2 H + k^2 H = -\hat{u}\delta(r - r_0) \tag{1.24}$$

where

$$\hat{u} = \hat{x},\ \hat{y},\ \hat{z}$$

The delta function in three dimensions is defined by the usual equations

$$\delta(r - r_0) = \infty \qquad \text{when } r = r_0$$

$$= 0 \qquad \text{when } r \neq r_0$$

subject to the integrability condition (in three dimensions)

$$\iiint_{-\infty}^{\infty} \delta(r - r_0)\, dv = 1$$

This three-dimensional impulse function has the following sifting property in three dimensions (see Appendix D):

$$f(r_0) = \iiint_{-\infty}^{\infty} f(r)\delta(r - r_0)\, dv$$

Now, the delta function term on the RHS of (1.24) will give rise to fields that vary as a function of radius only, so this is the type of solution we'll look for. And since we know that the delta function is zero everywhere except at the source point, that is, at $r = r_0$, we can first find the radially symmetric solution to the homogeneous wave equation (whose forcing function is identically zero) to get a solution that is at least valid everywhere except at the source point. We'll solve (1.22) for the three rectangular coordinates of magnetic field separately. So, we'll now solve the scalar wave equation

$$\nabla^2 H_u + k^2 H_u = -\delta(r - r_0) \tag{1.25}$$

where

$$u = x,\ y,\ z$$

Equation (1.25) was obtained thanks to the very simple form of the vector Laplacian in rectangular coordinates, that is,

$$\nabla^2 H = \nabla^2 H_x \hat{x} + \nabla^2 H_y \hat{y} + \nabla^2 H_z \hat{z}$$

We'll take r_0 at the origin, so that

$$|r - r_0| = r$$

where r is the radial distance from the origin. The form of the scalar Laplacian in spherical coordinates may be used (see Appendix A) to obtain an explicit expression for the radially symmetric scalar Laplacian. Hence, the radially symmetric form of (1.25) is

$$\frac{1}{r}\frac{\partial}{\partial r}\left(r\frac{\partial H_u}{\partial r}\right) + k^2 H_u = 0$$

With this, the radially symmetric solution to (1.25) for $r \neq 0$ may be verified to be

$$H_u(r) = A\,\frac{e^{-jkr}}{r} \tag{1.26}$$

where A is some unknown constant. The reader should verify this solution by substituting (1.26) into (1.25).

Equation (1.26) verifies that Huygens was indeed correct in taking a spherical wave as the field due to a point source radiator. (The phase fronts of this field are surfaces of constant radius, r; thus, the wave is referred to as a *spherical wave*.) This equation also shows, however, that this wave has not only spherical phase, but also an amplitude that varies inversely with distance from the source, due to the spreading of the spherical wave. This "one-over-r" dependence is the same as that of the static potential arising from a charged particle. In the static case, however, the potential varies inversely with radius, and the *field* varies inversely as the *square* of the radius. In this (dynamic) case, however, the actual field varies only inversely with distance, not inversely as the square of the distance. In the time-varying case, the fields extend outward much farther than in the static field case (i.e., they separate from their sources and are able to radiate to distant receivers).

It's worth taking a look at the solution to (1.25) at the origin. This is actually very easy. All we have to do is note that the scalar Laplacian in (1.25) is defined as the divergence of the gradient. So, if we integrate both sides of (1.25) over a small spherical ball (of radius, ϵ) about the origin, taking the incremental volume element in spherical coordinates as

$$dv = r^2 \sin\theta \, dr \, d\theta \, d\phi$$

we see that the second term on the LHS of (1.25) integrates to zero as ϵ tends to zero. The definition of the three-dimensional delta function above shows that the RHS integrates to -1. The first term on the LHS of (1.25) is integrated using the divergence theorem, to convert the integral over the spherical volume to an integral over the spherical surface. Then, taking the gradient of the Green's function in spherical coordinates (using the formula in Appendix A), along with the formula for the element of surface area,

$$ds = r^2 \sin\theta \, d\theta \, d\phi$$

it's readily shown that the integral of the first term is equal to -4π. So, the constant, A, in (1.27) is equal to $1/4\pi$.

The solution to (1.25),

$$\psi(r) = \frac{e^{-jkr}}{4\pi r} \tag{1.27}$$

is called the free-space *scalar Green's function*. To get the Green's function for the actual *field* (called the free-space *tensor Green's function*), we simply use the superposition principle (applied to all three Cartesian components of magnetic field) to get

$$H(r) = \iiint_{v'} \nabla' \times J(r') \, \frac{e^{-jk|r-r'|}}{4\pi|r-r'|} \, dv' \tag{1.28}$$

We obtained this equation by convolving the Green's function (1.27) with the forcing function from the RHS of (1.19). This is the solution to the magnetic field wave equation (1.19). It is also the tensor Green's function for the magnetic field. Using an entirely analogous procedure, we can show that the tensor Green's function for the electric field is

$$E(r) = -jk\eta \iiint_{-\infty}^{\infty} J(r') \, \frac{e^{-jk|r-r'|}}{4\pi|r-r'|} \, dv'$$

$$-\frac{1}{\epsilon} \iiint_{-\infty}^{\infty} \nabla'\rho(r') \, \frac{e^{-jk|r-r'|}}{4\pi|r-r'|} \, dv' \tag{1.29}$$

where

$$\eta = \sqrt{\frac{\mu}{\epsilon}}$$

is the characteristic impedance of the medium.

One small feature of these two equations may be unfamiliar to many readers. This is the concept of *source coordinates* and *field coordinates*. In Maxwell's equations, there is never any confusion as to the meaning of the vector differential operators. You just pick a point in space and—say, in the case of Ampère's law—you calculate the curl of the magnetic field at that point and relate it to the current and electric field at that same point. Things are a little more complicated in (1.28) and (1.29), however. Because of the convolution (superposition) nature of the solution, there are currents and charges distributed over the "source coordinates," denoted with primes in (1.28) and (1.29), and there are the resulting fields that appear at the "field coordinates," denoted by unprimed coordinates. When we derived the solution to (1.25), we arbitrarily placed the origin at r' and calculated the scalar Green's function with respect to that origin. Now we have the situation where both r and r' are variable and the origin is located at some fixed point elsewhere.

When we take differential operators now, they can be with respect either to the source coordinates or field coordinates. (This is much different than was the case with Maxwell's equations.) When the operator is taken with respect to the source coordinates, the field point is assumed to be fixed—it is the temporary origin. When the operator is taken with respect to the field coordinates, all the source currents are assumed to be fixed in space and only the field point is variable.

In Eqs. (1.28) and (1.29), the differential operators are taken with respect to the source (primed) coordinates. However, the tensor Green's functions are traditionally expressed in terms of field region (unprimed) differential operators. Appendix C shows how the differential operators in (1.28) and (1.29) can be shifted onto the field coordinates. The resulting equations for the fields are

$$H(r) = \nabla \times \iiint_{v'} J(r') \frac{e^{-jk|r-r'|}}{4\pi|r-r'|} \, dv' \tag{1.30}$$

$$E(r) = -jk\eta \iiint_{-\infty}^{\infty} J(r') \frac{e^{-jk|r-r'|}}{4\pi|r-r'|} \, dv'$$

$$\qquad - \frac{1}{\epsilon} \nabla \iiint_{-\infty}^{\infty} \rho(r') \frac{e^{-jk|r-r'|}}{4\pi|r-r'|} \, dv' \tag{1.31}$$

These two equations may be rewritten in terms of the scalar and vector potentials (mentioned at the outset of this chapter) as

$$H(r) = \nabla \times A(r) \tag{1.32}$$

$$E(r) = -jk\eta A(r) - \frac{1}{\epsilon} \Phi(r) \tag{1.33}$$

where

$$A(r) = \iiint_{v'} J(r') \frac{e^{-jk|r-r'|}}{4\pi|r-r'|} \, dv' \tag{1.34}$$

and

$$\Phi(r) = \iiint_{v'} \rho(r') \frac{e^{-jk|r-r'|}}{4\pi|r-r'|} \, dv' \tag{1.35}$$

By the earlier discussion on superposition, it's evident from the convolution form of A, Φ that they themselves satisfy wave equations of the form

$$\nabla^2 A + k^2 A = -J \tag{1.36}$$

$$\nabla^2 \Phi + k^2 \Phi = -\rho \tag{1.37}$$

The continuity equation (1.12) may be used to rewrite Eq. (1.33) entirely in terms of the vector potential, A as

$$E(r) = -jk\eta A(r) - j\frac{\eta}{k}\nabla[\nabla \cdot A(r)] \qquad (1.38)$$

The two equations (1.32) and (1.38) for the electric and magnetic fields represent the solutions to Maxwell's equations. (We've extracted E, H from under the vector differential operators and expressed them directly in terms of the currents.) A considerable amount of time was spent developing these equations, even though the principles behind their derivations weren't all that complicated. This development time reflects directly on the importance of these equations. They form the foundation of physical optics diffraction theory, one of the three main techniques we'll study in this book for analyzing optical phenomena. (In the next chapter, we'll derive the equations for the plane wave spectrum theory, and in Chapter 3 we'll derive the equations for geometrical optics theory.) The time spent on these equations was worthwhile for forming a good understanding of the origin of these equations and their meaning.

As it turns out, Eqs. (1.32) and (1.38) aren't quite complete. They only allow us to calculate the fields due to a current distribution when the currents radiate in an unbounded homogeneous dielectric medium having infinite extent. Such an assumption won't be general enough for later needs in this book, where we'll have lots of boundaries—between lenses and air. Therefore, we must add extra terms to (1.32) and (1.38), which take material boundaries (such as those that exist between a glass lens and the air) into account.

To see how the field equations are modified for material boundaries, consider Fig. 1.1, which shows a two-region problem. The reader may be familiar with the fact that a sheet of current causes a discontinuity in magnetic field (known as a *jump discontinuity*). This is readily seen from Ampère's law, for if

$$\nabla \times H = j\omega\epsilon E + K \qquad (1.39)$$

where K is a sheet current (measured in Amps/meter, and illustrated in Fig. 1.2), then integrating this equation over the surface shown (as ϵ tends to zero), using Stokes' theorem yields

$$H_{\text{tan}}(z = 0^+) - H_{\text{tan}}(z = 0^-) = K \times \hat{z} \qquad (1.40)$$

Thus, the electric surface current, K, supports a jump discontinuity in the tangential magnetic field, H. By analogy, what would happen if we were to modify Faraday's law (1.8) to read

$$\nabla \times E = -j\omega\mu H - M \qquad (1.41)$$

where M is a "sheet magnetic current," measured in volts/meter? Well, if M occupies the $z = 0$ plane, we may use the exact same logic as above to show that

$$E_{\text{tan}}(z = 0^+) - E_{\text{tan}}(z = 0^-) = -M \times \hat{z} \qquad (1.42)$$

In other words, this sheet magnetic current causes a jump discontinuity in the tangential electric field. Of course, all we're doing here is playing games with Maxwell's equations, just to see how they might be modified in order to yield step discontinuities in the tangential electric and magnetic fields. These games have important practical uses, however. For example, say we were to take the situation in Fig. 1.1 and change it to the one

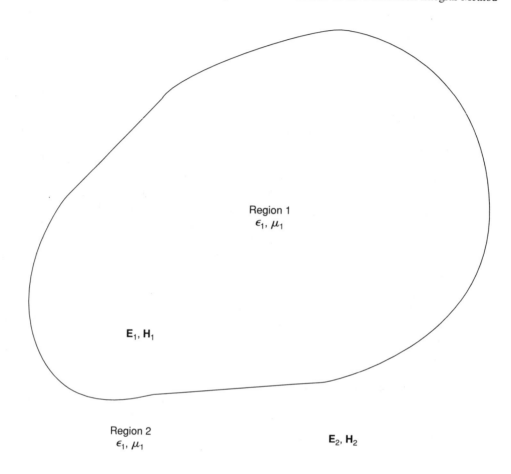

Region 1
ϵ_1, μ_1

$\mathbf{E}_1, \mathbf{H}_1$

Region 2
ϵ_1, μ_1

$\mathbf{E}_2, \mathbf{H}_2$

Figure 1.1 Field calculations in bounded dielectric media.

shown in Fig. 1.3. In Fig. 1.3, we have produced an infinite homogeneous medium [to which we can apply (1.30, 1.32)] from the mixed-medium situation shown in Fig. 1.1. The only difference now is that we've placed tangential electric and magnetic currents on the boundary surface between the two media, to support the discontinuities in the fields (from the original field values just inside the boundary to the null field values just outside the boundary). Both configurations will be equivalent (within medium 1, at least) if the following conditions hold at the boundary. These conditions are taken directly from the jump conditions above as

$$E_1 = \hat{n} \times M \tag{1.43a}$$

$$H_1 = K \times \hat{n} \tag{1.43b}$$

where the normal vector, $\hat{n}$ points into region 1.

All we have to do now is include the effects of K, M into (1.32) and (1.38). This isn't hard, because we've already done all the work. Noting that the magnetic current, M (in Faraday's law), has the same relationship to H as J has to E (in Ampère's law) and that M has the negative relationship to E that J has to H (with the roles of ϵ and μ

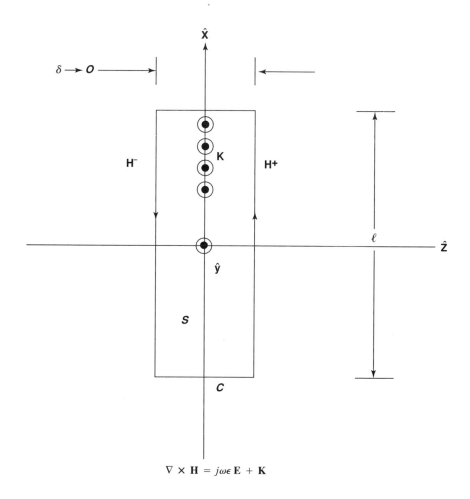

$$\nabla \times \mathbf{H} = j\omega\epsilon\, \mathbf{E} + \mathbf{K}$$

So,

$$\iint_s (\nabla \times \mathbf{H}) \cdot \hat{y}\, ds = \iint_s \mathbf{K} \cdot \hat{y}\, ds = K_y l$$

By Stokes' theorem however,

$$\iint_s (\nabla \times \mathbf{H}) \cdot \hat{y}\, ds = \oint_c \mathbf{H} \cdot d\mathbf{l} = (H_x^+ - H_x^-)l$$

So,

$$\mathbf{H}^+ - \mathbf{H}^- = \mathbf{K} \times \hat{z}$$

Figure 1.2 A jump discontinuity in the tangential magnetic field due to a sheet of current.

interchanged), we may immediately write the solutions to Maxwell's equations for bounded regions directly from (1.32) and (1.38). Thus,

$$E(r) = -j\omega\mu A(r) + \frac{1}{j\omega\epsilon} \nabla[\nabla \cdot A(r)] - \nabla \times F(r) \tag{1.44}$$

$$H(r) = -j\omega\epsilon F(r) + \frac{1}{j\omega\mu} \nabla[\nabla \cdot F(r)] + \nabla \times A(r) \tag{1.45}$$

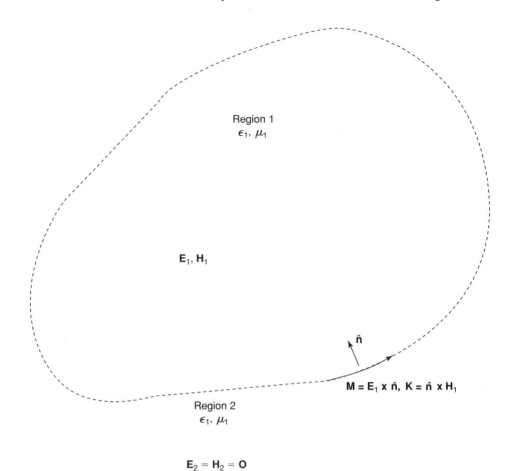

Figure 1.3 An electromagnetic problem that is equivalent (in region 1) to the two-region problem shown in Fig. 1.1.

where

$$F(r) = \iiint_{v'} M(r') \frac{e^{-jk|r - r'|}}{4\pi|r - r'|} \, dv' \tag{1.46}$$

These equations are written in terms of ω, ϵ, μ, rather than in terms of k and η, in order to show the dual nature of the equations. These are the equations we'll use in the diffraction integral analysis of optical systems, and they represent a mathematical statement of Huygens' principle. (Note: since M is a surface current, the integral in (1.46) is actually taken over a surface, not a volume.)

In the old days of optics, scalar equations similar to these were derived in various ad hoc ways. The names of numerous well-known researchers in classical optics have traditionally been associated with those earlier equations (usually hyphenations of names such as Huygens, Kirchhoff, Fresnel, Rayleigh, Sommerfeld, and Helmholtz). Coincidentally, the equations we've just derived also go by the hyphenation of two names, but they are the names of two twentieth-century researchers in electromagnetics, not optics. These are: J. A. Stratton and L. J. Chu, formerly of MIT, and their formulation of this solution to Maxwell's equations is known as the Stratton-Chu formulation [1].

1.4 ELECTROMAGNETIC POWER

The Poynting vector theorem relates stored and dissipated power within a reference volume to the integral of a "power flux" vector over the bounding surface of the volume. A very well-worn derivation of the Poynting vector theorem is generally used in electromagnetics [2], which, though rigorous from a mathematical standpoint, yields little physical understanding of the concept of electromagnetic power. Fortunately, it is a relatively simple matter to obtain a sound understanding of electromagnetic power from a simple intuitive approach. That is the approach we'll follow here.

Any concept of power must ultimately incorporate the idea of some type of mechanical motion. After all, power is a mechanical construct, being equal to work performed per unit time. When we think of ordinary electrical power, for example, what we're really thinking of is: how much mechanical power can this "electrical power" deliver? That is, can it drive a motor or cause a speaker to vibrate? The power available in an electromagnetic field can also be thought of in mechanical terms. For example, suppose we were to place a planar sheet of charged particles (charge/area) in the path of an electromagnetic wave. How much mechanical energy is available in the wave to expend on moving the charges in this sheet?

This question is readily answered. We know from elementary physics that the force exerted on a charge, q, by an electromagnetic field is given by

$$F = q(E + v \times B) \tag{1.47}$$

We also know from elementary physics that power is given as the dot product of force times velocity. (In the phasor domain, this is force times the *conjugate* of velocity; see Appendix B.) Thus, for an elemental square patch in the sheet,

$$P = F \cdot v^* = dq(E \cdot v^*) = E \cdot dK^* \tag{1.48}$$

where K is the surface current density, defined in the previous section. We know from the jump condition in the previous section that

$$H(z = 0^+) - H(z = 0^-) = K \times \hat{z} \tag{1.49}$$

So, say that all of the energy from the electromagnetic wave is absorbed by the sheet. That is, no energy gets past the sheet; hence,

$$H(z = 0^+) = 0$$

and

$$H(z = 0^-) = -K \times \hat{z}$$

or

$$K = H(z = 0^-) \times \hat{z}$$

Therefore,

$$P = E \cdot (H \times \hat{z})^* = \hat{z} \cdot (E \times H^*) \tag{1.50}$$

This leads naturally to the concept of the power vector (the so-called *Poynting vector*) as

$$P = E \times H^* \tag{1.51}$$

With this, the power dissipated (per unit area) in the sheet of charge is given by (1.50) as

$$P_z = \hat{z} \cdot P = \hat{z} \cdot (E \times H^*) \qquad (1.52)$$

This equation directly relates the mechanical power dissipated on a sheet of charge to the component of the Poynting vector *normal* to the sheet. So, the Poynting vector is a "flux" type of vector; it points in the direction of power flow, and its integral over a surface determines the amount of power crossing that surface.

1.5 IMAGE THEORY

In this section, we'll show how to make the diffraction integral method practical. If this method is applied hastily to the analysis of optical systems, it can result in many unnecessary calculations. *Image theory* can dramatically reduce the number of calculations involved whenever planar (or nearly planar) boundaries are involved.

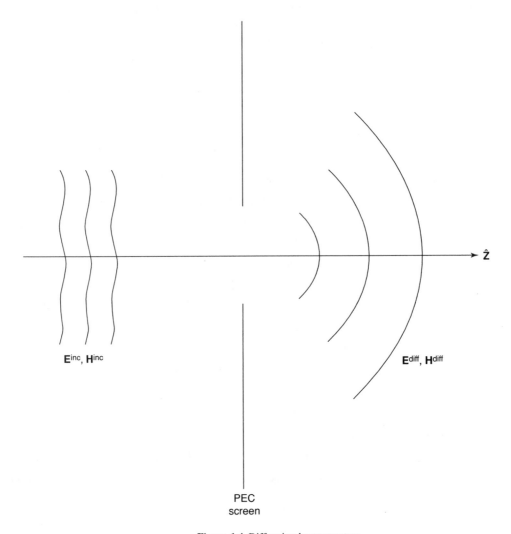

Figure 1.4 Diffraction by an aperture.

For example, consider the aperture in the planar wall shown in Fig. 1.4. We assume the wall to be perfectly electrically conducting (PEC); that is, the wall is so highly conductive that the tangential electric field can never take on a finite value. (We'll discuss perfect conductors in more detail below.) If we were to use the Stratton-Chu equations to calculate the electric field transmitted through the aperture to the $z > 0$ region, we'd have to integrate over the electric currents on the entire $z = 0$ plane, as well as integrate over the "equivalent electric and magnetic currents" on the aperture (which are expressed in terms of the tangential aperture fields by [1.43]). This is a lengthy calculation that is greatly simplified through the use of images.

We may verify from (1.44) and (1.45) that a planar sheet of "magnetic current" will radiate symmetric magnetic fields and antisymmetric electric fields. This is shown in Fig. 1.5. Since the tangential radiated electric field is antisymmetric, it must therefore be zero on those portions of the $z = 0$ plane where there are no magnetic currents. (Magnetic currents in the $z = 0$ plane will produce a step discontinuity, from a nonzero positive value on one side to the opposing negative value on the other.) Thus, magnetic currents in the $z = 0$ plane radiate fields that exactly satisfy the aperture and wall boundary conditions of the original aperture diffraction problem. That is, the tangential electric field is zero at $z = 0$, outside the aperture portions of the plane, and equal to the original aperture electric field in the aperture.

We may use these unique radiative properties of planar magnetic current sheets to advantage in analyzing the aperture diffraction problem. As shown in Fig. 1.6, the original aperture diffraction problem may be modified by setting the electric field on the left side

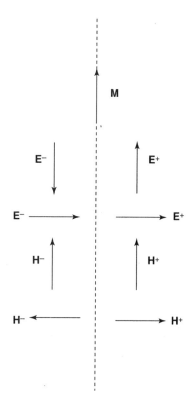

Figure 1.5 Fields produced by a planar magnetic current distribution.

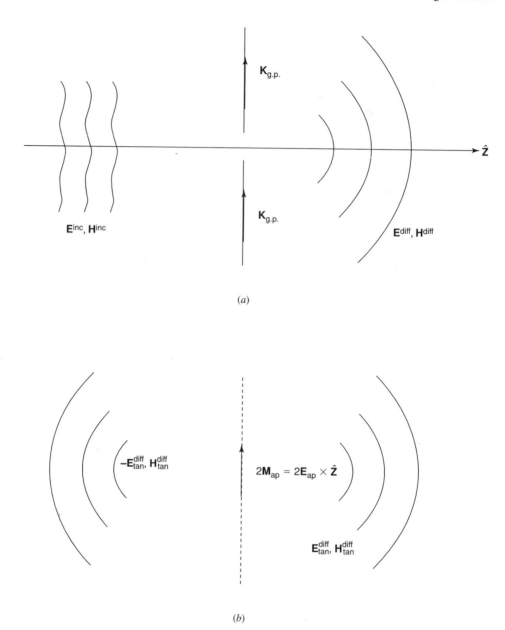

Figure 1.6 Simplification of the aperture diffraction problem: (a) Original aperture diffraction problem and (b) Transformed aperture diffraction problem equivalent to the original in the right-hand half space.

exactly equal to the negative of the diffracted electric field on the right side, and the magnetic field on the left side equal to the diffracted magnetic field on the right side. Thus, a step discontinuity has been created in the tangential electric field at $z = 0$ (equal to twice the aperture field), and the tangential magnetic field has been made continuous at $z = 0$. So, since the tangential magnetic field is now continuous everywhere at $z =$

0, we may remove all the electric currents from the $z = 0$ plane of our transformed problem. And since we've created a step discontinuity in the electric field (in the aperture), we must now add aperture magnetic currents to support this newly created discontinuity.

The original aperture diffraction problem shown in Fig. 1.6a now becomes the transformed aperture diffraction problem shown in Fig. 1.6b. The fields on the right-hand side of the aperture are exactly the same as those in the case of the original aperture diffraction problem—only the fields in the left half space have been altered. So, the price we've paid for obtaining a simplified mathematical problem is that we now have a problem that is equivalent to the original problem in the $z > 0$ region only.

Calculating the diffracted fields for the problem shown in Fig. 1.6c is relatively easy. We know the aperture electric field distribution. (In the optical limit, it's just equal to the incident field in the aperture.) And we can easily calculate the electric field radiated by the magnetic currents using (1.44). So, we've succeeded in making the initially intractable aperture diffraction problem tractable.

In electromagnetics, the construct of a PEC material is often used. This material is so highly conductive that an infinitesimal tangential electric field on its surface will result in a finite electrical current on the surface of the material. Materials that simulate PEC surfaces include metals such as silver, copper, and gold. Today's high-temperature super-conducting materials are (for all practical purposes) PEC materials, at least up to moderate microwave frequencies. The optical analogue of a PEC material is a reflective mirror. In many types of optical systems (cameras, telescopes, etc.), however, nonreflective (absorptive) black-colored materials are preferable to highly reflective PEC materials. These opaque materials absorb stray light and prevent it from propagating further through the system. However, the PEC construct is still used for aperture penetration problems. (It is the only possible way to "image away" the electric currents in the aperture and on the groundplane, and confine the integration to the magnetic currents in the aperture.) It is remarkable that this assumption produces accurate results for aperture-diffracted fields.

1.6 APPLICATIONS: FRAUNHOFER REGION FIELDS OF AN ILLUMINATED APERTURE

In the majority of calculations in optics, the diffraction integral formulation is applied in only two main regimes. One of these is the *Fraunhofer,* or far-field region, which is used primarily for analyzing the interaction between an optical instrument and its outside environment. For example, this situation would describe the field incident on a camera lens from some sort of environmental scene, or the field radiated by a laser beam. The second regime is the *Fresnel* regime, which is used primarily to describe the movement, or propagation, of optical energy from one transverse plane to another *within* an optical system. The diffraction integral formulation can be used to describe either regimes, and in this section we'll consider the Fraunhofer regime.

With reference to Fig. 1.7, we can obtain the far-zone expressions for the electric and magnetic fields by invoking the *parallel rays* approximation. In this approximation, the free-space scalar Green's function is approximated as

$$\frac{e^{-jk|r-r'|}}{4\pi|r-r'|} \cong \frac{e^{-jkr}}{4\pi r} e^{jk(\hat{r}\cdot r')} \tag{1.53}$$

Using this approximation, it is shown [3] that

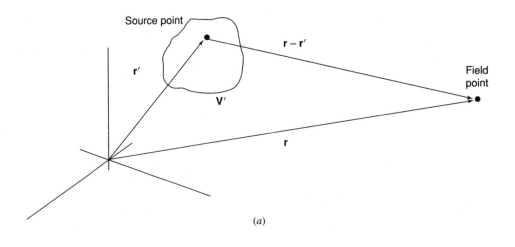

(a)

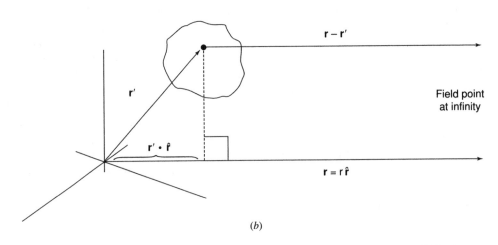

(b)

Figure 1.7 The parallel rays approximation for Fraunhofer region fields: (a) field point
at finite distance from source region and (b) field point infinitely far away
from sources.

$$E(\hat{r}) = -j\omega\mu[A - \hat{r}(\hat{r} \cdot A)] - \hat{r} \times F \qquad (1.54)$$

$$H(\hat{r}) = -j\omega\epsilon[F - \hat{r}(\hat{r} \cdot F)] + \hat{r} \times A \qquad (1.55)$$

These expressions show that in the far-zone region, both electric and magnetic fields
are transverse to each other and to the radial vector, r. They also show that in the far-
zone region, the electric and magnetic fields of a planar distribution are related to the
Fourier transform of the source fields and/or currents since

$$J(r') = J(x', y')$$

and

$$\hat{r} \cdot r' = (\alpha x + \beta y)$$

This is a very important result in the study of optical systems.

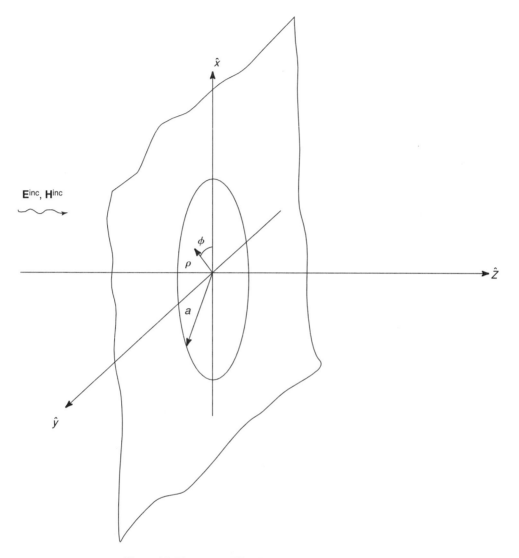

Figure 1.8 Plane wave diffraction by a circular aperture in a PEC screen.

EXAMPLE 1.1 THE AIRY PATTERN FOR CIRCULAR APERTURES

A uniformly illuminated circular aperture radiates a pattern that is often referred to as the *Airy pattern*. We may easily derive this Airy function pattern using (1.54) (with $A = 0$), in conjunction with an image transformation of the type described in Section 1.6. If the electric field incident on the aperture is a plane wave field, as shown in Fig. 1.8, then (from the results of Section 1.6),

$$M(x, y) = 2E^{\text{inc}}(x, y, z = 0) \times \hat{z}$$

$$\text{so, if}\quad E^{\text{inc}}(x, y, z = 0) = \frac{1}{2}\hat{y}$$

$$\text{then,}\quad M(x, y) = \hat{x}$$

(where we have approximated the true field in the aperture with the incident field in the aperture via the Physical Optics approximation).

The integral for the electric vector potential then becomes

$$F(r, \theta, \phi) = \frac{e^{-jkr}}{4\pi r}\,\hat{x}\int_0^{2\pi}\int_0^a e^{jk_\rho\rho\,\sin\theta\,\cos(\phi-\phi_k)}\,\rho\,d\rho\,d\phi$$

where

$$k_\rho\rho\,\sin\theta\,\cos(\phi - \phi_k) = \alpha x + \beta y$$

with,

$$x = \rho\,\cos\phi$$
$$y = \rho\,\sin\phi$$
$$\alpha = k_\rho\,\cos\phi_k$$
$$\beta = k_\rho\,\sin\phi_k$$

This integral expression is readily evaluated in closed form using the following two integral identities for Bessel functions (see ref. 16, Chapter 2):

$$J_0(x) = \frac{1}{2\pi}\int_0^{2\pi} e^{\pm jx\,\cos(\phi-\phi_0)}\,d\phi$$

and

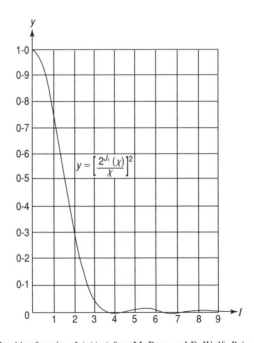

Figure 1.9 The Airy function $J_1(x)/x$ (after, M. Born and E. Wolf, *Principles of Optics*, 6th ed, Pergamon Press).

$$x \, J_1(x) = \int_0^x J_0(u) \, u \, du$$

to yield the electric vector potential as

$$F(r, \, \theta, \, \phi) = \frac{a^2}{2} \frac{e^{-jkr}}{4\pi r} \frac{J_1(ka \, \sin\theta)}{ka \, \sin\theta}$$

The function $J_1(x)/x$ is the classic Airy pattern, shown in Fig. 1.9.

EXAMPLE 1.2   THE FAR-FIELD CRITERION

In the text, it is shown how the parallel rays approximation may be used to simplify field calculations when the field point is infinitely distant from the source distribution. In practical calculations, the parallel ray approximation is valid when the *far-field criterion* is met. This criterion determines how far a field point must be away from a radiating aperture (for a given frequency and aperture size) in order for the parallel rays approximation to be valid. We may readily find the far-field criterion with the aid of Fig. 1.10.

Say a field point is located on-axis, a distance R away from the aperture. The aperture has a lateral extent, D, in the $x - z$ plane. Rays from the two extreme edges of the aperture are clearly not parallel, so the parallel rays approximation is not strictly satisfied. We can use the approximation, however, when the optical phase, kR_1 is not too different from the phase kR. So, say we require

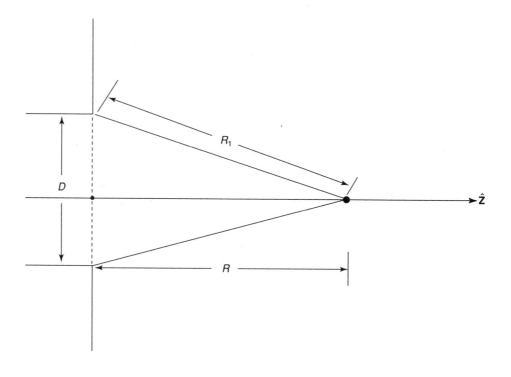

Figure 1.10 On the far-field criterion.

$$k(R_1 - R) < \frac{\pi}{8}$$

Then, since,

$$R_1 = \sqrt{(D/2)^2 + R^2} \cong R\left[1 - \frac{1}{8}(D/R)^2\right]$$

the phase criterion above becomes

$$R > \frac{2D^2}{\lambda}$$

According to this criterion, the far-field range increases as the aperture size increases and as the frequency increases.

EXAMPLE 1.3 FRESNEL ZONES

As we saw in the previous example, for field points located a finite distance from the aperture, the distance function does not necessarily satisfy the far-field criterion. This criterion is satisfied only when the wavefront curvature remains below an acceptable level. (We arbitrarily set that level equal to $\pi/8$ in the example above.) What happens when the wavefront curvature gets larger and perhaps even exceeds 2π? Well, with reference to Fig. 1.11a, we may approximately say that when kR_1 is between $-90°$ and $90°$ of kR, the field is basically in phase with the center of the aperture, and when the phase is outside this range, the field is out of phase. In-phase fields tend to add and out-of-phase fields tend to subtract, as shown in Fig. 1.11c. The alternate in-phase and out-of-phase zones on an aperture are known as *Fresnel zones*. We may readily find expressions for the Fresnel zones of a circular aperture. (*Note:* Zones similar to Fresnel zones are also created when (1.54) and (1.55) are evaluated for field points that lie off the main axis of the aperture; the computational difficulty involved in evaluating (1.54) and (1.55) in this case is directly proportional to the number of Fresnel zones in the aperture.)

So, since

$$k(R_1 - R) \cong kR\,\frac{1}{2}\,(x/R)^2$$

the in-phase Fresnel zones lie between

$$x = \sqrt{nR\lambda}$$

and

$$x = \sqrt{\left(n + \frac{1}{2}\right)R\lambda}$$

A Fresnel zone lens consists of a mask placed over an unfocused circular aperture, with the out-of-phase Fresnel zones covered. An example of such a masked aperture is shown in Fig. 1.12; the mask produces nearly the same focusing properties as a dielectric lens. It should be noted, however, that since the Fresnel zone mask is a *binary* mask (i.e., the zones are either 0% or 100% transmissive, and these zones coincide with the "in-phase" and "out-of-phase" regions), it may launch both converging and diverging spherical wave fields (the first constituting a real image and the second a virtual image, as we'll read about later). This property of a planar structure to produce multiple transmitted fields is characteristic of many types of devices we'll study in this book,

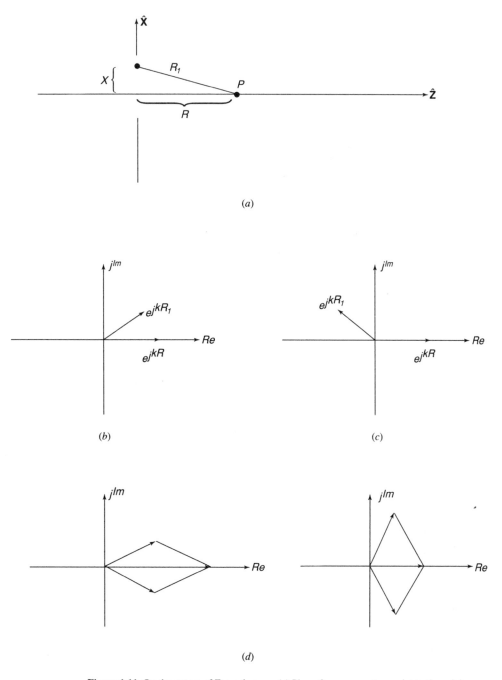

Figure 1.11 On the nature of Fresnel zones: (a) Phase from an aperture point to the axial field point, P; (b) Approximate in-phase condition: $|k(R_1 - R)| \leq \pi/2 \pmod{2\pi}$; (c) Approximate out-of-phase condition: $|k(R_1 - R)| \geq \pi/2 \pmod{2\pi}$; and (d) Addition of ''in-phase'' and ''out-of-phase'' complex numbers.

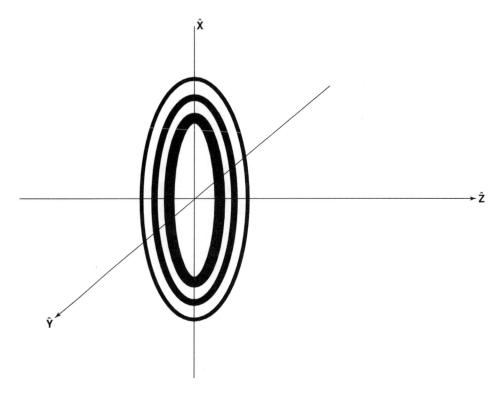

Figure 1.12 Fresnel zone mask.

including diffraction gratings, holograms, holographic lenses, and the like. In fact, the Fresnel zone concept has been used in the past both to understand the operation of holograms on optical fields [4] and to provide insight into the design of holographic devices such as laser beam scanners [5].

REFERENCES

[1] Stratton, J. A., *Electromagnetic Theory,* New York: McGraw-Hill, 1941.

[2] Scott, C. R., *Field Theory of Acousto-Optic Signal Processing Devices,* Norwood, MA: Artech House, 1992.

[3] Scott, C. R., *Modern Methods of Reflector Antenna Analysis and Design,* Norwood, MA: Artech House, 1992.

[4] Siemens-Wapniarski, W. J., and Givens, M. Parker, "The Experimental Production of Synthetic Holograms," *Applied Optics,* vol. 7, no. 3, March 1968.

[5] Lee, Wai-Hon, "Holographic Grating Scaners with Aberration Corrections," *Applied Optics,* vol. 16, no. 5, May 1977.

2

Foundations of the Plane Wave Spectrum Method

In the hundred years since Maxwell first organized the equations of classical electromagnetics into the set of relations that now bear his name, an almost endless array of techniques has been devised for solving these four differential equations. While the equations may be simple enough to look at, they are often quite difficult to apply to problems of practical interest. As a result, many specialized techniques have been proposed through the years for applying Maxwell's equations to different types of problems in electromagnetics. In this book, we'll consider three such solution methods that are of particular value in optics: the ray optic method, the diffraction integral method, and the plane wave spectrum method. In this chapter, we'll concentrate on the plane wave spectrum method.

One important means of classifying the different approaches to solving Maxwell's equations is by the frequency of the fields under consideration. For example, when material bodies are very small in comparison with the electromagnetic wavelength (the "low-frequency" or "quasi-static" regime), the time dependence of the fields can often be neglected and static field solutions to Maxwell's equations can be employed. When material bodies are comparable in linear dimension with the electromagnetic wavelength (the "resonance region"), highly accurate analytical and numerical methods may be used to solve Maxwell's equations in their exact form. These methods include: the eigenfunction expansion method, the spatial domain method of moments (MoM), the spectral domain method of moments (based on the plane wave spectrum approach to be described in this chapter), the finite element method (FEM), and the finite-difference time domain method (FDTD). And when the dimensions of material bodies are very large in comparison to the wavelength, the diffraction integral method (called the physical optics, or PO method, in electromagnetics) and geometrical optics (GO) methods (to be presented in later chapters) may be used to calculate the fields.

When antennas, scatterers, or refracting bodies in the resonance region conform to one of the *separable coordinate systems* (rectangular, cylindrical, and spherical coordinates are the only practical separable coordinate systems), field calculations can be performed in analytic form using *eigenfunction* techniques. In rectangular coordinates, the eigenfunctions of Maxwell's equations are the plane wave functions and—as we'll show in this chapter—these eigenfunctions form the basis for the plane wave spectrum method. Later in this chapter, we'll adapt the plane wave spectrum method to high-frequency (ray optical) calculations of the type commonly encountered in connection with typical optical systems. The plane wave spectrum method is very versatile and has many applications beyond what we'll study in this book. It may also be used in conjunction with the method of moments to obtain accurate solutions to Maxwell's equations in the resonance region [1]).

In cylindrical coordinates, the eigenfunctions for Maxwell's equations are the cylindrical wave functions (Bessel functions in the radial direction and sinusoids in the azimuthal and longitudinal directions) [2]. In spherical coordinates, the eigenfunctions are the spherical wave functions (spherical Bessel functions in the radial direction, sinusoids in the azimuthal direction, and associated Legendre functions in the elevation direction) [3]. Eigenfunction techniques for cylindrical coordinates are used primarily in the quasi-static and resonance regions. However, they can also be adapted to the high-frequency regime via asymptotic techniques similar to those presented in this chapter for rectangular coordinates [4–6].

Eigenfunction techniques have both historical and contemporary value. For example, the plane wave spectrum technique was used historically for calculating the fields of a dipole above a conducting ground [7] and more recently, for analyzing a variety of microwave printed circuit structures [8–10]. The cylindrical wave expansion method has been used to find the guided modes in optical fibers [11]; the spherical wave expansion technique (known in optics as the Mie series technique) has been used historically to determine light scattered by raindrops [12], and in recent times has been adapted for calculating the scattering from body-of-revolution scatterers [13].

The plane wave spectrum technique was invented twice in two completely different contexts. First described in qualitative fashion by Ernst Abbe in connection with optical imaging systems [14], this method was also used by Sommerfeld [7] in connection with the scattering of radio waves by a conducting ground. It is not known whether the two researchers were aware of each other's work. In any event, this technique (known today among microwave engineers as the spectral domain method) represents an important tool in modern electromagnetic analysis. In this chapter, we'll introduce this technique and show how it may be applied to optical problems in the high-frequency regime.

2.1 PLANE WAVE SOLUTIONS OF THE WAVE EQUATION

In the previous chapter, we solved the *nonhomogeneous* wave equations for the electric and magnetic field vectors. We did this using the superposition integral technique, by convolving the forcing function against the impulse response (Green's function) of the equation. In this chapter, we'll examine solutions to a set of *homogeneous* wave equations, that is, wave equations for which the right-hand side is zero. Clearly, we can't use the superposition principle this time, because there is now no forcing vector to convolve with—it's zero. So we have to come up with a new approach for solving homogeneous equations. The approach we'll use is the eigenfunction technique discussed at the top of this chapter.

The term *eigenfunction* is a big word that just means a function which, when operated on by some differential operator, is either transformed back into itself or zero. For example,

$\sin x$ is an eigenfunction of the differential operator d^2/dx^2, since the second derivative of a sine function is again a sine. The scalar Green's function, introduced in Chapter 1, is almost an eigenfunction of the scalar wave equation, since it yields a zero value everywhere except at the origin, where it produces a delta function.

In this chapter, we'll obtain the eigenfunctions to the wave equation in rectangular coordinates. It will be as easy as solving the simple differential equation above for the sine function.

We begin with the wave equations for the magnetic and electric fields in a source-free region, obtained in Chapter 1 as

$$\nabla^2 E + k^2 E = 0 \tag{2.1a}$$

$$\nabla^2 H + k^2 H = 0 \tag{2.1b}$$

These equations take on a very nice form in rectangular coordinates, owing to the simplified form of the vector Laplacian in these coordinates (as discussed in Chapter 1). So, since

$$\nabla^2 a = \hat{x}\nabla^2 a_x + \hat{y}\nabla^2 a_y + \hat{z}\nabla^2 a_z \tag{2.2}$$

we have on a component-by-component basis,

$$\nabla^2 E_j + k^2 E_j = 0 \tag{2.3a}$$

$$\nabla^2 H_j + k^2 H_j = 0 \tag{2.3b}$$

for $j = x, y, z$. That is, in rectangular coordinates, the vector Laplacian reduces to the scalar Laplacian on a component-by-component basis.

Expanding the scalar Laplacian in rectangular coordinates, (2.3a, b) become simply

$$\frac{\partial^2}{\partial x^2}\begin{pmatrix} E_x \\ E_y \\ E_z \end{pmatrix} + \frac{\partial^2}{\partial y^2}\begin{pmatrix} E_x \\ E_y \\ E_z \end{pmatrix} + \frac{\partial^2}{\partial z^2}\begin{pmatrix} E_x \\ E_y \\ E_z \end{pmatrix} + k^2\begin{pmatrix} E_x \\ E_y \\ E_z \end{pmatrix} = 0 \tag{2.4}$$

$$\frac{\partial^2}{\partial x^2}\begin{pmatrix} H_x \\ H_y \\ H_z \end{pmatrix} + \frac{\partial^2}{\partial y^2}\begin{pmatrix} H_x \\ H_y \\ H_z \end{pmatrix} + \frac{\partial^2}{\partial z^2}\begin{pmatrix} H_x \\ H_y \\ H_z \end{pmatrix} + k^2\begin{pmatrix} H_x \\ H_y \\ H_z \end{pmatrix} = 0 \tag{2.5}$$

These two scalar partial differential equations are now transformed into ordinary differential equations using an argument known as the principle of *separation of parameters*. This principle applies *only* to *separable* coordinate systems of the type mentioned at the outset of this chapter; it is not of general utility. On the other hand, the separation of parameters concept is the vehicle for obtaining the *eigenfunctions* for Maxwell's equations in rectangular, cylindrical, and spherical coordinates. (It is also used for finding the eigenfunctions of other equations in mathematical physics, such as Schröedinger's equation.) In the fields of optical imagining and optical image processing, the eigenfunctions of Maxwell's equations in rectangular coordinates are of paramount importance.

To apply the separation of parameters principle to rectangular coordinates, we make the following postulate. We suppose that in rectangular coordinates, a general solution to the wave equation can be represented as a sum over terms having a *separated form*. That is, the individual terms in the sum take the form of the product of a function of x times a function of y times a function of z. The solution is (not surprisingly) termed a *product solution* and physically represents a plane wave field. The various plane wave solutions to the wave equations constitute the eigenfunctions for Maxwell's equations in rectangular coordinates.

So, the electric and magnetic vector fields take the form

$$\begin{pmatrix} A_z \\ F_z \end{pmatrix} = \psi_x(x)\psi_y(y)\psi_z(z) \tag{2.6}$$

With this, Eqs. (2.4) and (2.5) now take the form

$$\frac{1}{\psi_x}\frac{d^2\psi_x}{dx^2} + \frac{1}{\psi_y}\frac{d^2\psi_y}{dy^2} + \frac{1}{\psi_z}\frac{d^2\psi_z}{dz^2} + k^2 = 0 \tag{2.7}$$

If the first term on the LHS of this equation is moved over to the RHS (and given a minus sign), then the RHS of the equation becomes a function of the x-coordinate variable only, whereas the LHS is a function of y and z. The only way that both of these conditions may be satisfied simultaneously is if both sides of the equation are constant. Thus, it is shown that all four terms on the LHS of (2.7) must be constant. If we denote the first three constants by $-\alpha$, $-\beta$, $-\gamma$, respectively, then we obtain three ordinary differential equations, plus one *separation condition*,

$$\frac{d^2\psi_x}{dx^2} + \alpha^2\psi_x = 0 \tag{2.8a}$$

$$\frac{d^2\psi_y}{dy^2} + \beta^2\psi_y = 0 \tag{2.8b}$$

$$\frac{d^2\psi_z}{dz^2} + \gamma^2\psi_z = 0 \tag{2.8c}$$

$$\alpha^2 + \beta^2 + \gamma^2 = k^2 \tag{2.9}$$

According to the separation condition, α, β, and γ are not all independent. It is customary to regard α and β as the independent parameters and γ as the dependent parameter. In fact, the character γ is not even generally used to represent the propagation constant in the z-direction. The dependence of γ on α and β is usually emphasized by writing

$$\gamma = k_z = \sqrt{k^2 - \alpha^2 - \beta^2} \tag{2.10}$$

In the last equation, the root is chosen which has a nonpositive imaginary part, corresponding to an attenuated (or, at least nonamplified) plane wave. Note that a negative real part is, in principle, physically permissible and is sometimes encountered with dielectric materials having resonant frequencies in the optical regime. As noted at the beginning of this section, the solutions to Eqs. (2.8) are sinusoidal functions. Therefore, the elementary product solutions to the wave equations are

$$\begin{pmatrix} E_x \\ E_y \\ E_z \\ H_x \\ H_y \\ H_z \end{pmatrix} = e^{j(\alpha x + \beta y)} e^{\pm j\sqrt{k^2 - \alpha^2 - \beta^2}z} \tag{2.11}$$

The independent parameters α and β are completely arbitrary; either (or both) may have magnitudes greater than that of k itself. When

$$k^2 > \alpha^2 + \beta^2$$

the waves are ordinary plane waves propagating in a direction normal to the vector

$$\hat{n} = -\alpha\hat{x} - \beta\hat{y} \mp k_z\hat{z}$$

These propagating plane waves lie in the *visible* part of the spectrum. When

$$k^2 < \alpha^2 + \beta^2$$

the waves still propagate in the *x*, *y* plane, but they decay exponentially in the direction normal to the plane. These waves are known as *evanescent waves* and are important in the study of surface wave phenomena. They will not figure significantly in the remainder of this book, however, since we'll be exclusively concerned with propagating wave phenomena.

The existence of exponentially decaying evanescent waves has great significance in the area of optical imaging systems. A plane wave will not propagate whenever α, β, or both are greater than k. Thus, that portion of an object's spectral content that lies outside the visible range will not propagate and cannot reach the image space. (We say in this case that the plane wave spectrum is *bandlimited*.) As a result, the finest detail that can be observed in an optical image is roughly equal to the wavelength of the light used to produce that image. No finer detail can be observed, because the plane waves that carry the more detailed information cannot propagate. The propagation medium therefore acts as a low-pass filter, passing only those (spatial) frequency components that lie in the visible region.

Note. New scanning near-field optical microscopy (SNOM) technology [15] strives to overcome the low-pass property of propagation media by placing optical probes very close to the object being probed (i.e., within a wavelength of the surface). Such close placement allows the probe to capture some of the slowly decaying evanescent energy from the surface, allowing the resultant image of the surface to have a bandwidth that is somewhat broader than an image produced by ordinary optical means (i.e., by simply capturing the propagating energy). This increased bandwidth allows for increased resolution of the object surface.

A general expression for the fields will involve all possible combinations of sinusoidal solutions of the form given in (2.11). That is,

$$E_x(x, y, z) = \iint E_x(\alpha, \beta)_{z=0} \, e^{j(\alpha x + \beta y)} e^{\pm j\sqrt{k^2 - \alpha^2 - \beta^2}\,z} \, d\alpha \, d\beta \qquad (2.12)$$

Similar equations hold for the other field components. This equation is very important for the rest of the chapter. It indicates that a general vector field in rectangular coordinates may be expressed in the form of a Fourier integral over an infinite set of plane wave functions (i.e., over an infinite *spectrum* of plane waves, hence the name: plane wave spectrum). This representation is entirely analogous to the Fourier spectrum representation of time-varying signals. Readers familiar with the Fourier spectrum representation of time signals will find that background extremely valuable in this context.

The Fourier spectrum representation of optical fields is important in the analysis of optical systems. If the optical field is known at one transverse plane of an optical system, it may be Fourier analyzed and then decomposed into plane wave components. Then, the total field may be calculated at any other transverse plane by merely propagating each individual plane wave component to the new plane and then reassembling all of the individual plane wave components according to (2.12). We'll examine this process in more detail in later sections.

There are many different ways in which an optical field on a two-dimensional plane may be decomposed into a functional spectrum and analyzed. Many well-known transforms such as the Karhunen-Loeve transform, the Walsh-Hadamard transform, the Hartley transform, and the discrete cosine transform are widely used in image analysis to represent a 2-D image in terms of a discrete set of basis functions. These differ from the ordinary Fourier transform only in terms of the *kernel* used. The Fourier kernel is the complex exponential; hence, the Fourier transform of a function is obtained by multiplying the function by the exponential Fourier kernel and then integrating the product over the domain of the function. The exponential Fourier kernel sifts out sinusoidal features in the original function. The other transforms use kernels that have other functional forms, which highlight different features in the original function.

One of the most popular current trends in this area involves decomposing fields according to various *wavelet transforms.* We'll study this type of signal decomposition in a later chapter in connection with optical image processing. However, the wavelet transform expansion of fields does not lend itself well to the problem of transforming fields from one transverse plane to another. Each wavelet in the transverse plane is composed of an infinite spectrum of plane waves, unlike the simple transverse plane sinusoid, which corresponds to one and only one plane wave field.

With Eq. (2.12) in hand, expressing the magnetic and electric fields in terms of the plane wave spectrum expansion, it's possible to calculate the impedance relationship that exists between tangential components of the two fields. This is a straightforward process that is accomplished by merely substituting the representations for the fields from (2.12) into Faraday's law for source-free media,

$$\nabla \times E = -j\omega\mu H \tag{2.13}$$

Using Gauss' law in rectangular coordinates, we obtain

$$E_z = \frac{\alpha}{k_z} E_x + \frac{\beta}{K_z} E_y \tag{2.14a}$$

$$H_z = \frac{\alpha}{k_z} H_x + \frac{\beta}{k_z} H_y \tag{2.14b}$$

for plane wave fields propagating/decaying in the positive z-direction. For waves propagating/decaying in the negative z-direction, the signs of the terms on the LHS are reversed.

Substituting (2.14) into (2.13) and equating the x, y components of the fields on either side of Faraday's law yield the following impedance relation between tangential components of the electric and magnetic fields,

$$H_{\tan}(\alpha, \beta) = \pm Y(\alpha, \beta) E_{\tan}(\alpha, \beta) \tag{2.15}$$

where

$$Y(\alpha, \beta) = \frac{1}{k\eta} \cdot \frac{1}{k_z} \begin{bmatrix} -\alpha\beta & \alpha^2 - k^2 \\ k^2 - \beta^2 & \alpha\beta \end{bmatrix} \tag{2.16}$$

This is an impedance relation between *transverse* components of the field. The matrix Y is an admittance matrix. The plus sign applies for waves propagating or decaying in the positive z-direction, and the negative sign applies for waves propagating or decaying in the negative z-direction. This relation is useful for computing the reflection and transmission from a planar dielectric interface.

EXAMPLE 2.1   PLANE WAVE REFLECTION AT A DIELECTRIC BOUNDARY

Consider the problem of calculating plane wave reflection and transmission at the dielectric interface shown in Fig. 2.1. The incident fields are given in the form

$$E^{\text{inc}}(x, y) = E^{\text{inc}}(\alpha, \beta)e^{j(\alpha x + \beta y)}e^{-j\sqrt{k_1^2 - \alpha^2 - \beta^2}z}$$

$$H^{\text{inc}}(x, y) = Y_1(\alpha, \beta)E^{\text{inc}}(\alpha, \beta)e^{j(\alpha x + \beta y)}e^{-j\sqrt{k_1^2 - \alpha^2 - \beta^2}z}$$

The reflected fields are given as

$$E^{\text{refl}}(x, y) = E^{\text{refl}}(\alpha, \beta)e^{j(\alpha x + \beta y)}e^{j\sqrt{k_1^2 - \alpha^2 - \beta^2}z}$$

$$H^{\text{refl}}(x, y) = -Y_1(\alpha, \beta)E^{\text{refl}}(\alpha, \beta)e^{j(\alpha x + \beta y)}e^{j\sqrt{k_1^2 - \alpha^2 - \beta^2}z}$$

and the transmitted fields are given as

$$E^{\text{trans}}(x, y) = E^{\text{trans}}(\alpha, \beta)e^{j(\alpha x + \beta y)}e^{-j\sqrt{k_2^2 - \alpha^2 - \beta^2}z}$$

$$H^{\text{trans}}(x, y) = Y_2(\alpha, \beta)E^{\text{trans}}(\alpha, \beta)e^{j(\alpha x + \beta y)}e^{-j\sqrt{k_2^2 - \alpha^2 - \beta^2}z}$$

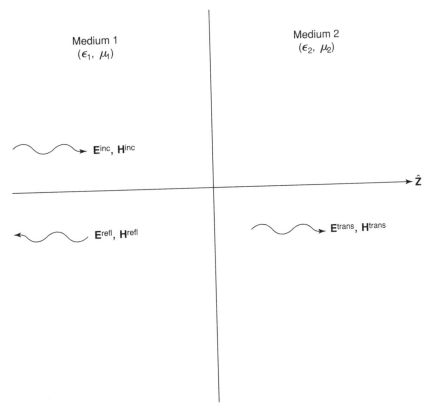

Figure 2.1 Plane wave reflection and transmission at a planar dielectric interface.

where the subscripts on the admittance matrices refer to the medium. Enforcing continuity of the tangential E and H fields at $z = 0$ yields the following *matrix* reflection and transmission coefficients:

$$R = (Y_1 + Y_2)^{-1}(Y_1 - Y_2)$$

$$T = 2(Y_1 + Y_2)^{-1}Y_1$$

These matrix (tensor) reflection and transmission coefficients are useful not only for planar scattering problems but also for ray optic scattering problems involving curved surfaces, as we'll see in Chapter 3. *Note:* In the example, we assumed α, β to be the same for all fields (incident, reflected, and transmitted). This is known as *phase matching*. Whenever boundary conditions are imposed on the tangential components of (phasor) fields, the fields must be matched in terms of amplitude, phase, and vector direction. The matrix reflection and transmission coefficients ensure that the tangential components of the amplitude and vector directions of the fields are matched at the boundary.

2.2 ASYMPTOTIC EVALUATION OF THE SPECTRAL INTEGRALS IN THE FRAUNHOFER REGION: INTRODUCTION TO THE STATIONARY PHASE METHOD

In this section, we'll re-derive the equations obtained in Chapter 1 for the Fraunhofer region fields. In so doing, we'll first begin to explore a concept that is central to all of optics. This concept is known by a variety of names—Fermat's principle, the principle of least time, the principle of minimum optical path length, and the principle of stationary phase. We'll study this principle in more detail later, in connection with ray optical theories of light. In this section, however, we'll use this principle of stationary phase to obtain the Fraunhofer region fields and to see how the ray theory of light can be obtained from the plane wave expansion of an aperture field (fulfilling a promise made at the beginning of Chapter 1). To do this, we'll derive the form of the plane wave expansion in the far field of the radiating aperture.

With Eq. (2.12) in hand, expressing the electromagnetic field in terms of its value on a given reference plane, we can now use this equation to perform a number of different kinds of field calculations. An example might be the calculation of the Fraunhofer region field emitted by a laser, for the purpose of estimating beam divergence. For example, say the field in the exit pupil of the laser is given by

$$E(x, y) = \iint_{-\infty}^{\infty} E(\alpha, \beta)_{z=0}\, e^{j(\alpha x + \beta y)} d\alpha\, d\beta \tag{2.17}$$

We assume that the electric field: $E(x, y, z = 0)$ in the exit pupil is known. In fact, the mathematical form of the aperture fields of many lasers is quite well known, many having radially symmetric Gaussian amplitude and quadratic phase distributions. (We'll discuss this special type of distribution in more detail in a later chapter.)

In order to calculate the far-zone radiation field of the aperture distribution (2.17), we employ (2.12) in the limit as the radial distance to the observation point (x, y, z) tends to infinity. This produces what is known as an *asymptotic* form of the integral. Under this special limiting condition, the exponential terms in (2.12) act to perform a *sifting function* on the integrand, in much the same way that the unit impulse function performs a sifting function when integrated against another function. In the case of the unit impulse function, the sifting takes place because the amplitude of the impulse is zero everywhere except at one point, where it tends to infinity, in an integrable fashion. In the case of the exponential functions in (2.12), however, the phase of the exponential oscillates with a frequency that tends to infinity as the radial distance tends to infinity (except at one point, however, where the phase remains *stationary* for all values of r). This stationary phase point is determined by the condition that the *wavenumber vector,*

$$k = k(\alpha\hat{x} + \beta\hat{y} + k_z\hat{z})$$

is parallel to the far-field vector,

$$r = x\hat{x} + y\hat{y} + z\hat{z} = r(\sin\theta_0\cos\phi_0\hat{x} + \sin\theta_0\sin\phi_0\hat{y} + \cos\theta_0\hat{z})$$

because it is when

$$\psi(\alpha, \beta) = k \cdot r = kr$$

that

$$\frac{\partial\psi}{\partial\alpha} = \frac{\partial\psi}{\partial\beta} = 0$$

That is, the phase (dot product—related to the cosine of the angle between the two vectors) is stationary with respect to the integration variables α and β. The reader should try to verify this mathematically.

Since the oscillation frequency of the exponential tends to infinity away from the stationary phase point, the rapid positive and negative excursions of its real and imaginary parts tend to cancel each other out when multiplied by the relatively slowly varying aperture function and integrated. The only portion of the range of integration that then contributes to the final integral is that portion located in the vicinity of the stationary phase point. Thus, a sifting function is achieved.

One more comment on sifting is worth mentioning. In Appendix D, it's shown that the unit impulse function is an infinitesimally narrow version of a Gaussian (quadratic amplitude) distribution, and this function performs an *amplitude* sifting function. A quadratic phase function (a Gaussian distribution with a purely imaginary exponent) also performs a *phase* sifting function in the limit as the coefficient of the quadratic exponent tends to infinity.

The characteristics of the phase function near the stationary phase point will determine the nature of the final asymptotic integral, so it is only necessary to approximate the phase function in the immediate vicinity of this stationary point. Therefore, a Taylor series expansion can be used to represent the phase locally about this point. Since the phase is defined to be stationary with respect to the spatial frequency variables α, β, its first derivative with respect to both of these integration variables will be zero. So, in order to evaluate the integral, we'll only need the constant term plus the quadratic term in the Taylor series expansion for the phase function. The choice of approximating function for the phase is governed only by two criteria, that is, that the approximating function has the same value and second derivative as the real function at the stationary phase point, and that the approximating function itself has only one stationary phase point (namely, at the same point as the original function).

To show how the stationary phase integration process works mathematically, we'll use (2.17) in cylindrical coordinates. So, let

$$x = \rho \cos\phi$$

$$y = \rho \sin\phi$$

$$z = \sqrt{r^2 - \rho^2}$$

and

$$\alpha = k_\rho \cos\phi_k$$

$$\beta = k_\rho \sin\phi_k$$

$$k_z = \sqrt{k^2 - k_\rho^2}$$

It may at first seem odd that spectral space (or in the jargon of physics, *k-space*) may assume the same geometrical properties as ordinary "spatial space." Yet, we may freely use geometrical intuition to move between coordinate systems in both the spatial and spectral domains. Such transformations between coordinate systems are often helpful in obtaining identities between the eigenfunctions of the three major separable coordinate systems [2].

Substituting these last six relations into (2.17) yields

$$E(r, \theta, \phi) = \int_0^\infty \int_0^{2\pi} E(\alpha = k_\rho\cos\phi_k, \beta = k_\rho\sin\phi_k)\ e^{jk_\rho\rho\cos(\phi_k - \phi)}e^{-jk_z z}\ k_\rho\, dk_\rho\, d\phi_k$$

(2.18)

The phi-integral is evaluated in closed form using the relation [16]

$$J_0(x) = \frac{1}{2\pi} \int_0^{2\pi} e^{jx\cos\phi}\, d\phi$$

(2.19)

where J_0 is the Bessel function of order zero [14]. Thus,

$$E(r, \theta, \phi) = 2\pi E(\alpha_0, \beta_0) \int_0^\infty J_0(k_\rho\rho)\, e^{-j\sqrt{k^2 - k(2/\rho)}z}k_\rho\, dk_\rho$$

(2.20)

The electric field at the stationary phase point has been pulled outside the integral sign, in accordance with the discussion on sifting above. In the equation, α_0 and β_0 are the spectral wavenumbers in the direction of the far-field stationary phase point; that is,

$$\alpha_0 = k \sin\theta \cos\phi$$

$$\beta_0 = k \sin\theta \sin\phi$$

The integral above is to be evaluated in the limit as r approaches ∞. Now, as long as α_0 and β_0 are not both zero, the argument of the Bessel function will approach ∞ along with the radial distance, r. Thus, we may employ the large-argument expansions for the Bessel function [16] to simplify the integral. The large-argument approximation for the Bessel function of order zero is given as

$$J_0(x) = \sqrt{\frac{2}{\pi x}} \cos\left(x - \frac{\pi}{4}\right)$$

(2.21)

The cosine function in (2.21) can be expressed as the sum of exponentials as

$$\cos x = \frac{1}{2}(e^{jx} + e^{-jx})$$

However, only one of these exponentials (i.e., the negative one) will combine with the exponential in (2.20) to produce a stationary phase point in the range of integration. So, for the purposes of evaluating the integral, we can ignore the positive exponential in the expression for the cosine to obtain

$$E(r,\ \theta,\ \phi) = 2\pi E(\alpha_0,\ \beta_0)\ \sqrt{\frac{1}{2\pi k_{\rho,0}\rho}}\ k_{\rho,0}\ e^{j\pi/4} \int_0^\infty e^{-jk_\rho\rho}e^{-j\sqrt{k^2-k_\rho^2}z}\ dk_\rho$$

(2.22)

where

$$k_{\rho,0} = k\sin\theta$$

Now all we need is the Taylor series expansion for the phase function about the stationary phase point. The Taylor series for the phase is

$$\phi(k_\rho) = \phi(k_{\rho,0}) + (k_\rho - k_{\rho,0})\frac{d\phi}{dk_\rho} + \frac{1}{2}(k_\rho - k_{\rho,0})^2\frac{d^2\phi}{dk_\rho^2}$$

where

$$k_{\rho,0} = k\sin\theta$$

and

$$
\begin{aligned}
\phi(k_\rho = k_{\rho,0}) &= -(jk\sin\theta)\rho - (jk\cos\theta)\,z \\
&= -jk(\rho\sin\theta + z\cos\theta) \\
&= -jkr(\sin^2\theta + \cos^2\theta) \\
&= -jkr
\end{aligned}
$$

$$\frac{d\phi}{dk_\rho}(k_{\rho,0}) = 0$$

$$\frac{d^2\phi}{dk_\rho^2}(k_{\rho,0}) = j\frac{r}{k\cos^2\theta}$$

so, the expression for the far-zone field becomes

$$E(r,\ \theta,\ \phi) = E(\alpha_0,\ \beta_0)\ \sqrt{\frac{2\pi k}{r}}\ e^{j(\pi/4)}e^{-jkr} \int_0^\infty e^{j(r/2k\cos^2\theta)(k_\rho - k_{\rho,0})^2}\ dk_\rho \quad (2.23)$$

The integral is evaluated by making the change of variables

$$u = k_\rho - k_{\rho,0}$$

and using the identity

$$\int_{-\infty}^\infty e^{jAu^2}du = \sqrt{\frac{\pi}{A}}\ e^{j\pi/4}$$

to yield finally the asymptotic value of the far-field integral (2.17) as

$$E_{\text{tan}}(r,\ \theta,\ \phi) = 2\pi j(k\cos\theta)\ \frac{e^{-jkr}}{r}\ E_{\text{tan}}(\alpha_0,\ \beta_0)$$

(2.24)

NOTE: The last change of variables transformed the range of integration from 0 → ∞ into $-k_{\rho,0}$ → ∞. But since only that portion of the integrand near the stationary phase point contributes to the integral (in the limit as $r \to \infty$), we may go ahead and extend the lower limit of integration all the way out to $-\infty$, and then use the integration formula above.

This last result shows that out of the entire continuous Fourier spectrum of plane waves comprising the aperture field, only that component propagating in the direction of the far-field point contributes to the far-field pattern at a distant point. (That particular plane wave component has been sifted out.) In other words, we've performed a Fourier transform (FT) operation on the aperture field distribution. (This type of FT operation is known as a *lensless Fourier transform.*) The term on the LHS of (2.24) is the transverse plane field; the z-component of electric field is obtained from Gauss' law. Note that the stationary phase integral can also be readily evaluated in rectangular coordinates [17].

To the Reader. In obtaining this asymptotic far-zone expression for the field, we've employed some relations involving Bessel functions that may be unfamiliar. These relations (and a few others involving Bessel functions) pop up again and again in optical analysis, especially in calculations involving circular apertures. (The famous *Airy function* for circular apertures is derived using equations such as these). If you're not already familiar with Bessel functions and their properties, it will be worthwhile to get a text or handbook on the functions of mathematical physics. (Abramowitz and Stegun [16] is almost indispensable.) Mathematical handbooks are like dictionaries—you don't need to know the derivations of the formulas any more than you need to know the history of a particular word. But in the same way that you'd use a dictionary to find the meaning of a word and learn how to use it in a sentence, you would use a mathematical handbook to learn how to manipulate certain standard mathematical expressions, in order to get simplified results. These results are described in terms of mathematical functions that have been studied before—in many cases by people who spent their entire lives in that study—and are now available for engineers and scientists to use freely. So we might as well use them. Other useful books on this general subject include Watson [18], Magnus and Oberhettinger [19], and Erdèlyi [20]. In addition, most any junior-level text on differential equations will have a good discussion on higher functions. One good example is Simmons [21].

EXAMPLE 2.2   UNIFORMLY ILLUMINATED RECTANGULAR APERTURE

Consider the uniformly illuminated rectangular aperture shown in Fig. 2.2, for which

$$E_{ap}(x, y) = \hat{x}$$

In this case, the Fourier transform of the aperture distribution is given by inverting the relation (2.17) as

$$E(\alpha, \beta) = \frac{1}{4\pi^2} \iint_{-\infty}^{\infty} E(x, y) e^{-j(\alpha x + \beta y)} dx \, dy$$

to yield

$$E(\alpha, \beta) = \frac{ab}{4\pi^2} \operatorname{sinc}(\alpha a/2) \operatorname{sinc}(\beta b/2)$$

The Fourier transform for this aperture distribution is a purely real function of α and β. Therefore,

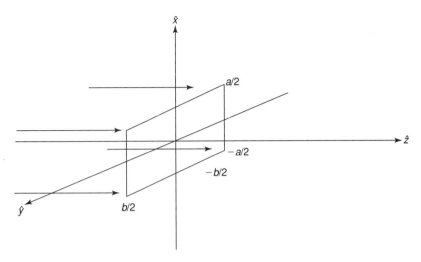

Figure 2.2 Uniformly illuminated rectangular aperture.

by (2.24), the far-zone electric field due to the uniformly illuminated aperture is a purely spherical phase function, whose phase center is at the center of the rectangular aperture (i.e., the spherical wave appears to emanate from the center of the rectangular aperture).

EXAMPLE 2.3 QUADRATIC PHASE ILLUMINATION ON RECTANGULAR APERTURE

Consider again the rectangular aperture of Fig. 2.2, illuminated as shown in Fig. 2.3. This time, the illumination function is due to a spherical wave source located behind the screen, given by

$$E(x, y) = \hat{x}e^{-jk(x^2+y^2)2R}$$

We'll evaluate the Fourier transform of this distribution in the region for which $ka \gg \alpha R, \beta R$ (large aperture, with α and β lying near the z-axis).

Using the asymptotic form of the Fresnel integral for large arguments,

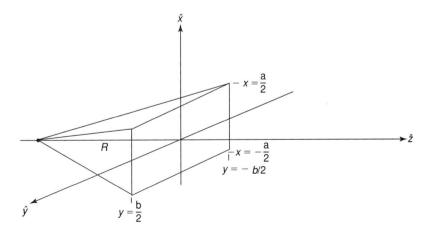

Figure 2.3 Rectangular aperture with spherical wave (quadratic phase) illumination.

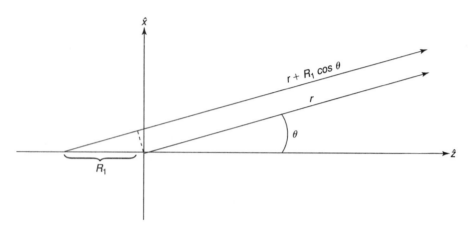

Phase $= e^{-jk\,(r\,+\,R_1\,\cos\,\theta\,)}$

$\cong e^{-jkr}\,e^{-jkR_1}\,e^{-jkR_1\,\theta^2/2}$

(Quadratic phase for $\theta \ll 1$)

Figure 2.4 Ray description of fields radiated by point source located behind aperture.

$$\int_0^x e^{ju^2}du = \frac{\sqrt{\pi}}{2}\,e^{j\pi/4} - \frac{j}{2x}\,e^{jx^2}$$

(where x is greater than zero in the equation above, and the sign of the integral is reversed for x less than zero), we obtain the Fourier transform of the aperture distribution as

$$E(\alpha,\,\beta) = \hat{x}e^{j(R/2k)(\alpha^2\,+\,\beta^2)} \cdot f(\alpha,\,\beta)$$

where f is a purely real function. Thus, the Fourier transform of the aperture center distribution is now no longer purely real; it has a quadratic phase term that is a function of the radiation direction determined by $\alpha,\,\beta$. The reason for this quadratic phase term is that the aperture center is not the phase center for the far-field wave. The phase center is located behind the aperture plane a distance R. This is shown in Fig. 2.4.

2.3 GEOMETRICAL RAY OPTICS REPRESENTATION OF FRAUNHOFER REGION FIELDS

The results of the previous section lead directly into the ray theory of optics. The ray theory of light is a far-field (Fraunhofer region) concept that allows for tremendous simplification of many types of optical calculations. The concept of light rays is probably familiar to most readers of this book—many having first seen it in high school—and the purpose of this short section is to derive the concept from the far-field equations of the previous section.

To begin, we can calculate the magnetic field associated with the electric field given in (2.24). To do that, we may show that an equation similar to (2.24) holds for the magnetic field. That is,

$$H_{\tan}(r, \theta, \phi) = 2\pi j(k\cos\theta)\frac{e^{-jkr}}{r}H_{\tan}(\alpha_0, \beta_0) \qquad (2.25)$$

Now, by (2.15),

$$H_{\tan}(\alpha, \beta) = \pm Y(\alpha, \beta)E_{\tan}(\alpha, \beta) \qquad (2.26)$$

Thus,

$$H_{\tan}(r, \theta, \phi) = 2\pi j(k\cos\theta)\frac{e^{-jkr}}{r}YE_{\tan}(\alpha_0, \beta_0) \qquad (2.27)$$

where all field quantities lie in the transverse (x, y) plane. Now, in source-free media, both E and H satisfy Gauss' law, that is,

$$\nabla \cdot E = 0, \qquad \nabla \cdot H = 0$$

and this enables the z-component of the two fields to be evaluated as

$$\vec{k} \cdot \vec{E} = 0$$

$$E_z(\alpha, \beta) = \frac{\alpha}{k_z}E_x(\alpha, \beta) + \frac{\beta}{k_z}E_y(\alpha, \beta)$$

$$H_z(\alpha, \beta) = \frac{\alpha}{k_z}H_x(\alpha, \beta) + \frac{\beta}{k_z}H_y(\alpha, \beta)$$

for propagation in the positive z-direction (the signs of the z-components are reversed for propagation in the negative z-direction).

The two (E, H) fields are perpendicular to each other, as well as to the propagation direction given by the propagation vector,

$$k = \frac{2\pi}{\lambda}(\sin\theta\cos\phi\,\hat{x} + \sin\theta\sin\phi\,\hat{y} + \cos\theta\,\hat{z})$$

(The reader should verify this by taking the dot products of E, H, and k). The fact that the field amplitude in (2.24) decays in a $1/r$ fashion implies that the electromagnetic power given by

$$P = E \times H*$$

decays in a $1/r^2$ fashion. This leads to the ray type of field description shown in Fig. 2.5. In this model, electromagnetic power flows through "ray tubes," which (in homogeneous media) expand in a spherical fashion, with all light rays propagating along straight-line paths. Since the Poynting vector is parallel to the ray tubes, no power crosses the sides of the tubes; it only emerges from the end caps of the tubes, to which the Poynting vector is perpendicular. The electric and magnetic fields are perpendicular to the direction of propagation, to the direction of power flow, and to each other.

2.4 THE EFFECT OF APERTURES AND LENSES ON PLANE WAVE SPECTRA: ABBE'S THEORY

The action of optical systems and components on optical fields may be understood in a variety of ways, and we'll try to explore a number of these in this book. For example, it is well known that a convex lens converts an incident plane wave into a converging

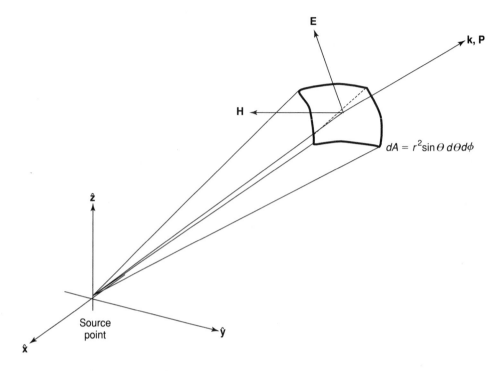

Figure 2.5 Ray model for Fraunhofer region fields.

spherical wave, which converges on the focal point of the lens. (We'll examine lens properties more thoroughly in Chapters 4–6.) So, one way of viewing lens action is to say that the lens converts an incident plane wave into a spherical wave (as shown in Fig. 2.6).

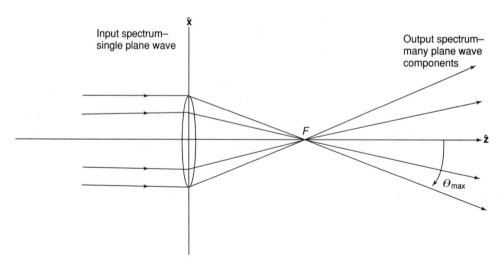

Figure 2.6 An ordinary lens creates new spectral components not present in the incident field.

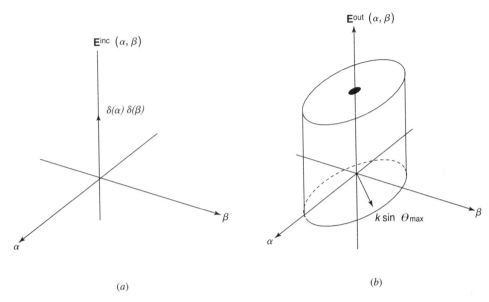

Figure 2.7 Lens action in the spectral domain: (a) Incident field plane wave spectrum and (b) Output field plane wave spectrum (according to geometric optics). (Note: the output spectrum is actually not perfectly flat.)

By the ray-tracing sketch in the figure, we see that after passing through the focal point, the rays diverge to infinity, producing the far-zone field shown. From the results of Section 2.2, in which we demonstrated the equivalence of the far-zone field and the plane wave spectral decomposition (Fourier transform) of a field, we can see that the spectrum of the field on the RHS of the lens is as shown in Fig. 2.7. Thus, the lens has converted a single plane wave field into an entire spectrum of plane waves. Of course, a second lens can also transform the spectrum of plane waves back into a single plane wave component (there is actually a device, known as a *beam expander,* which uses just such a combination of lenses to transform a thin collimated beam from a laser into a larger-diameter, collimated beam for use in an optical system).

In the language of electrical engineering, the lens acts as a nonlinear element, creating additional (spatial) frequencies on output which were not present in the input waveform. We'll examine the implications of these "nonlinear" effects in Chapter 13.

2.5 APPLICATION OF THE STATIONARY PHASE METHOD IN THE ZERO-WAVELENGTH LIMIT

Equation (2.24) shows that the Fraunhofer region field of an aperture distribution is equal to the Fourier transform of the field distribution. This result was obtained by integrating the plane wave spectrum integrals in the limit as r tends to ∞. In the same way, the integral for the Fourier transform may also be evaluated using the stationary phase method, in the limit as k tends to ∞ (wavelength approaches zero). This "zero-wavelength" limit defines the geometrical optics (GO) limit in optics.

The stationary phase integration is accomplished using basically the same process that was used in Section 2.2. The Fourier transform of the aperture distribution is given as

$$E(\alpha, \beta) = \frac{1}{4\pi^2} \iint_{\text{aper}} E(x, y)e^{-j(\alpha x + \beta y)}dx \, dy$$

Say the aperture field has a spherical wave phase shown in Fig. 2.3 and given mathematically by

$$E(\rho, \phi) = E_0(\rho, \phi) \frac{e^{-jkR}}{R} e^{-j(k\rho^2/2R)}$$

where

$$\rho = \sqrt{x^2 + y^2}$$

where E_0 is a slowly varying function over the aperture. If we rewrite the Fourier transform relation in cylindrical coordinates as in Section 2.2, then

$$E(\alpha = k_\rho \cos\phi_k, \beta = k_\rho \sin\phi_k)$$

$$= \frac{1}{4\pi^2} \frac{e^{-jkR}}{R} \iint_{\text{aper}} E_0(\rho, \phi)e^{-j[k_\rho \rho \cos(\phi - \phi_k) + (k\rho^2/2R)]}\rho \, d\rho \, d\phi$$

This time, the stationary phase point is pretty easy to spot. The phi-derivative of the exponent is clearly zero when

$$\phi - \phi_k = 0 \quad \text{or} \quad \pi$$

(only the first condition corresponds to a *minimum* phase condition), and the radial derivative is zero when

$$k_\rho = k \frac{\rho}{R}$$

So, knowing where the stationary phase point of this integral is, we can now evaluate (2.24) in the limit as k approaches infinity. We do this as in Section 2.2 (i.e., we factor out the value of E_0 at the stationary phase point, evaluate the phi-integral in closed form to get a Bessel function, employ the large-argument expression for the Bessel function, approximate the exponential phase by a quadratic function near the stationary phase point, and evaluate the remaining integral from 0 to infinity in the radial direction). The result (which is obtained as in Section 2.2) is

$$E(\alpha = k_\rho \cos\phi_k, \beta = k_\rho \sin\phi_k) = -\frac{j}{2\pi k} E\left(\rho = R\frac{k_\rho}{k}, \phi = \phi_k\right) e^{j(1/2)(k_\rho R)(k_\rho/k)}$$

Combining this last equation with (2.24) yields

$$E_{\text{tan}}(r, \theta, \phi) = \cos\theta \, E_{\text{tan}}\left(\rho = R\frac{k_\rho}{k}, \phi = \phi_k\right) \frac{e^{-jk(r+R)}}{r} e^{j(1/2)(k_\rho R)(k_\rho/k)}$$

This result says that in the zero-wavelength limit, the plane wave field in the direction of α_0, β_0 is due entirely to the aperture field at the stationary phase point. So, in the zero-wavelength limit, the far-zone field in the direction of θ, ϕ is due to the one point in the aperture that lies on the straight-line path connecting from the spherical wave phase center and the far-field point. This straight line is called a *ray,* and this straight-line propagation property of geometrical optics fields gives rise to the so-called ray-tracing method of calculating geometrical optics fields. In the next chapter, we'll study these ray fields in much more detail.

This result also has usefulness in the numerical evaluation of Fourier transforms when the wavelength is small but not infinitesimal. It says that the major contribution to the transform integral is due to those portions of the function that lie near any stationary phase points (should any exist).

The magnitude of the geometrical optics field decreases in a $1/r$ fashion, and the phase of the field is taken from the spherical wave center. This is how fields propagate in the geometrical optics regime, which is in the limit of (i) infinite distance from the source and (ii) zero wavelength.

Figure 2.8 illustrates the aperture diffraction problem in the geometrical optics (GO) limit. Far-field points that lie on line-of-sight paths through the aperture to the source will have a GO field. (The aperture stationary phase point for these far-field points will lie within the aperture.) Far-field points that do not lie on line-of-sight paths through the aperture to the source will not have a GO field (though they will still have diffracted fields). The GO shadow boundary shown in Fig. 2.8 separates those far-field points having aperture stationary phase points from those that do not.

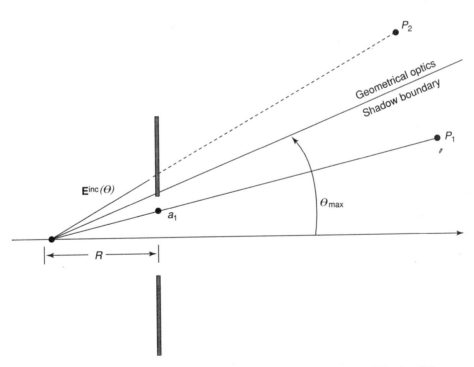

Figure 2.8 Geometrical optics field transmitted through an aperture. Point P_1 will have a GO field since there is a stationary point at the aperture point a_1. Point P_2 will not have a GO field since the stationary point for P_2 lies outside the aperture range of integration.

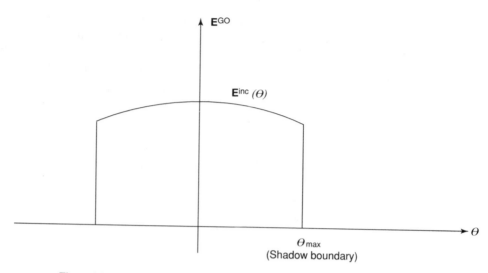

Figure 2.9 Plot of Fraunhofer region GO field showing the discontinuity in the field at the shadow boundary.

Figure 2.9 illustrates the "cookie cutter" effect that the aperture has on the geometrical optics field transmitted through the aperture. At the shadow boundary, there is a discontinuity in the GO field. (Note that the GO field only applies in the far-field/zero-wavelength limit.) In reality, such step discontinuities in the field distribution do not exist because the GO field is only an approximation to the actual field. However, this GO picture often gives a quick, approximate understanding of many types of optics problems. Note that if the aperture in Fig. 2.8 were illuminated by a plane wave (rather than spherical wave) field, there would be no stationary phase points in the aperture and GO could not be used to determine the transmitted fields. In that case, we say that the aperture is *diffraction limited* since diffraction effects account entirely for the far-field pattern. Other diffraction limited fields include the fields in the vicinity of a perfect focal point.

2.6 A PLANE WAVE RESONATOR: THE FABRY-PEROT INTERFEROMETER

The Fabry-Perot interferometer is a device that acts on plane wave fields. Although called an interferometer, it is nothing more than a one-dimensional cavity that is used as an optical bandpass filter, passing plane waves as a function of frequency for a given angle of incidence, or as an angular filter, passing plane waves as a function of angle for a given frequency (see Fig. 2.10).

We can readily find an expression for the filter response of the Fabry-Perot shown in Fig. 2.10. The basic Fabry-Perot consists of two parallel reflecting planes separated by some distance, d. The reflecting planes could consist of partially silvered mirrors (in which case the reflecting planes would be slightly lossy, due to the current flow in the silver); they could consist of very thin, high-epsilon, lossless dielectric films (in which case, the

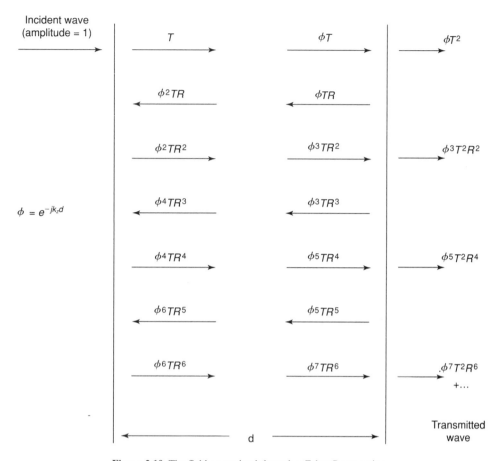

Figure 2.10 The field transmitted through a Fabry-Perot cavity.

reflecting planes would be lossless); or they could simply consist of the interfaces between an external dielectric medium (air) and the optically denser dielectric medium inside the Fabry-Perot. (In this case, the cavity simply consists of a slab of dielectric material.) Our analysis won't be concerned with the actual nature of the reflecting planes; we'll just assume that each plane can be characterized by a reflection coefficient, R, and a transmission coefficient, T, each assumed to be the same at both ends of the cavity.

As shown in Fig. 2.10, the field transmitted by the resonator can be expressed in terms of an infinite geometrical series of the form

$$S = 1 + x + x^2 + x^3 + \ldots$$

where

$$x = (Re^{-jk_z d})^2$$

This series is easily summed using the formula

$$S = \frac{1}{1 - x^2}$$

when

$$|x| < 1$$

Therefore, the total transmission through the Fabry-Perot cavity is

$$T^{\text{Tot}} = \frac{T^2 e^{-jk_z d}}{1 - R^2 e^{-j2k_z d}}$$

and we see that whenever

$$2k_z d = 2n\pi + \pi,$$

where

$$n = 0, 1, 2, \ldots$$

then the transmission equals

$$|T^{\text{Tot}}| = \frac{T^2}{1 + R^2}$$

and the transmission through the cavity is a minimum. On the other hand, when

$$2k_z d = 2n\pi$$

where

$$n = 0, 1, 2, \ldots$$

the transmission through the cavity is a maximum (this is the passband of the cavity), and

$$|T^{\text{Tot}}| = \frac{T^2}{1 - R^2}$$

the exponential term varies as a function of

$$k_z = \sqrt{k^2 - \alpha^2 - \beta^2}$$

where

$$k = \omega\sqrt{\mu\epsilon}$$

so, the exponent can vary either as a function of frequency (ω) or as a function of the inclination angle of the plane wave (α, β). In this way, the Fabry-Perot can either selectively filter waves on the basis of frequency or inclination angle (or both). It is a plane wave filter.

2.7 UNCERTAINTY IN FIELDS AND TRANSFORMS: HEISENBERG'S PRINCIPLE

Most readers probably remember Heisenberg's principle from freshman physics. It states (roughly speaking) that it is impossible to measure both the position and the momentum of a particle with perfect accuracy. If the position is measured with high accuracy, the momentum will be measured with poor accuracy, and conversely. A remarkably similar principle applies for functions and their Fourier transforms.

In the case of optical signals, for example, this duality between the time-domain viewpoint and the Fourier transform domain viewpoint arises all the time, especially in connection with today's new fiberoptic communication systems. For example, a primary goal in fiberoptic cable design is to maximize the number of data bits passing a given transverse plane in a given time. Since the propagation velocity of the cable is fixed, the only way to increase the data rate of the cable is to reduce the time window occupied by any one bit. Bits are sent as optical pulses, so a major goal in fiberoptic cable design is to minimize the width of each pulse that can be transmitted down the fiber. Unfortunately, when the pulse width becomes too short, it can become distorted as it propagates along the fiber (possibly changing shape or even broadening enough to extend outside its designated ''time window'' and into that occupied by an adjacent pulse.

At present, picosecond (10^{-12} second) pulses are typical for many fibers. In order for a fiber to carry picosecond pulse, it needs terahertz (10^{12} cycles per second) bandwidth in the frequency domain. That is, the frequency bandwidth of a pulse is inversely proportional to its width. Stated another way, the narrower the pulse, the broader its Fourier transform spectrum, and conversely. This is why *broadband* optical fibers are so important in modern communications systems.

We can readily see the inverse relation between time domain and frequency domain by noting that the Fourier transform of a rectangular pulse of duration T is given as

$$P(\omega) = \frac{T}{2\pi} \frac{\sin(\omega T/2)}{\omega T/2}$$

$$= \frac{T}{2\pi} \operatorname{sinc}(\omega T/2)$$

which is shown in Fig. 2.11. Clearly, the smaller the pulse width T, the broader the

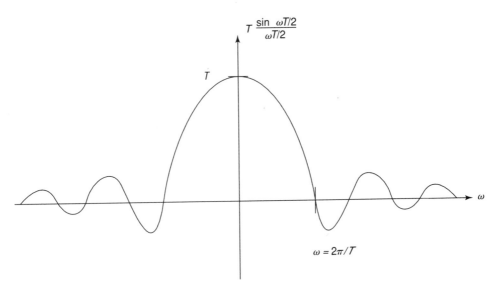

Figure 2.11 Plot of the frequency content (Fourier transform) of a rectangular pulse of duration, T.

sin x/x spectrum is in the frequency domain. This inverse relation between the width of a function and its spectrum is universal for deterministic functions. (It doesn't hold for random noise functions, which are generally considered to have infinite extent in both the time and frequency domains.)

The same Heisenberg principle that relates the time and frequency domains also relates the space and spatial frequency domains. If an optical field is highly concentrated in the spatial domain, it will have a very broad plane wave spectrum (and the field is termed broadband).

One measure of the information content of an optical signal is the time-bandwidth product. A signal with a high time-bandwidth product might consist of a two-hour-long data stream consisting of picosecond bits. A signal with a low time-bandwidth product might consist of a one-minute-long data stream of microsecond bits. The product of the duration of the data stream and the bandwidth of the data stream (or, in a more common-sense view, dividing the total signal duration by the width of each pulse) gives the total number of pulses/bits transmitted, and that is clearly related to the information content of the signal.

In the spatial/spatial frequency domains, the space-bandwidth product is used to measure the complexity of an optical image. A mural covered with microfilm-size print will have a higher space-bandwidth product than a business card printed with headline-size print.

In optical systems, the "space side" of the space-bandwidth product can be increased by using larger aperture lenses. As we've mentioned before, however, the "bandwidth side" is always limited to the visible portion of the plane wave spectrum, and hence can only be increased by decreasing the frequency of the light used. In fact, one current trend for increasing the data storage capacity of optical disks involves the use laser light at the highest possible (blue) frequency [22].

We can also speak about the time-bandwidth product of a single pulse. For example, if we define the bandwidth of the sinc function above as the distance between the first two nulls, then this distance in the ω domain is

$$BW \cdot T/2 = 2\pi$$

or

$$BW = 4\pi/T$$

Since the duration of the pulse in the time domain is T, the product of the time duration of the pulse and the bandwidth of the pulse is

$$T \cdot BW = 4\pi$$

So, according to this definition of the bandwidth of the sinc function, the time-bandwidth product for a single pulse is simply equal to 4π—a constant independent of the pulse duration, T. In other words, the time-bandwidth product is a function of the *shape* of the pulse, not the width of the pulse.

From the foregoing, we see that the concept of the time-bandwidth product is somewhat inexact. While the rectangular pulse function has a well-defined temporal duration, its frequency bandwidth is not nearly so well-defined. Thus, the exact value of the time-bandwidth product depends on how the "bandwidth" of the pulse is defined.

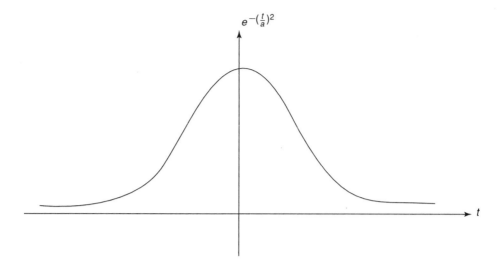

Figure 2.12 Gaussian pulse.

The Gaussian pulse shown in Fig. 2.12 is an ideal pulse function for studying the concept of time-bandwidth product, since the Fourier transform of a Gaussian pulse is again a Gaussian pulse. We can readily calculate the Fourier transform of the Gaussian pulse and thereby calculate its time-bandwidth product.

As it turns out, it is easier to calculate the two-dimensional Fourier transform of a 2-D Gaussian pulse function, so we'll do that calculation first and then show how the 1-D Fourier transform of the 1-D pulse may be obtained from the 2-D result.

So, consider the 2-D Gaussian pulse function of the form

$$e^{-(x^2+y^2)/a^2} = e^{-r^2/a^2}$$

The Fourier transform of this 2-D Gaussian pulse is given by

$$FT(\alpha, \beta) = \frac{1}{4\pi^2} \iint_{-\infty}^{\infty} e^{-(r/a)^2} e^{-j(\alpha x + \beta y)} dx\, dy$$

$$= \frac{1}{4\pi^2} \int_0^{\infty} \int_0^{2\pi} e^{-(r/a)^2} e^{-j\eta r \cos(\gamma - \phi)}\, r\, dr\, d\phi$$

$$= \frac{1}{2\pi} \int_0^{\infty} J_0(\eta r) e^{-r^2/a^2}\, r\, dr$$

where

$$\alpha = \eta \cos\gamma, \qquad \beta = \eta \sin\gamma$$

$$x = r \cos\phi, \qquad y = r \sin\phi$$

We may use a standard integration formula (11.4.29 from [16]) to show finally that

$$FT(\alpha, \beta) = \frac{a^2}{4\pi} e^{-a^2\eta^2/4}$$

We see right away that the width of the Gaussian pulse in the spatial domain is inversely proportional to the width of the pulse in the spectral domain. If we consider the width of the Gaussian pulse to be the width of the pulse at the $1/e$ point, then the width of the Gaussian pulse in the spatial domain is a^2 and the width of the pulse in the spectral domain is $4/a^2$. Thus, the product of the spatial bandwidth and the spectral bandwidth for the 2-D pulse is

$$\text{space-bandwidth product} = 4$$

We may now find the time-bandwidth product of the 1-D Gaussian pulse by setting $\alpha = \beta$, so that

$$\eta^2 = 2\alpha^2$$

In this case,

$$FT_{2D} = \frac{a^2}{4\pi} e^{-a^2\alpha^2/2}$$

Now, noting that

$$FT_{2D} = FT_{1D}^2$$

we obtain

$$FT_{1D} = \frac{a}{2\sqrt{\pi}} e^{-a^2\alpha^2/4}$$

and we see that the space-bandwidth product of the 1-D Gaussian pulse is the same as for the 2-D Gaussian pulse.

REFERENCES

[1] Scott, C. R., *The Spectral Domain Method in Electromagnetics*, Norwood, MA: Artech, 1989.

[2] Harrington, R. F., *Time-Harmonic Electromagnetic Fields*, New York, McGraw-Hill, 1961, Ch. 5.

[3] Ibid., Ch. 6.

[4] Keller, J. B., "Diffraction by an Aperture," *J. Appl. Physics*, vol. 28, pp. 426–444, April 1957.

[5] Kouyoumjian, R. G., "Asymptotic High-Frequency Methods," *Proc. IEEE*, vol. 53, August 1965, pp. 864–876.

[6] Pathak, P. H., and Kouyoumjian, R. G., "The Dyadic Diffraction Coefficient for a Perfectly Conducting Wedge," ElectroScience Lab., Dept. Elec. Eng., Ohio State University, Columbus, Rep. 2183-4, June 5, 1970.

[7] Sommerfeld, A., *Partial Differential Equations in Physics*, New York: Academic Press, 1949.

[8] Itoh, T., and Mittra, R., "Spectral Domain Approach for Calculating the Dispersion Characteristics of Microstrip Lines," *IEEE Trans. Microwave Theory Tech.*, vol. MTT-21, July 1973, pp. 496–499.

[9] Pozar, D. M., and Voda, S. M., "A Rigorous Analysis of a Microstripline Fed Patch

Antenna,'' *IEEE Trans. Antennas Propagat.,* vol. AP-35, no. 12, December 1987, pp. 1343–1349.

[10] Rana, I. E., and Alexopoulos, N. G., "Current Distribution and Input Impedance of Printed Dipoles," *IEEE Trans. Antennas Propagat.,* vol. AP-29, no. 1, January 1981, pp. 99–105.

[11] Snitzer, E., "Cylindrical Dielectric Waveguide Modes," *J. Opt. Soc. Am.,* vol. 51, no. 5, May 1961, pp. 491–498.

[12] Born, M., and Wolf, E., *Principles of Optics,* 6th ed. New York: Pergamon Press, 1980, p. 650.

[13] Andreason, M. G., "Scattering from Bodies of Revolution," *IEEE Trans. Antennas Propagat.,* vol. AP-13, March 1965, pp. 303–310.

[14] Born, M., and Wolf, E., *Principles of Optics,* 6th ed. New York: Pergamon Press, 1980, pp. 419–424.

[15] Moyer, P., and Van Slambrouck, T., "Near-Field Optical Microscopes Break the Diffraction Limit," *Laser Focus World,* October 1993, pp. 105–109.

[16] Abramowitz, M., and Stegun, I. A., *Handbook of Mathematical Functions,* New York, Dover, 1965.

[17] Scott, C. R., *Modern Methods of Reflector Antenna Analysis and Design,* Norwood, MA: Artech House, 1990.

[18] Watson, G. N., *A Treatise on the Theory of Bessel Functions,* Cambridge (UK): Cambridge University Press, 1980.

[19] Magnus, W., and Oberhettinger, F., *Formulas and Theorems for the Special Functions of Mathematical Physics,* New York: Chelsea, 1949.

[20] Erdèlyi, A., et al., *Higher Transcendental Functions,* New York: McGraw-Hill, 1953.

[21] Simmons, G. F., *Differential Equations,* New York: McGraw-Hill, 1972.

[22] Higgins, T. V., Technologies Merge to Create High-Density Data Storage, *Laser Focus World,* August 1993, pp. 57–65.

3

Foundations
of Geometrical Optics

The ray theory of light was obtained in Chapter 2 as an asymptotic far-field/high-frequency limit of Maxwell's equations. In that derivation, the diffraction integral was integrated in the limit as the radial distance tended to infinity and the wavelength tended to zero. Then we evaluated the diffraction integral using the stationary phase method. In this chapter, we'll look more in-depth at the fields that arise in this geometrical optics limit.

3.1 RAY PROPAGATION IN INHOMOGENEOUS MEDIA: THE EIKONAL EQUATION AND FERMAT'S PRINCIPLE

Let's broaden our viewpoint a little from that used in Section 2.3. We'll presume the existence of high-frequency vector fields of the form

$$E = e(r)e^{-jk_0\phi(r)} \tag{3.1}$$

$$H = h(r)e^{-jk_0\phi(r)} \tag{3.2}$$

where

$$k_0 = \omega\sqrt{\epsilon_0\mu_0}, \text{ the free-space propagation constant}$$

and

$$\phi(r) \text{ is some scalar function of position.}$$

This form is slightly more general than the ''one-over-r'' type of amplitude dependence derived in Section 2.3. This broader representation is used in order to allow for the greater variety of ray fields that can arise in inhomogeneous media. One main reason for developing this general ray theory is in fact to enable us to analyze optical fields in inhomogeneous media. The exponents in (3.1) and (3.2) involve the free-space propagation constant; therefore, the spatially dependent part of the phase will involve *relative* dielectric constant values.

We may substitute (3.1) and (3.2) into Maxwell's equations (1.8)–(1.9) to obtain the following equations

$$e \times \nabla \phi(r) = -\sqrt{\epsilon_r}\, \eta_0 h \tag{3.3}$$

$$h \times \nabla \phi(r) = \frac{\sqrt{\epsilon_r}}{\eta_0}\, e \tag{3.4}$$

$$e \cdot \nabla \phi(r) = 0 \tag{3.5}$$

$$h \cdot \nabla \phi(r) = 0 \tag{3.6}$$

where

$$\eta_0 = \sqrt{\frac{\mu_0}{\epsilon_0}} = \text{the free-space impedance}$$

and

$$\epsilon_r = \text{relative dielectric constant}$$

In obtaining these equations, we've used the following chain rule for the gradient of a composite function,

$$\nabla f[g(r)] = f'[g(r)]\nabla g(r) \tag{3.7}$$

where f, g are real functions (f being an ordinary real function of a real variable and g being a real function of a vector variable). In obtaining (3.3) and (3.4), (3.7) was used to produce the equation

$$\nabla e^{-jk_0 \phi(r)} = -jk_0 e^{-jk_0 \phi(r)} \nabla \phi(r)$$

In addition, we've assumed that as k_0 tends to infinity, only those terms that are multiplied by k_0 are significant in the final expressions (3.3)–(3.6).

From (3.5) and (3.6) it's clear that the electric and magnetic field vectors are perpendicular to the phase fronts defined by

$$\phi(r) = \text{const.}$$

similar to the results derived in Section 2.5. So, as in Section 2.5, we may associate the ray direction with the direction of this normal gradient vector. In Section 2.5, wherein we studied geometrical optics in homogeneous dielectric media, the ray paths were determined to be straight lines. In this chapter, we'll study inhomogeneous media (in which the dielectric constant varies as a function of position in the medium) and see how ray trajectories distort to follow curvilinear paths.

Equations (3.3)–(3.6) lead to the fields shown in Fig. 2.5. In the figure, the electric and magnetic field vectors lie in the plane tangent to the constant-phase surface. The vector normal to the surface determines the ray direction.

We may substitute (3.4) into (3.3) to obtain the equation,

$$(h \times \nabla \phi) \times \nabla \phi = -\epsilon_r h \tag{3.8}$$

Using a vector identity, we can rearrange this equation in the form

$$(\nabla \phi \cdot \nabla \phi) = \epsilon_r = n^2 \tag{3.9}$$

where n is the index of refraction of the medium.

Thus,

$$|\nabla \phi|^2 = n^2 \qquad (3.10)$$

that is, the magnitude of the gradient vector is equal to the value of the relative dielectric constant at any given point in the medium. Thus,

$$\nabla \phi = n\hat{\imath} \qquad (3.11)$$

where $\hat{\imath}$ is the unit tangent to the ray path.

In differential geometry, the unit vector tangent to a curve $r(s)$ is given as

$$\hat{\imath} = \frac{dr}{ds} \qquad (3.12)$$

where s is distance, measured along the length of the curve. So, the equation for the ray trajectory is immediately obtained as

$$\frac{dr}{ds} = \frac{1}{n} \nabla \phi \qquad (3.13)$$

We know from vector calculus that a gradient vector field is *conservative;* that is, the line integral of this field from one point to another is independent of the path taken. This is shown in Fig. 3.1. Thus,

$$\int_a^b \nabla \phi \cdot dr = \phi(b) - \phi(a) \qquad (3.14)$$

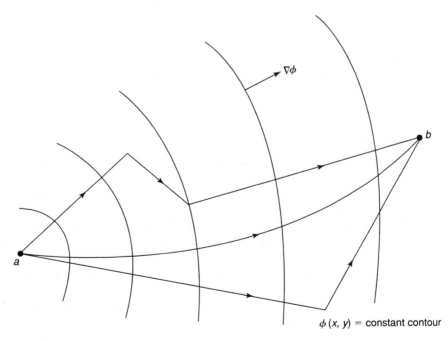

Figure 3.1 Level contours of a function $\phi(x, y)$. Line integrals along all three paths yield $\phi(b) - \phi(a)$.

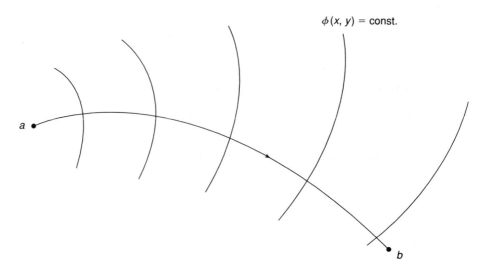

Figure 3.2 A ray trajectory in the field $\phi(x, y)$.

Now let's modify the line integral a little bit. Let's say that both a and b lie at two points along a given ray trajectory and that the path connecting them is a ray path. Notice that we're now placing strict restrictions on these two points (as well as on the path connecting them), as shown in Fig. 3.2. In this case, the two vectors in the dot product in the LHS of (3.14) are always parallel, by (3.13) above. So,

$$\nabla \phi \cdot dr = \frac{1}{n} |\nabla \phi|^2 \, ds = n \, ds \qquad (3.15)$$

Hence, for two points that lie on the same ray trajectory, we can say

$$\int_a^b n \, ds = \phi(b) - \phi(a) \qquad (3.16)$$

where the integral is evaluated along the ray path joining points a, b.

When the two points a, b are connected by a path that is *not* a ray path, the integral on the LHS of (3.16) will be *greater* than $\phi(b) - \phi(a)$. So what we've shown is that the line integral (3.16) is minimized when the two endpoints lie on the same ray path. This is Fermat's principle of minimum optical path length. The integral on the LHS of (3.16) is known as the *optical path length (OPL)* between the two points a, b. We can see that this integral clearly depends on the path chosen because n can never be negative. So the longer the path between a and b, the bigger the integral becomes; the integral becomes monotonically larger as the path becomes longer. The integral is a minimum for two points lying on the same ray trajectory. *Note:* The OPL between the two points is a minimum, and it is also *stationary* at that minimum value. (Once again, the stationary phase concept comes into play, now from a "path length" perspective.) The stationary phase concept is central to so many phenomena in optics, and this is why it was given so much attention in chapter 2.

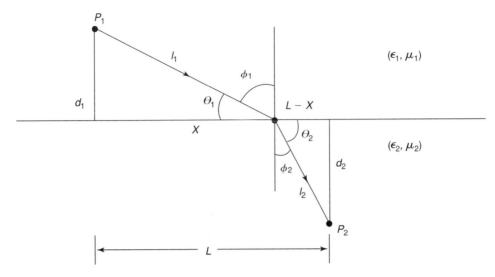

Figure 3.3 On Fermat's principle and phase matching.

EXAMPLE 3.1 PLANE WAVE REFRACTION AT A PLANAR DIELECTRIC INTERFACE

In Example 2.2, we mentioned the importance of matching the phase (as well as amplitude and vector direction) of tangential fields along a dielectric boundary. Fermat's principle gives us another viewpoint for looking at phase matching in planar refraction problems.

With reference to Fig. 3.3, the optical path length between P_1 and P_2 is given by the integral,

$$\phi(x) = \int_{P_1}^{P_2} n \, dl = n_1 l + n_2 l_2$$

$$= n_1 \sqrt{d_1^2 + x^2} + n_2 \sqrt{d_2^2 + (L - x)^2}$$

Setting

$$\frac{d\phi}{dx} = 0$$

yields the equation

$$n_1 \sin\phi_1 = n_2 \sin\phi_2$$

This, of course, is Snell's law. The analysis performed in Example 2.2 is a little more general than this one, since the plane wave spectrum approach there allowed for nonpropagating evanescent waves. (Snell's law can be generalized to allow for evanescent wave fields, by permitting the angle ϕ to take on purely imaginary values; in that case, the trigonometric sine function becomes a hyperbolic sine.) Note that the vector refracted field (in the plane of the interface) may be calculated using the results of Example 2.2.

EXAMPLE 3.2 TOTAL REFLECTION AT A DIELECTRIC BOUNDARY: RAY CONFINEMENT IN A DIELECTRIC SLAB

From the way Fig. 3.3 has been drawn, we've assumed medium 2 to be the optically denser medium, that is,

$$k_2 > k_1$$

(ϵ_1, μ_1)

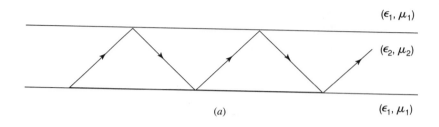

(ϵ_2, μ_2)

(a) (ϵ_1, μ_1)

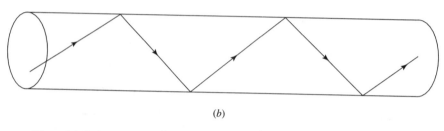

(b)

Figure 3.4 Optical transmission systems based on the principle of total internal reflection:
(a) Slab waveguide showing a propagating (totally reflected) ray and (b) Optical fiber waveguide.

In this case, the ray in medium 2 will always have the steeper angle with respect to the interface. Assuming now ray propagation from point P_2 to point P_1, we may set

$$\sin\phi_1 = 1$$

that is,

$$\phi_1 = \pi/2$$

in Snell's law above, to obtain the value of ϕ_1 for which no field can be transmitted into medium 1. This angle is known as the *critical angle,* and it is important in the design of optical transmission systems that confine and guide light rays, such as the planar dielectric slab and the circular optical fiber, shown in Fig. 3.4. In the case of the slab, all propagating rays in the slab will be tilted at angles greater than the critical angle. *Note:* In the case of the circular fiber waveguide, this ray-based analysis is an oversimplification; the actual fields are obtained by solving a *waveguide eigenvalue problem,* which, for most graded index fibers, is a fairly complex mathematical problem.

EXAMPLE 3.3 RAY PROPAGATION IN A MULTILAYER DIELECTRIC SLAB STRUCTURE

In the slab waveguide discussed in the previous example, the rays follow straight-line paths, as we know they do in homogeneous media. In the five-layer dielectric slab shown in Fig. 3.5, the ray paths, while straight lines in each medium, now take on a more curved appearance. (In this structure, the center slab has the highest dielectric constant.) Say we want the piecewise-linear path of the rays to approximate the sine-wave path shown in Fig. 3.6. How should the index be graded in order to approximate this sinusoidal propagation path?

Well, let the assumed slab be two units high (one unit from center line to the upper edge, as shown in Fig. 3.6) and divided into five equal-height sections, with two and a half sections lying

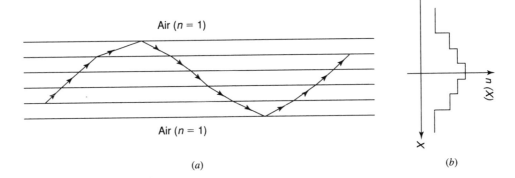

Figure 3.5 Trajectories of a confined ray in a five-layer dielectric slab waveguide having index profile as shown: (a) Ray trajectory and (b) Index profile.

between the center line and the upper edge. We lay out the intended sine-wave ray path on top of the sketch of the slab waveguide. Next, we mark off the intersection points of the intended ray path with the slab interfaces to get the piecewise linear approximation to the continuous sine wave path. Then we compute the various ϕ angles in each slab by taking inverse tangents as shown. Once the ray path is laid out, the index gradient is calculated via repeated application of Snell's law, starting from the outermost interface and working inward toward the center of the slab. We see from the figure that the slab has roughly a quadratic index gradient.

3.2 THE DIFFERENTIAL EQUATION OF LIGHT RAYS

In this section, we continue looking at ray propagation in inhomogeneous media. We'll obtain a differential equation for the ray trajectory as a function of the dielectric constant of the medium. This equation has great utility in the study of GRIN (GRadient INdex) lenses.

To begin, let's write (3.13) in the component-by-component form as (we'll only write out the x-component of the equation)

$$n \frac{d}{ds} x(s) = \frac{\partial \phi}{\partial x}$$

We can differentiate both sides of this equation with respect to distance along the ray to get

$$\frac{d}{ds} \left\{ n(s) \frac{dx}{ds} \right\} = \frac{d}{ds} \left(\frac{\partial \phi}{\partial x} \right)$$

$$= \nabla \left(\frac{\partial \phi}{\partial x} \right) \cdot \hat{t}$$

$$= \nabla \left(\frac{\partial \phi}{\partial x} \right) \cdot \frac{\partial r}{\partial s}$$

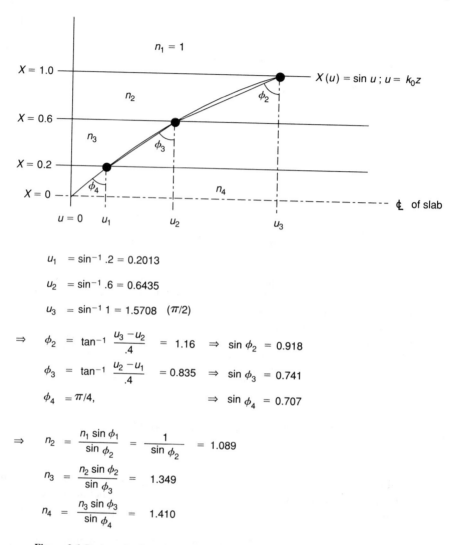

$$u_1 = \sin^{-1} .2 = 0.2013$$

$$u_2 = \sin^{-1} .6 = 0.6435$$

$$u_3 = \sin^{-1} 1 = 1.5708 \quad (\pi/2)$$

$$\Rightarrow \quad \phi_2 = \tan^{-1} \frac{u_3 - u_2}{.4} = 1.16 \quad \Rightarrow \quad \sin \phi_2 = 0.918$$

$$\phi_3 = \tan^{-1} \frac{u_2 - u_1}{.4} = 0.835 \quad \Rightarrow \quad \sin \phi_3 = 0.741$$

$$\phi_4 = \pi/4, \qquad\qquad \Rightarrow \quad \sin \phi_4 = 0.707$$

$$\Rightarrow \quad n_2 = \frac{n_1 \sin \phi_1}{\sin \phi_2} = \frac{1}{\sin \phi_2} = 1.089$$

$$n_3 = \frac{n_2 \sin \phi_2}{\sin \phi_3} = 1.349$$

$$n_4 = \frac{n_3 \sin \phi_3}{\sin \phi_4} = 1.410$$

Figure 3.6 Design of a five-layer graded index slab for a nearly sinusoidal propagation path.

$$= \frac{1}{n} \nabla \left(\frac{\partial \phi}{\partial x} \right) \cdot \nabla \phi$$

$$= \frac{1}{2n} \frac{\partial}{\partial x} (\nabla \phi \cdot \nabla \phi)$$

$$= \frac{1}{2n} \frac{\partial}{\partial x} n^2 = \frac{\partial n}{\partial x}$$

So, combining the other two equations for the y-, z-components of the equation, we obtain the *differential equation for light rays* as

$$\frac{d}{ds}\left(n\frac{dr}{ds}\right) = \nabla n \qquad (3.17)$$

As mentioned earlier, this equation is very important in calculating the fields in GRIN rod lenses. This equation directly relates various arc-length derivatives of the position vector to the index of refraction of the medium. It's important to note that all three spatial variables x, y, z in (3.17) are functions of the single independent parameter s, the arc length along the curve. So even though the index n is not directly a function of the arc length s, it is still a function of s since n is a function of x, y, z, which in turn are functions of s, the distance along the curve. Thus, when evaluating the derivative of n with respect to s, we'd use the chain rule, that is,

$$\frac{d}{ds}n(x) = \frac{d}{dx}n(x)\frac{dx}{ds}$$

If the index, n, is a function of both transverse variables (x, y), then,

$$\frac{d}{ds}n(x, y) = \frac{\partial}{\partial x}n(x, y)\frac{dx}{ds} + \frac{\partial}{\partial y}n(x, y)\frac{dy}{dz}$$

Since x, y, z are only functions of the single variable s, the derivatives of x, y, z with respect to s are ordinary derivatives.

To solve (3.17) for an arbitrary inhomogeneous medium, one would assume initial values for the quantities:

$$r(s = 0) \qquad \text{(initial position)}$$

and

$$\frac{dr}{ds}(s = 0) = \hat{\imath}(s = 0) \qquad \text{(initial unit tangent vector)}$$

Then, using (3.18) below, the curvature vector

$$\frac{d^2r}{ds^2}(s = 0)$$

can be calculated, allowing the position vector and unit tangent vector to be calculated at some incremental distance Δs along the ray path. In this way, the solution for the ray path may be obtained.

Note. For the purpose of performing this calculation, it may be easier to expand the derivative on the LHS of (3.17) using the product rule. This gives the equation,

$$n\frac{d^2r}{ds^2} = \nabla n - \frac{dn}{ds}\frac{dr}{ds} \qquad (3.18)$$

3.3 RAY PROPAGATION IN A GRADED INDEX SLAB

In Example 3.3, we saw how a continuously graded slab can be approximated by a discretely graded slab. It's also possible to analyze the continuous case. In fact, any kind of discrete numerical solution to the differential equation essentially amounts to approximating the continuous medium with a finely graded, discrete medium.

An often-used solution to (3.17), (3.18) for slabs of the type shown in Figs. 3.5 and 3.6 involves an assumption that the ray trajectories never tilt too far from the center line of the slab. This "small-angle" assumption allows the arc length derivative to be replaced by a derivative with respect to z. Thus, (3.17) becomes

$$\frac{d}{dz}\left\{n(x)\frac{d}{dz}[x(z)\hat{x} + z\hat{z}]\right\} = \frac{\partial n}{\partial x}\hat{x} \tag{3.19}$$

In this case, we've replaced the vector

$$r(s) = x(s)\hat{x} + y(s)\hat{y} + z(s)\hat{z}$$

by

$$r(z) = x(z)\hat{x} + z\hat{z}$$

Now, the longitudinal distance, z, is the single independent variable in the equation. Evaluating the derivatives in (3.19) yields the equation

$$\frac{dn}{dx}\cdot\frac{dx}{dz}\left(\frac{dx}{dz}\hat{x} + \hat{z}\right) + n\frac{d^2x}{dz^2}\hat{x} = \frac{\partial n}{\partial x}\hat{x} \tag{3.20}$$

Equating x-, z-components of this equation gives

$$n\frac{d^2x}{dz^2} - \frac{\partial n}{\partial x} = 0 \tag{3.21}$$

If, motivated by the results of Example 3.3, we choose a quadratic index distribution in the slab of the form

$$n(x) = n_0\left[1 - \left(\frac{x}{x_0}\right)^2\right] \tag{3.22}$$

then the differential equation for the light rays becomes

$$\frac{d^2x}{dz^2} + \frac{2}{x_0^2}x = 0 \tag{3.23}$$

(where we've neglected the quadratic term in the expression for n, multiplying the second derivative of x with respect to z). This equation is clearly the differential equation for the sine function

$$x(z) = \sin\sqrt{2}\frac{z}{x_0} \tag{3.24}$$

This equation says, then, that in the *paraxial limit* of small tilt angles, the ray trajectory is a sine (or cosine, or combination) function. Since the ray trajectories continually pass through the axis of the slab, they produce real images periodically along the length of the slab. Circular cylinder versions of our graded index slab are called graded index (GRIN) lenses and may be used for imaging, exactly the same as ordinary lenses. The GRIN lens, however, is somewhat unique in the sense that it is a type of image transmission system. We'll see in Chapter 5 how periodic systems of ordinary lenses may also be used to transmit images.

There are some very good descriptions of techniques for tracing rays through GRIN media, and these are described in the literature [1–7].

3.4 THE THEOREM OF MALUS

The theorem of Malus probably doesn't need an entire section in order to be described; the fact is, it can be stated immediately. However, the concept is very significant, and the reader should be aware of it. The theorem states that a family of rays remains perpendicular to its associated wavefront after any number of reflections or refractions. A proof of this theorem may be found in Born and Wolf [8]. However, the alert reader will recall that we initially defined rays as being the vectors normal to the phase fronts of an optical wave. So naturally, these vectors are always going to remain perpendicular to the phase fronts, since that's how we defined them in the first place.

The theorem of Malus, though seemingly trivial, has some important and useful consequences. The most important involves the design of coherent optical systems in which the optical phase is important—for example, in synthesizing a planar phase front. The theorem of Malus says that you don't have to trace thousands of rays through a complicated optical system to make sure they all have the same path length to the plane of interest where you want to produce a planar phase front. All you have to do is make sure all the rays are parallel to each other when they finally do reach the plane of interest (and are perpendicular to the plane itself). As long as you ensure this "parallel ray" condition, all the rays will automatically be in phase. Without a doubt, it is much easier to design an optical system to a parallel ray condition than to compute the lengths of numerous paths and then try to set them all to some constant value.

REFERENCES

[1] Montagnino, L., "Ray Tracing in Inhomogeneous Media," *J. Opt. Soc. Am.,* vol. 58, no. 12, December 1968, pp. 1667–1668.

[2] Marchand, E. W., "Ray Tracing in Gradient-Index Media," *J. Opt. Soc. Am.,* vol. 60, no. 1, January 1970, pp. 1–7.

[3] Kapron, F. P., "Geometrical Optics of Parabolic Index-Gradient Cylindrical Lenses," *J. Opt. Soc. Am.,* vol. 60, no. 11, November 1970, pp. 1433–1436.

[4] Marchand, E. W., "Ray Tracing in Cylindrical Gradient-Index Media," *Applied Optics,* vol. 11, no. 5, May 1972, pp. 1104–1106.

[5] Moore, D. T., "Ray Tracing in Gradient Index Media," *J. Opt. Soc. Am.,* vol. 65, no. 4, April 1975, pp. 451–455.

[6] Marchand, E. W., "Third Order Aberrations of the Photographic Wood Lens," *J. Opt. Soc. Am.,* vol. 66, no. 12, December 1976, pp. 1326–1330.

[7] Sharma, A., Vizia Kumar, D., and Ghatak, A. K., "Tracing Rays Through Graded-Index Media: A New Method," *Applied Optics,* vol. 21, no. 6, March 1982, pp. 984–987.

[8] Born, M. and Wolf, E., *Principles of Optics,* 6th ed. New York: Pergamon Press, 1980.

PART II
LENS ACTION FROM THREE VIEWPOINTS

4

Focusing and Imaging Properties of Lenses: Ray Optical Viewpoint

So far, the properties of optical fields have been studied from the diffraction integral viewpoint, the plane wave spectrum viewpoint, and the geometrical optics viewpoint. Now, some of this knowledge of optical fields will be applied to various practical optical imaging instruments. In this chapter, we'll look at the properties of lens systems from the geometrical optics perspective. The point here is not to present a laundry list of lens types and lens analysis techniques, but rather to present the concepts surrounding lens action in a coherent and (hopefully) interesting way.

4.1 GENERAL PROPERTIES OF OPTICAL IMAGING INSTRUMENTS: MAXWELL'S THEOREM AND THE SINE AND COSINE CONDITIONS

Whenever optical fields are used for the purpose of creating an image of some object scene, there are certain fundamental physical limitations on what can be done in terms of magnification, image fidelity, and so on. Optical imaging is considerably different from electronic imaging, in which images can be manipulated at will via software, to create ''morphed'' or colorized images. In optics, Maxwell's equations place restrictions on the image manipulation that can be done, and some of these restrictions are described in this section.

There is a well-known theorem in optics—known as Maxwell's theorem—which places the most basic limits on image fidelity. It states in effect that the image must be an imperfect replica of the object whenever the magnification of the imaging instrument is different from unity. Even though this chapter deals with ray fields and their properties, we'll prove this theorem using the plane wave theory from Chapter 2.

71

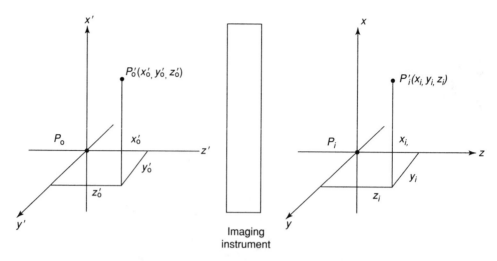

Figure 4.1 On Maxwell's theorem: $x_i = x'_0$, $y_i = y'_0$, $z_i = z'_0$

Consider Fig. 4.1, which illustrates an optical imaging instrument, along with its associated object and image spaces. Say that the point P_0 in the object space is imaged perfectly onto the point P_i in the image space. Now, since P_0 is a perfect point, it will have a Fourier transform representation of the form (neglecting a factor of $4\pi^2$)

$$\delta_0(x', y') = \iint_{-\infty}^{\infty} e^{j(\alpha x' + \beta y')}\, d\alpha\, d\beta \tag{4.1}$$

In order for the image of point P_0 at P_i to be perfect, it must have exactly the same plane wave spectrum expansion as the source distribution, namely,

$$\delta_i(x, y) = \iint_{-\infty}^{\infty} e^{j(\alpha x + \beta y)}\, d\alpha\, d\beta \tag{4.2}$$

In other words, by some incredible process, all of the plane waves (propagating and evanescent) in the expansion for the object point have to pass through the imaging system and reach the plane containing P_i having exactly the same amplitude and phase relationship (i.e., uniform amplitude and phase) that they had back in the object plane containing P_0.

Now, let's look at the image of the object point P'_0. As in the case of P_0, the point field at P'_0, is represented in terms of a plane wave spectrum in the $z' = z_0$ plane, of the form given by (4.1), namely, via a spectrum having uniform amplitude and phase. We can propagate this plane wave spectrum back to the $z' = 0$ plane to get the following representation for the object point P'_0 at $z' = 0$

$$P'_0(x' - x'_0, y' - y'_0, z' = 0) = \iint_{-\infty}^{\infty} e^{j\sqrt{k^2 - \alpha^2 - \beta^2}z'_0} e^{j[\alpha(x' - x'_0) + \beta(y' - y'_0)]}\, d\alpha\, d\beta \tag{4.3}$$

From what we've already seen in connection with the imaging of point P_0, however, we know that the field in the $z' = 0$ plane in the object space will be reconstituted in the plane containing P_i, with exactly the same spectrum that it had in the $z' = 0$ plane. So what we can do then is to propagate the plane wave spectrum for P'_0 from the $z = z'_0$ plane, backwards to the $z' = 0$ plane, then forward through the optical system to the point

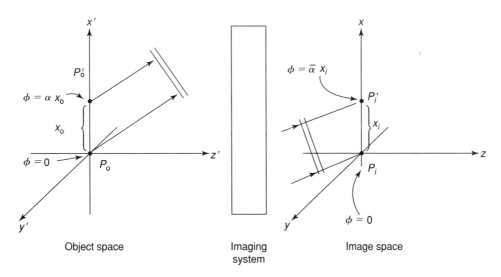

Figure 4.2 On the Abbe sine condition.

P_i, where it once again assumes the spectrum it had in the $z' = 0$ plane, which is given by (4.3). From there, we simply propagate this spectrum forward to the point P_i', where the perfect uniform amplitude and phase spectrum are recovered, and the delta function produced at $x_i = x_0'$, $y_i = y_0'$, $z_i = z_0'$. So, we see that perfect (optical) imaging is possible only when the imaging system exactly replicates the object, that is, with no magnification. Imaging for which a point in the object plane is transformed into a corresponding perfect point in the image plane is called *stigmatic*.

By Maxwell's theorem, it's impossible to get stigmatic imaging everywhere in the object space unless that imaging involves no magnification. In real optical systems (possibly containing magnification), there are two conditions, known as the (Abbe) sine and (Herschel) cosine conditions, which set criteria for approximate stigmatic imaging in the transverse and longitudinal directions, respectively. These conditions determine the design of an optical system for stigmatic imaging. (*Note:* The Abbe condition is generally satisfied to a first order whenever the lens law is satisfied.)

With reference to Fig. 4.2, suppose that the imaging system images point P_0 perfectly onto point P_i. How shall the imaging system be designed so that P_i' is a perfect image of point P_0'? Well, say that both object points P_0, P_0' in the $z' = 0$ plane have uniform amplitude and phase spectral expansions, as shown in (4.1). Thus, for points P_0, P_0':

$$\delta_{P_0}(x', y') = \iint_{-\infty}^{\infty} e^{j(\alpha x' + \beta y')} \, d\alpha \, d\beta \qquad (4.4)$$

$$\delta_{P_0'}(x', y') = \iint_{-\infty}^{\infty} e^{j[\alpha(x' - x_0) + \beta y']} \, d\alpha \, d\beta \qquad (4.5)$$

and we presume (by the assumption of perfect imaging) that

$$\delta_{P_i}(x, y) = \iint_{-\infty}^{\infty} e^{j(\bar{\alpha} x + \bar{\beta} y)} \, d\bar{\alpha} \, d\bar{\beta} \qquad (4.6)$$

$$\delta_{P_i'}(x, y) = \iint_{-\infty}^{\infty} e^{j[\bar{\alpha}(x - x_i) + \bar{\beta} y]} \, d\bar{\alpha} \, d\bar{\beta} \qquad (4.7)$$

Different notations have been used for α, β in the object and image spaces. This is to allow for the fact that α, β in the object and image spaces may not be the same, as shown in Fig. 4.2.

Now, by comparing (4.5) and (4.7), we see that perfect imaging at P_0' is obtained when

$$\alpha x_0 = \overline{\alpha} x_i \qquad (4.8a)$$

or

$$k_0 x_0 \sin \theta_0 = k_i x_i \sin \theta_i \qquad (4.8b)$$

This is the well-known *Abbe sine condition* required for stigmatic imaging in the transverse plane. Now, since this condition (in principle) holds for all plane waves, this condition gives the *transformation relation* between α and $\overline{\alpha}$, the spectral wavenumbers in the object and image planes.

The sine condition is not particularly profound, as it's sometimes made out to be. In fact, it's nothing more than a restatement of Heisenberg's principle, that is, the broader/narrower a function is in the spatial domain, the narrower/broader it is in the spectral domain. In the case of pure magnification, the function doesn't change shape, only its scaling. The spatial and spectral domain scaling takes place in such a way that the products

$$\alpha x, \ \beta y$$

remain constant.

The cosine condition is similar to the sine condition and may be derived in exactly the same way. It states that longitudinal object points are imaged stigmatically when the condition

$$\sqrt{k_0^2 - \alpha^2 - \beta^2} \ z_0 = \sqrt{k_i^2 - \overline{\alpha}^2 - \beta^2} \ z_i \qquad (4.9)$$

is met. Since this equation gives a different transformation relation between object and image plane wavenumbers from what the sine condition gives, it is clear that the sine and cosine conditions cannot be simultaneously met (unless, of course, the imaging system produces unit magnification, which is in accordance with Maxwell's theorem).

4.2 FOCUSING AND IMAGING BY CURVED DIELECTRIC INTERFACES

The simplist possible type of focusing device is the curved interface between two dielectric media. Such an imaging device is generally of little practical use (at least for forming real images) since it would be impossible to access an image formed within a solid dielectric medium. On the other hand, such a focusing surface is useful in studying the focusing action of lenses, since a lens is nothing more than two of these dielectric boundary interfaces placed back to back.

In Example 2.2, we looked at the problem of plane wave reflection and transmission at a planar dielectric boundary. In that example, we noted that the three quantities that must always be matched (when no surface currents flow on the boundary) are the phase,

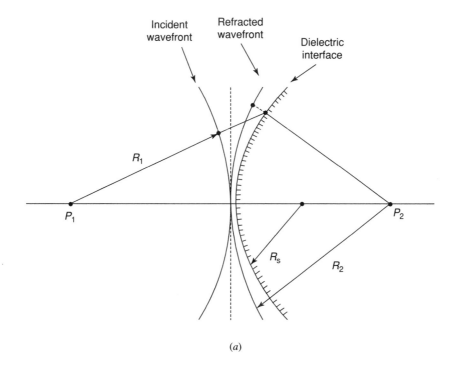

(a)

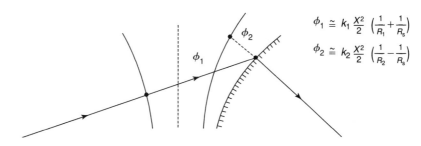

$$\phi_1 \cong k_1 \frac{x^2}{2} \left(\frac{1}{R_1} + \frac{1}{R_s} \right)$$

$$\phi_2 \cong k_2 \frac{x^2}{2} \left(\frac{1}{R_2} - \frac{1}{R_s} \right)$$

Figure 4.3 Refraction at a curved dielectric interface: (a) Overall view and (b) Close-up view.

amplitude, and vector direction of transverse fields. In that example, matching the phase was almost trivial, since it merely consisted of equating the wavenumbers α, β of all fields at the planar interface. In the context of curved dielectric interfaces, the phase matching problem is somewhat more complicated.

We can see the effect of a quadratic surface on a quadratic incident field, with the aid of Fig. 4.3. We'll look at what happens when an axial point source radiates a field that is incident on a curved surface. Note that the problem of a nonaxial source is a little more complicated mathematically [1,2], so we won't go into that solution here.

For the purpose of matching the incident field to the refracted field, we're going to use Fermat's principle from Chapter 3 to note that the optical phase from P_1 to P_2 is the same for all ray paths. In particular, choosing the path along the z-axis, we have

$$\phi = k_1 R_1 + k_2 R_2 \qquad (4.10)$$

Now let's look at the tilted ray path. Along this path, the phase (as shown in Fig. 4.3) is

$$\phi \cong k_1 R_1 + k_1 \frac{x^2}{2}\left(\frac{1}{R_1} + \frac{1}{R_s}\right) + k_2 R_2 - k_2 \frac{x^2}{2}\left(\frac{1}{R_2} - \frac{1}{R_s}\right) \tag{4.11}$$

Equating this phase to the phase along the axial path yields the equation relating the incident wavefront curvature, the surface curvature, and the refracted wavefront curvature, given as

$$\frac{k_1}{R_1} + \frac{k_2}{R_2} = \frac{k_2 - k_1}{R_s} \tag{4.12}$$

or, in terms of indices of refraction,

$$\frac{n_1}{R_1} + \frac{n_2}{R_2} = \frac{n_2 - n_1}{R_s} \tag{4.13}$$

where

$$n = \sqrt{\mu\epsilon}$$

So we see that, indeed, a simple curved dielectric boundary can act—to a first order—as a stigmatic imaging surface. Note that the refracted field vector (tangent to the plane of the interface) may be obtained from the planar analysis of Example 2.2.

We may also look at imaging by curved surfaces from a "phase-matching" point of view. As we'll see in Chapter 5, this viewpoint is somewhat more general than the ray viewpoint, since rays are a far-field/zero-wavelength construct, whereas phase fronts can apply to any optical or electromagnetic field. The alert reader may note that we *defined* rays to be the normal vectors to the constant-phase fronts; therefore, the two concepts (ray versus phase front) should be one and the same. However, we made this definition only in the context of the far-field/zero-wavelength limit. As we'll show in Chapter 5, extending this definition outside the specified limiting regime leads to the concept of *curved* ray propagation (as in the case of Gaussian beam propagation), which is nonphysical in homogeneous media.

The method of phase matching is much more general than what we'll be presenting in this section. Here, we'll use phase matching to analyze an entire lens, but the concept may be used to analyze the refractive and reflective properties of locally quadratic patches on lens surfaces, to produce highly accurate analyses of the fields reflected and transmitted by dielectric lenses [2].

So, with reference to Fig. 4.4, we'll look at the phase on the front face of the dielectric interface due to a point source radiator as shown, at $z = 0$. The incident field phase is given by

$$\phi_{\text{inc}} = e^{-jk_1 r} = e^{-jk_1\sqrt{z^2 + \rho^2}} = e^{-jk_1 z\sqrt{1 + (\rho/z)^2}} \cong e^{-jk_1 z[1 + 1/2(\rho/z)^2]} \tag{4.14}$$

Now, the surface, S_1, is defined by the equation

$$z = d_0 + \frac{\rho^2}{2R_{S,1}} \tag{4.15}$$

so, plugging (4.15) into (4.14) yields the phase *over the surface* S_1, due to the point source. This is given as

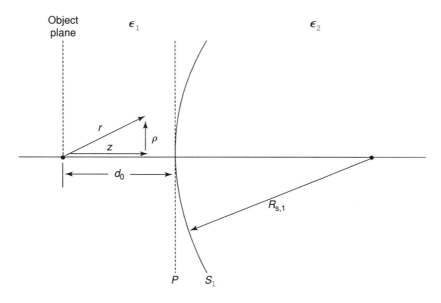

Figure 4.4 Wavefront focusing by a curved dielectric interface.

$$\phi_S = e^{-jk_1 d_0} \, e^{-jk_1 (\rho^2/2)(1/R_{S,1} + 1/d_0)} \tag{4.16}$$

NOTE: We're only going to be worried about the part of this expression that varies as the radial distance, ρ; the constant-phase term won't affect our results. Now we're going to track this field back to the plane P, assuming for the moment that the plane P lies inside medium 2. We'll trace the fields backwards, assuming a paraxial field approximation, in which the fields are propagating substantially parallel to the z-axis, along with an approximation that the curvature of the interface surface is large. With these considerations in mind, we obtain the field *on the plane P* as

$$\phi_P = e^{jk_2(\rho^2/2)[(1/R_{S,1})(k_2 - k_1)/k_2 - k_1/(k_2 d_0)]} \tag{4.17}$$

This represents a field in region 2 having curvature

$$\frac{1}{R_2} = \frac{1}{R_{S,1}} \left(\frac{k_2 - k_1}{k_2} \right) - \frac{k_1}{k_2 d_0} \tag{4.18}$$

When R_2 is positive, the wave converges, and when R_2 is negative, the wave diverges in medium 2. We see from the previous expression that as the dielectric constant (refractive index) of medium 2 increases (and as the radius of curvature of the surface decreases), the field in region 2 eventually becomes convergent.

4.3 FOCUSING AND IMAGING BY THIN LENSES

We may continue with the phase-matching analysis begun in the previous section and obtain an expression for the field transmitted through a thin lens. All we have to do is place a second curved dielectric interface in the path of the optical field already calculated

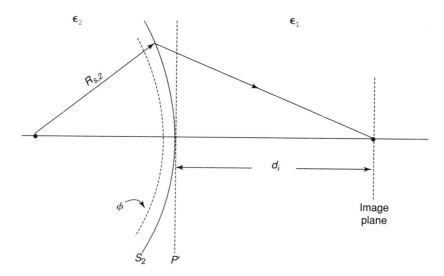

Figure 4.5 On the analysis of a thin lens.

in the previous section. First, we want to calculate the phase along the inside portion of the interface, shown by dashed lines in Fig. 4.5. Once again, assuming the paraxial approximation, the phase *on the inside surface of* S_2 is given as

$$\phi_S = e^{jk_2(\rho^2/2R_2)}\, e^{jk_2(\rho^2/2R_{S,2})} \tag{4.19}$$

and on the plane P' just outside the lens,

$$\phi_{P'} = \phi_S\, e^{-jk_1(\rho^2/2R_{S,2})} = e^{jk_1(\rho^2/2)[(k_2/k_1)(1/R_2 + 1/R_{S,2}) - (1/R_{S,2})]} \tag{4.20}$$

and from this, we readily obtain the image distance from the quantity in brackets. That is,

$$\frac{1}{d_i} = \frac{k_2}{k_1}\left(\frac{1}{R_2} + \frac{1}{R_{S,2}}\right) - \frac{1}{R_{S,2}} \tag{4.21}$$

In the terminology of classical optics, the object and image points of an optical system are often referred to as *conjugate points*. Substituting (4.18) into (4.21), we obtain the lens law as

$$\frac{1}{d_o} + \frac{1}{d_i} = \frac{1}{f}$$

where

$$\frac{1}{f} = \left(\frac{k_2 - k_1}{k_1}\right)\frac{1}{R_{S,1}} + \left(\frac{k_2 - k_1}{k_1}\right)\frac{1}{R_{S,2}} \tag{4.22}$$

This last equation gives the focal length as a function of the lens dielectric constant and the two radii of curvature.

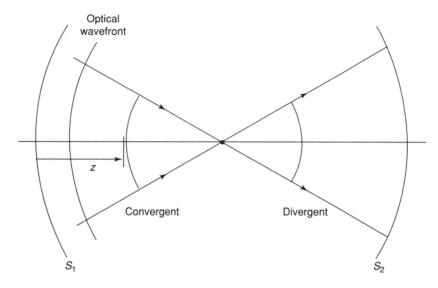

Figure 4.6 Variation of optical phase front inside a thick lens.

4.4 FOCUSING AND IMAGING BY THICK LENSES

Just a simple change in the formulas allows the previous expressions to be generalized to thick lenses. In the last section, we assumed the lenses to be so thin and to have such large radii of curvature that the radii of curvature of all wavefronts were everywhere the same along both faces of the lens. When the lens is thicker, the curvature of the wavefront inside the lens changes from the front face to the back. This is shown in Fig. 4.6.

From the figure, it's evident that the curvature of the wave inside the lens varies as a function of distance from the front face. The wave may even pass through a focal point, as shown, and begin to diverge again before it encounters the back surface of the lens. So, the phase of the wave a distance z from the front face is

$$\phi = e^{jk_2(\rho^2/2)[1/(R_2-z)]} \tag{4.23}$$

Note that when $z > R_2$, the sign of the denominator in the exponent is negative and the wave is divergent.

In any event, the focal length of the thick lens may be found as in the case of the thin lens above, except that we now replace R_2 by $R_2 - z$.

4.5 RAY-TRACING ANALYSIS OF THIN LENSES

The analyses we've given up until now have been somewhat nonstandard in comparison with the traditional (ray-tracing) way of looking at imaging by lenses. The above viewpoint of a lens as a device that transforms the curvature of an optical wavefront has been given for two reasons. First, even though our discussion is not accurate for analyzing entire lenses, it is very accurate for analyzing quadratic patches of lenses. Using the method

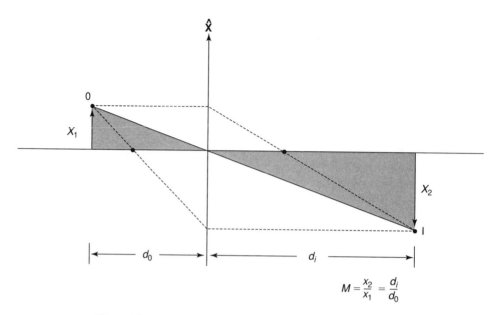

$$M = \frac{x_2}{x_1} = \frac{d_i}{d_0}$$

Figure 4.7 Derivation of the magnification law using similar triangles.

given above, we may accurately analyze a whole lens patch-by-patch. Second, if care is not taken, it is easy to become confused by a ray-tracing analysis. The tilt of a ray gives an indication of ray divergence only when the ray is considered in connection with its near neighbors (the other rays in the pencil). A "wavefront-based" analysis of the type presented in the preceding sections gives pencil divergence information directly, whereas a ray-tracing approach doesn't always do that.

In this section, we'll rejoin the rest of the optics world and present a ray-based analysis of thin lenses. With the wavefront background behind us, the ray approach will hopefully be more meaningful.

Let's first look at imaging by a thin lens, as shown in Fig. 4.7. The object and image planes—and lens focal planes—are as shown. The ray-tracing method is based on three simple postulates. (1) A ray passing through the front focal point is bent parallel to the symmetry axis of the lens. (2) An incident ray parallel to the lens axis is bent so as to pass through the back focal point of the lens. (3) A ray passing through the center of the lens is unchanged in direction.

These postulates do not tell us anything about the divergence of any of the ray bundles. Their main benefit is in enabling us to geometrically locate the image points of an imaging system. For example, in Fig. 4.7, the three rays shown emanating from the object point O all converge—according to our rules—on the same image point I. The location of the image point is determined using the simple geometric relations shown in Fig. 4.7. (By the way, this geometrical mapping from the object space to the image space defines an affine function mapping [3]).

By similar triangles, the magnification of the lens is easily seen to be

$$M = \frac{d_i}{d_0} \qquad (4.24)$$

This is a simple relation depending only on the object and image distances.

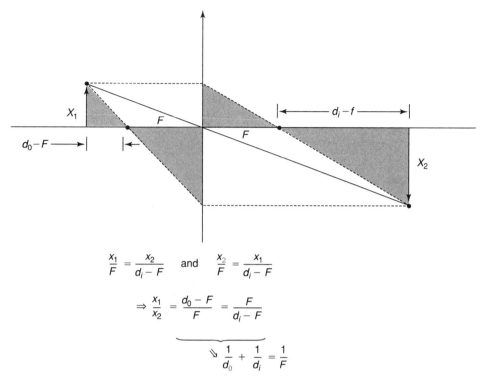

$$\frac{x_1}{F} = \frac{x_2}{d_i - F} \quad \text{and} \quad \frac{x_2}{F} = \frac{x_1}{d_i - F}$$

$$\Rightarrow \frac{x_1}{x_2} = \frac{d_0 - F}{F} = \frac{F}{d_i - F}$$

$$\searrow \frac{1}{d_0} + \frac{1}{d_i} = \frac{1}{F}$$

Figure 4.8 Derivation of the lens law, $1/d_0 + 1/d_i = 1/f$, using similar triangles.

Using a similar argument based on similar triangles, we may derive the lens law as in Fig. 4.8.

4.6 CAUSTICS AND FOCAL SURFACES

A focal point of an ideal lens is defined in the following way. If the lens is illuminated by a plane wave arising from a point source at infinity, the focal point is that point common to every ray transmitted by the lens. All transmitted rays pass through the focal point. Since the lens aperture is a two-dimensional area function, a doubly infinite number of rays will pass through the focal point (an infinitude of rays in each linear direction). And the focal point has a dimensionality of zero; it is a point (as opposed to a line, a surface, or a volume).

A caustic is similar to a focal point, but instead of having zero dimensionality, a caustic can extend in one or two linear directions. That is, instead of being a point, it is a curve or surface in space. And instead of arising from rays passing through a two-dimensional transverse area, a caustic can arise from a one-dimensional or zero-dimensional set of rays, and the intersection of those rays with the lens forms a one-dimensional curve or a zero-dimensional point in the plane of the lens.

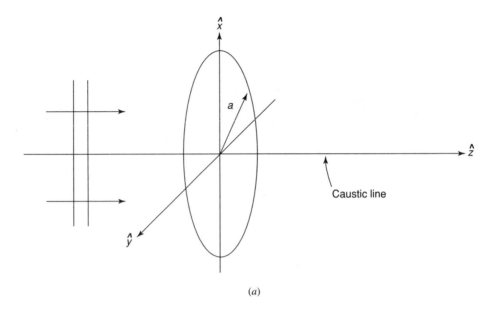

(a)

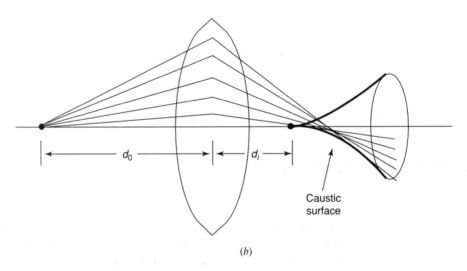

(b)

Figure 4.9 Examples of caustics: (a) Circular aperture under plane wave illumination and
(b) Imperfect lens under point source illumination.

We can look at these ideas in more detail with the help of Fig. 4.9. In Fig. 4.9a, a circular aperture is illuminated by an axially incident plane wave. Now, for points along the z-axis, every point on the rim of the aperture will give rise to an edge-diffracted field. So, in the case of the unfocused aperture, the z-axis is a caustic line, and the circular rim of the aperture is the locus of points (i.e., the one-dimensional curve in the aperture plane) which gives rise to the caustic curve.

A caustic can also be a surface, as shown in Fig. 4.9b. When a lens is imperfect and has no well-defined focal point through which all rays pass, the focal *point* can turn into a caustic *surface,* a cross section of which is shown in Fig. 4.9*b*. If the lens is assumed to be rotationally symmetric, the caustic surface will also be symmetric, as shown. This caustic surface is topologically the inverse of the situation for a perfect lens. In the case of a perfect lens, all rays passing through the transverse *area* of the lens converge to a single *point.* In this case, the ray passing through each *point* in the transverse plane of the lens defines the tangent to the 2-D caustic *surface.*

We've looked at the kinds of caustics that can arise from rotationally symmetric lenses/apertures which are illuminated by rotationally symmetric fields. Now we'll look at the caustic surfaces that arise when either the lens/aperture or the incident field is not rotationally symmetric. For example, say the incident field has two different radii of curvature in the two orthogonal transverse planes, as shown in Fig. 4.10. This field is termed *astigmatic.* This field is only slightly more general mathematically than the symmetric fields we've worked with up until now, its form being given by the equation

$$E(x, y, z) = e^{-jkz} \frac{e^{-jk(x^2/R_1 + y^2/R_2)}}{\sqrt{(z - R_1)(z - R_2)}} \tag{4.25}$$

This field may be made incident on a spherical lens, and, using the mathematical methods of the previous sections, it may be shown to focus on one focal line in the x-direction and on a second focal line in the y-direction. When $R_1 = R_2$, the two focal lines coalesce to a single focal point. These two focal lines are again caustic lines.

When the astigmatic field is incident on a rotationally symmetric lens from an off-axis direction, as shown in Fig. 4.11, each focal line will focus onto a separate focal line in the image plane, as shown. We'll assume to be viewing the plane containing the symmetry axis of the lens and the axis of the incident field; this plane is known as the *meridional plane.* We'll also assume that one focal line is perpendicular to this plane and that one lies in this plane, as shown in the figure. In the object space, the focal line perpendicular to the plane of incidence is called the primary focal line, and the other is called the secondary focal line. These are also called the *tangential* and *sagittal* foci. Note that, in general, the two focal lines won't lie in the meridional plane. The paper by Deschamps [2] gives an excellent discussion of how to analyze optical reflection and refraction from

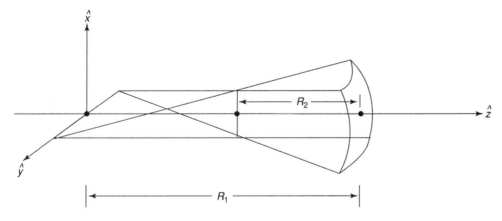

Figure 4.10 An astigmatic field.

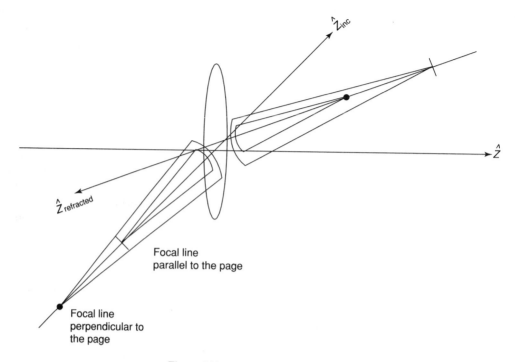

Figure 4.11 A tilted astigmatic field.

general quadratic dielectric and conductive surfaces under illumination by a general astigmatic ray pencil. This paper is highly recommended to the reader interested in this subject.

4.7 TRANSFORMATION MATRICES

In the *paraxial* or *Gaussian* regime wherein all rays are tilted with respect to the optic axis at angles small enough so that

$$\tan\theta \cong \sin\theta \cong \theta \tag{4.26}$$

we may use a matrix formulation to understand the operation of an optical system. Since this is only an approximation to the actual operation of the system—it neglects diffraction, aberration, and aperture stop effects—it is useful primarily in the initial layout of an optical system. The exact performance of the system would have to be determined through a more exact analysis using ray-tracing or diffraction analysis. However, for the purpose of getting a good initial design, the transformation matrix approach is very quick (it can easily be combined with optimization software) and can often give good insight to the physical effects governing the performance of the system. In this section, we'll look at the transformation matrix method.

In its most simplified form, the transformation matrix approach is capable of analyzing optical systems consisting of two basic components—ideal thin lenses and propagation paths (spacers) between the lenses. This matrix method is based on two simple properties of lens systems containing these elements. The first property is that the propagation path (in a homogeneous medium) is a straight line, so that the slope of the line does not change. Only the *distance* of the ray from the axis of the system changes (linearly as a function of axial distance through the system). The second property is the fact that an ideal thin lens does not change the position of the ray in the transverse plane of the lens; it only changes the *slope* of the ray. So it's evident that lenses and propagation media perform two dual functions: one changes the slope of the ray without changing its position, and the other changes the position of the ray without changing its slope.

From this discussion, then, the ray at any point is determined by its position in the transverse plane and its slope with respect to the z-axis. In ray matrix analysis, it is always assumed that the ray lies in a meridional plane, so when we refer to the position of the ray in the transverse plane, we're really talking about the distance of the ray from the symmetry axis. So, a ray is determined by two variables—its distance from the optic axis and its slope with respect to that axis. For a given ray emanating from a point in the image plane, these two variables will vary from point to point along the axis of the system. Since our attention is restricted to meriodinal rays only, we're limited to object points that lie on the axis of the system. Magnification from the object plane to the image plane will be determined by the relationship between the slope of a given ray in the object and image spaces. As we saw in the discussion on the Abbe sine condition, magnification relates to the amount of "stretching" that takes place between the plane wave spectrum in the object plane and in the image plane. This stretching in α, β space relates to the slopes of the plane waves in the two spaces. Magnification is proportional to the ratio of the ray or plane wave slopes in the object space and the image space. Mathematically,

$$M = \frac{\text{ray slope in object space}}{\text{ray slope in image space}} \qquad (4.27)$$

This relationship follows immediately from (4.8b).

The preceding remarks may be quantified with reference to Fig. 4.12. Since we're dealing with paraxial rays, the ray slope and angle to the system axis are considered equal. For the ideal thin lens, the object distance, the image distance, and the focal length are related through the usual lens law:

$$\frac{1}{d_0} + \frac{1}{d_i} = \frac{1}{f} \qquad (4.28)$$

Multiplying both sides by x_{in} (which is the same for both input and output planes on the front and back faces of the lens) gives

$$\frac{x_{\text{in}}}{d_0} + \frac{x_{\text{in}}}{d_i} = \frac{x_{\text{in}}}{f}$$

or

$$x'_{\text{in}} - x'_{\text{out}} = \frac{1}{f} x_{\text{in}}$$

which is one of our matrix transformation equations for the ideal thin lens. The other is implied by the discussion above; that is,

$$x_{\text{out}} = x_{\text{in}}$$

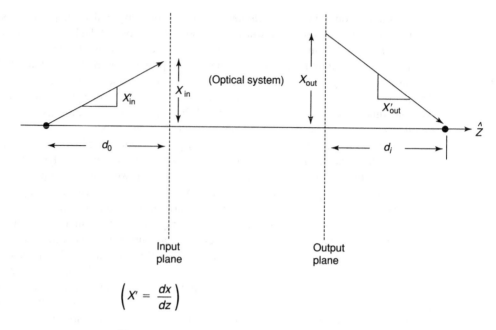

Figure 4.12 The transformation matrix of an optical system.

Rewriting these last two equations in matrix form gives the transformation matrix for the ideal thin lens as

$$\begin{pmatrix} x_{out} \\ x'_{out} \end{pmatrix} = \begin{pmatrix} 1 & 0 \\ -\dfrac{1}{f} & 1 \end{pmatrix} \begin{pmatrix} x_{in} \\ x'_{in} \end{pmatrix} \tag{4.29}$$

A similar matrix equation may be obtained for the propagation medium. A propagating ray will diverge from the symmetry axis (linearly with distance along the axis) by an amount proportional to the slope of the ray. Mathematically, this becomes

$$x_{out} = x_{in} + z \cdot x'_{in}$$

Combining this with the fact that the slope of the ray is unchanged as it propagates, we obtain the following matrix transformation equation for the propagation medium as

$$\begin{pmatrix} x_{out} \\ x'_{out} \end{pmatrix} = \begin{pmatrix} 1 & z \\ 0 & 1 \end{pmatrix} \begin{pmatrix} x_{in} \\ x'_{in} \end{pmatrix} \tag{4.30}$$

So, with these two matrix transfer equations, we may simply multiply them together to cascade elements to form a simple optical system and quickly find its magnification properties.

In general, the transformation matrix is represented in the generic form

$$\begin{pmatrix} x_{out} \\ x'_{out} \end{pmatrix} = \begin{pmatrix} A & B \\ C & D \end{pmatrix} \begin{pmatrix} x_{in} \\ x'_{in} \end{pmatrix} \tag{4.31}$$

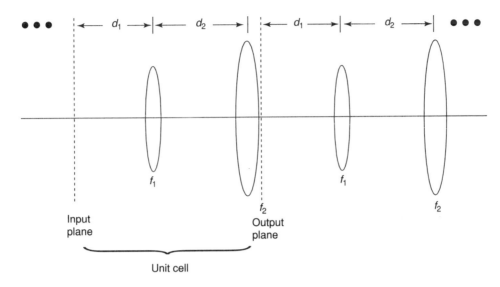

Figure 4.13 A lens waveguide.

4.8 INTRODUCTION TO MULTIPLE-LENS IMAGING SYSTEMS

Many types of optical systems employ multiple lenses, including telescopes, submarine periscopes, and lens waveguides. One of the more interesting multiple-lens systems is the "biperiodic" lens waveguide, shown in Fig. 4.13. This lens waveguide possesses certain periodicity characteristics similar to those found with diffraction gratings. In this case, however, the periodicity is in the axial direction, not the transverse direction. The lens waveguide has certain properties in common with other periodic electromagnetic waveguides, such as the corrugated waveguide [4]. The primary property of interest here is the fact that for propagating modes, the fields in the transverse plane of the waveguide are periodic, and they have the same periodicity of the waveguide itself. This is known as *Floquet's theorem* [5]. We can use Floquet's theorem, along with ray transfer matrices, to obtain the stable modes for a periodic lens waveguide.

The ray matrix of the unit cell segment of the infinite lens waveguide is easily found by multiplying the individual matrices of the various components together. The result is [6].

$$A = 1 - \frac{d_2}{f} \tag{4.32a}$$

$$B = d_1 + d_2 - \frac{d_1 d_2}{f_1} \tag{4.32b}$$

$$C = -\frac{1}{f_1} - \frac{1}{f_2} + \frac{d_2}{f_1 f_2} \tag{4.32c}$$

$$D = 1 - \frac{d_1}{f_1} - \frac{d_2}{f_2} - \frac{d_1}{f_2} + \frac{d_1 d_2}{f_1 f_2} \tag{4.32d}$$

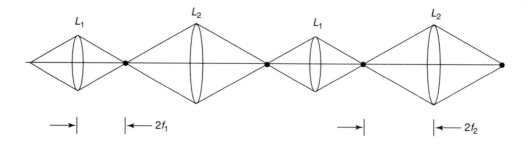

Figure 4.14 "Type 1" biperiodic lens waveguide.

For the field at the back plane of the unit cell to be equal to the field at the front plane, one of two conditions must be satisfied. Either we must have

$$x_{out} = x_{in}$$

and

$$x'_{out} = x'_{in}$$

or we must have

$$x_{out} = -x_{in}$$

and

$$x'_{out} = -x'_{in}$$

In the first case, we have

$$r_2 = Ar_1 + Br'_1 = r_1$$
$$r'_2 = Cr_1 + Dr'_1 = r'_1$$

and in the second case, we have

$$r_2 = Ar_1 + Br'_1 = -r_1$$
$$r'_2 = Cr_1 + Dr'_1 = -r'_1$$

The first equation leads to the following relation between the *ABCD* coefficients

$$BC = (1 - A)(1 - D)$$

and the second equation yields

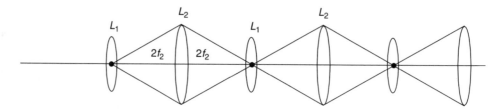

Figure 4.15 "Type 2" lens waveguide.

$$BC = (1 + A)(1 + D)$$

for propagation in an infinite periodic lens waveguide.

Substituting for the *ABCD* parameters for the biperiodic lens waveguide, we see in the first case that

$$f_1 + f_2 = \frac{d}{2} \tag{4.33}$$

which is shown in Fig. 4.14. For the second type of lens waveguide, we find

$$d = 2f_2; \quad f_1 \text{ is arbitrary} \tag{4.34}$$

which is shown in Fig. 4.15. For this waveguide, a focal plane exists at the plane of lens L_1. This lens waveguide is then degenerate and can be replaced by a singly periodic structure involving only lens L_2.

Whenever a focal plane coincides with the plane of a lens, a physical optics procedure (of the type used in the following chapter) may be used to analyze the system. When this situation occurs, we may (to a first order) regard the lens as modifying the phase of the image by adding a quadratic phase taper to it. As we'll see in the next chapter, adding a quadratic phase taper to an image plane distribution will cause the Fourier transform of the image to be formed at the plane located at the center of curvature of the quadratic wavefront.

4.9 GENERAL SUMMARY ON RAY-TRACING TECHNIQUES

In Chapter 2, we derived the basic ray concept as a far-field/zero-wavelength limit of a plane wave spectrum formulation. In that way, we obtained the equivalence between a ray and a plane wave on the infinite sphere. This means that ray-tracing analyses of lenses and mirrors proceed according to the tangent plane approach. That is, for the purpose of obtaining the refracted ray direction at a dielectric interface, the curved lens is locally taken to be infinite and planar, and is illuminated by an infinite plane wave whose polarization and angle of incidence are given by the ray at the given point on the interface. (Note that the radii of curvature of the reflected and refracted ray fields cannot be determined using the infinite plane approach—only the reflected and refracted ray *directions* and *polarizations*.) As mentioned previously, the reflected and refracted field radii of curvature are calculated as in [1,2].

So, it is in this context that the infinite plane wave problems from Chapter 2 become useful in practical lens analysis. This situation is shown in Fig. 4.16.

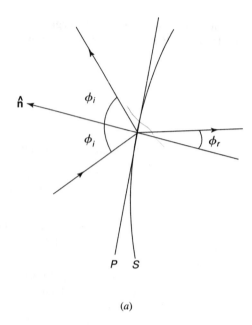

(a)

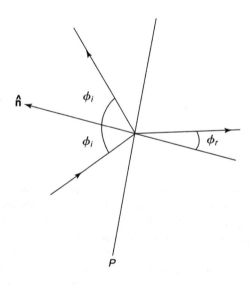

(b)

Figure 4.16 Tangent plane problem for calculating reflected and refracted fields: (a) Ray incident on curved dialectric boundary and (b) Equivalent tangent plane problem involving infinite plane wave fields.

REFERENCES

[1] Scott, C. R., *Modern Methods of Reflector Antenna Analysis and Design,* Norwood, MA: Artech House, 1990.

[2] Deschamps, G. A., ''Ray Techniques in Electromagnetics,'' *Proc. IEEE,* vol. 60, September 1972, pp. 1022–1035.

[3] Born, M., and Wolf, E., *Principles of Optics,* 6th ed. New York: Pergamon Press, 1980.

[4] Scott, C. R., *Field Theory of Acousto-Optic Signal Processing Devices,* Norwood, MA: Artech House, 1992.

[5] Scott, C. R., *The Spectral Domain Method in Electromagnetics,* Norwood, MA: Artech House, 1989.

[6] Kogelnik, H., and Li, T., Laser Beams and Resonators, *Applied Optics,* vol. 5, no. 10, October 1966, pp. 1550–1567.

5

Focusing and Imaging Properties of Lenses: Diffraction Integral Viewpoint

The last chapter dealt with ray analysis of lenses and imaging systems. In this chapter, the diffraction integral approach is applied to the problem of optical system analysis. This approach, of course, reveals the phenomenon of diffraction, which is not described by geometrical optics.

5.1 FRESNEL REGION FIELDS

In this section, we'll derive the diffraction integral equations for the Fresnel region fields. As mentioned in Chapter 1, the Fresnel region approximation is generally used for calculating fields at various transverse planes in an optical system. It may also be used to understand Gaussian beam propagation, since this is also basically a Fresnel region effect.

With reference to Fig. 5.1, assume we want to calculate the field at plane P_2 due to a set of magnetic currents on plane P_1. This is done using (1.44), in conjunction with (1.46). That is,

$$E(r) = -\nabla \times \iiint_{v'} M(r') \frac{e^{-jk|r-r'|}}{4\pi|r-r'|} \, d\,v' \qquad (5.1)$$

In the Fresnel region approximation, we approximate the distance function using the following quadratic approximation:

$$|r - r'| = \sqrt{(x-x')^2 + (y-y')^2 + (x-z')^2} \cong z\left[1 + \frac{(x-x')^2}{2z^2} + \frac{(y-y')^2}{2z^2}\right]$$

$$(5.2)$$

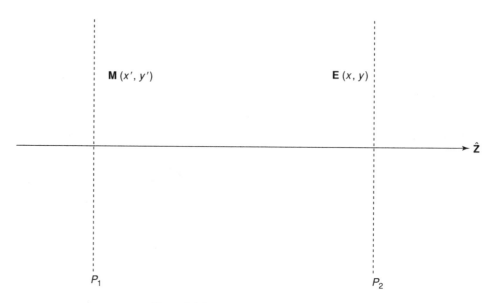

Figure 5.1 Fresnel region field calculations.

This equation is consistent with all of the analyses we'll be doing in this book. That is, whenever we describe lenses (except perhaps when we explicitly talk about aberrations), we'll assume that all fields obey the *paraxial rays* approximation. According to the paraxial rays approximation, all wavefronts consist of surfaces that may adequately be described in terms of quadratic functions in x and y. In point of fact, such wavefronts have parabolic, not spherical, shape. But to our degree of approximation (i.e., for paraxial rays that never tilt appreciably far away from the axis of the optical instrument), a parabolic (quadratic phase) wavefront is equivalent to the real (spherical wave) phasefront radiated by a Huygens point source. This quadratic phase approximation defines what is known as *Gaussian optics*—the optics of paraxial rays.

So, by (5.2), a Huygens point source in plane P_1 radiates a spherical wave, which we may approximate by a quadratic function in x and y. Thus, the equation for the electric field becomes

$$E(r) = \iint_P M(x', y')\, \nabla \left(\frac{e^{-jkz}}{4\pi z}\, e^{-(jk/2z)[(x-x')^2+(y-y')^2]} \right) dx'\, dy' \tag{5.3}$$

where we've used (A.14) in obtaining this equation, along with the fact that the source magnetic currents are constant with respect to the field region differential operator.

The gradient in (5.3) is easily evaluated, and keeping terms proportional to $1/z$, we obtain for the Fresnel region field,

$$E(r) = \frac{-jk\, e^{-j(kz)}}{4\pi z} \iint_{P_1} [M(x', y') \times \hat{z}]\, e^{-(jk/2z)[(x-x')^2+(y-y')^2]}\, dx'\, dy'$$

$$\tag{5.4}$$

Finally, recalling the discussion on equivalent magnetic currents in Chapter 1, which indicated that they are related to the tangential aperture electric field according to the cross-product relation,

$$M(x', y') = 2\, E^{\tan}(x', y') \times \hat{z} \tag{5.5}$$

we finally obtain the expression for the Fresnel region radiated field of an aperture distribution as

$$E(x, y) = jk \frac{e^{-jkz}}{2\pi z} \iint_{P_1} E^{\tan}(x', y')\, e^{-j(k/2z)[(x-x')^2+(y-y')^2]}\, dx'\, dy' \tag{5.6}$$

In this case, the relationship between the aperture field and the radiated field is not so simple as in the Fraunhofer region case. That is, instead of having a nice Fourier transform (linear phase) relationship between the two sets of fields, we now have a Fresnel transform (quadratic phase) relationship.

NOTE: For those readers so inclined, the linear phase term in the Fourier transform is known as the Fourier transform *kernel*. Similarly, the quadratic phase term in (3.6) is known as the Fresnel transform kernel. Any type of linear functional transform is defined by its kernel as well as the range over which the kernel is defined. Transform kernels may be defined over infinite ranges (as in the case of the Fourier and Fresnel transforms), or they may be defined over finite ranges, as in the case of wavelet and Gabor transforms. Functions defined over a finite range are said to have *finite support*, while those defined over an infinite range are said to have infinite support. Other transforms of use in optics include the Hartley, Lebedev, Hadamard-Walsh, cosine, and Mellin transforms.

5.2 THE THIN LENS AS A QUADRATIC PHASE PLATE

Within the context of the paraxial rays approximation, it is possible to give a very simple and intuitive explanation of thin lens operation; that is, for paraxial rays, the thin lens acts as a quadratic phase plate. Consider Fig. 5.2 in which an axially propagating plane wave field impinges on a dielectric lens.

Assuming a quadratic approximation to the spherical lens surface, we may give the various phase terms shown in the figure as follows, as a function of radius:

$$\phi_{1-2} = -k_0 z_1 = -\frac{k_0 x^2}{2R_1} \tag{5.7a}$$

$$\phi_{3-4} = -k_0 z_2 = -\frac{k_0 x^2}{2R_2} \tag{5.7b}$$

$$\phi_{2-3} = -k[(d_1 - z_1) + (d_2 - z_2)]$$
$$= -k\left[\frac{(D/2)^2 - x^2}{2R_1} + \frac{(D/2)^2 - x^2}{2R_2}\right] \tag{5.7c}$$

Combining these terms together, we obtain the total phase through the lens (neglecting constant terms) as

$$\phi(r) = \frac{1}{2}\left(\frac{1}{R_1} + \frac{1}{R_2}\right)(k - k_0)\, r^2 \tag{5.8}$$

In these equations, k is the propagation constant in the dielectric lens medium and k_0 is the free-space propagation constant.

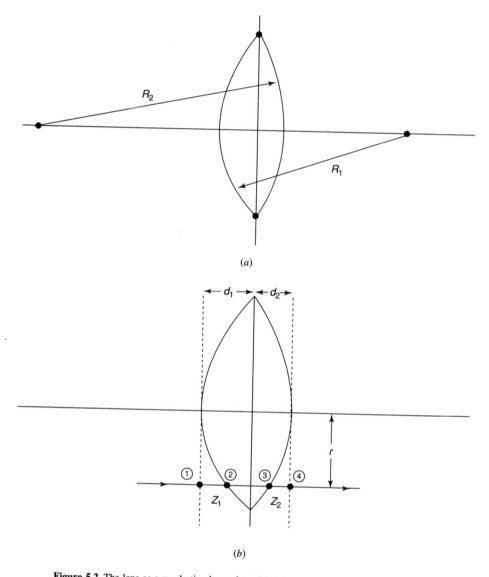

(a)

(b)

Figure 5.2 The lens as a quadratic phase plate: (a) A lens generated from two spherical surfaces; (b) Paraxial ray or plane wave propagating through the lens.

This is clearly a quadratic phase distribution. The positive sign indicates a converging wave. If this phase distribution is to represent a phase front that converges on the focal point of the lens, then we must have

$$\frac{1}{2}\left(\frac{1}{R_1} + \frac{1}{R_2}\right)(k - k_0)\, r^2 = \frac{k_0 r^2}{2f}$$

or

$$\frac{1}{f} = \left(\frac{1}{R_1} + \frac{1}{R_2}\right)(n - 1) \qquad\qquad (5.9)$$

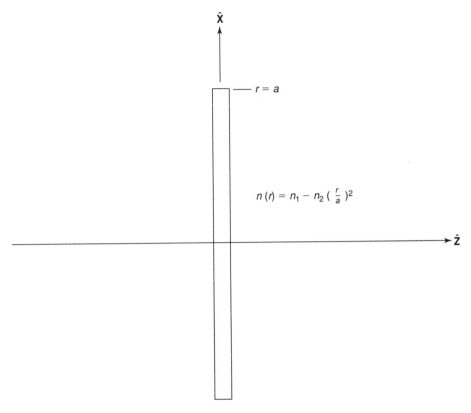

Figure 5.3 The Wood lens.

where n is the index of refraction of the dielectric medium, given by

$$n = \sqrt{\frac{\epsilon\mu}{\epsilon_0\mu_0}}$$

This equation gives the focal length of a lens in terms of its geometrical properties (radii of curvature) and its material properties (index of refraction). Equation (5.9) is equivalent to (4.23).

The concept of the lens as a quadratic phase plate lies behind the design of the Wood lens, which is topologically similar to the GRIN lens introduced in Chapter 3. The Wood lens is shown in Fig. 5.3. The lens is in the shape of a disk, and the index is graded in a quadratic fashion as shown. This lens has substantially the same imaging properties as the spherical lens.

We can also look at what happens when a tilted plane wave impinges on a thin lens. With reference to Fig. 5.4, we take the incident wave phase as

$$E^{\text{inc}} = e^{-j\alpha_0 x}$$

From the previous discussion on spherical lenses, we may regard the lens as a plate having a quadratic phase transmittance of the form

$$T = e^{j(k/2f)(x^2 + y^2)}$$

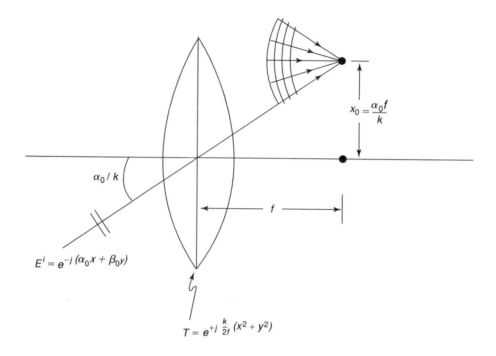

Figure 5.4 Tilted plane wave incident on a spherical lens.

Multiplying the incident field by the quadratic phase transmittance of the lens gives the transmitted field as

$$E^{\text{trans}} = e^{j(k/2f)[(x - \alpha_0 f/k)^2 + y^2]}$$

(up to a constant-phase factor).

This is a convergent spherical wave field, as shown in Fig. 5.4. Note that by the previous equation, this convergent spherical wave field is *not tilted,* as the incident plane wave field is. This converging spherical wave is *symmetric* about the horizontal line, $x = \alpha_0 f/k$; it is merely displaced in the transverse plane. This is a very important characteristic of laterally defocused spots that arise from tilted incident plane waves.

Lastly, we may look at the action of the spherical lens on an incident spherical wave field. This is shown in Fig. 5.5.

The incident field is given as

$$E^{\text{inc}} = e^{-j(k/2d_0)(x^2 + y^2)}$$

The transmission function of the lens is given (as before) as

$$T = e^{j(k/2f)(x^2 + y^2)}$$

The product of the incident field and the transmittance gives a transmitted field of the form

$$E^{\text{trans}} = e^{j(k/2)(1/f - 1/d_0)(x^2 + y^2)}$$

which represents a converging spherical wave of curvature

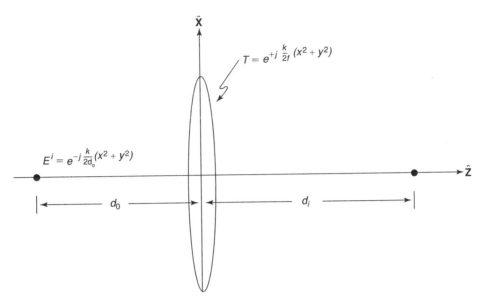

Figure 5.5 The effect of a lens on an incident spherical wave.

$$\frac{1}{d_i} = \frac{1}{f} - \frac{1}{d_0} \qquad (5.10)$$

(when $d_0 > f$) and again, we see the lens law popping out.

5.3 COHERENT IMAGING WITH LENSES: THE POINT SPREAD FUNCTION

The term *physical optics* is often used to describe the application of the diffraction integral formulation, especially when the source electric fields have been calculated using geometrical optic ray-tracing methods. For example, ray-tracing methods might be used to calculate the propagation of optical energy through an optical instrument to a plane just past the final lens. Then, the diffraction integral formulation can be used to calculate the electric field radiated/diffracted beyond this final aperture plane.

Since diffracted electromagnetic fields are stationary with respect to their source currents, whether those sources are electric or magnetic currents, it is not crucial that the source currents be known exactly. (Geometrical optics fields, on the other hand, are *not* stationary with respect to their sources because, as we saw in Chapter 2, geometrical optics fields are *directly proportional* to their source fields, being altered from them only in magnitude and phase.) Diffracted fields, however, *are* stationary, and this property is due to the integral representation of the diffracted fields. The integral tends to "smooth out" errors in the current distribution, making the diffracted field relatively insensitive to the exact form of the sources. This means that first-order errors in the representation of the currents will result in second-order errors in the diffracted field.

Ray-tracing methods are generally quite adequate for calculating source fields as long as these ray-based calculations only involve converging or diverging (i.e., uncollimated) spherical waves. As we saw in Chapter 2, converging and diverging spherical waves give rise to stationary phase points in the zero-wavelength (geometrical optics) limit. This stationary phase/ray optic term is the dominant contributor to the true field, with the diffracted field term representing only a secondary effect. When fields are collimated, however, as they often are in the exit pupil of many optical instruments, there is no stationary phase point in the zero-wavelength limit (except on-axis at infinity, in which case all aperture points are stationary phase points, as we saw in Chapter 2). Therefore, for collimated fields, there is *no* geometrical optics field, and the entire radiated field is due to the formerly second-order effect of diffraction, which is now a first-order effect since no other effects exist.

One other situation in which geometrical optics breaks down is near focal points. The problem with focal points is that the rays associated with a converging spherical wave all converge on a single point. This is another way of saying that the converging spherical wavefront has not just one, but an infinite number of stationary phase points. (This is the same situation mentioned in the previous paragraph for on-axis fields at infinity, due to collimated aperture fields.) This situation produces what is known as a *caustic point,* and geometrical optics breaks down at such points. Geometrical optics predicts an infinite optical field at caustics, and hence is of no use. Caustics and focal points that lie interior to the optical system generally do not present a problem because we always propagate the ray fields *through* the caustics; we never actually have to calculate the fields there.

So, these are some of the reasons why ray tracing is often adequate for calculating the fields interior to an optical system, whereas the diffraction integral method is generally better suited to calculating the fields radiated by such systems.

Let's look again more closely at the situation shown in Fig. 5.6, in which a thin lens is illuminated by an axial point source of light, given by

$$E^{\text{obj}} = \hat{x}\delta(x)\delta(y)\delta(z + d_0) \tag{5.11}$$

Any field a finite distance from a point source is in the Fraunhofer region of that source. This comes from our analysis of the Fraunhofer region from Chapter 1. Therefore, by (1.54), the equation for the Fraunhofer region field of the point source incident on the lens is given by

$$E^{\text{inc}}(x', y') = \hat{\theta}^{\text{obj}} e^{-j(k/2d_0)(x^2+y^2)}$$

where $\hat{\theta}^{\text{obj}}$ is taken with respect to a coordinate system centered at the object point, as shown in Fig. 5.6. The field transmitted through the thin lens is given by

$$E^{\text{trans}} = \theta^{\text{trans}} e^{j(k/2)(1/f - 1/d_0)(x^2+y^2)}$$

where θ^{trans} is taken with respect to a coordinate system centered at the image point.

Here we've used Malus's theorem and symmetry arguments to deduce that the electric field vector still lies in the plane of the wavefront even after refraction through the lens. (The field polarization would fall out automatically from a ray-tracing analysis.) The field in the image plane may now be calculated using the Fresnel region formula (5.6), as

$$E(x, y) = jk \, \frac{e^{-jkd_i}}{2\pi d_i} \, \hat{x} \iint_{P_1} e^{j(k/2)(1/f - 1/d_0)(x'^2+y'^2)} \, e^{-j(k/2d_i)[(x-x')^2+(y-y')^2]} \, dx' \, dy' \tag{5.12}$$

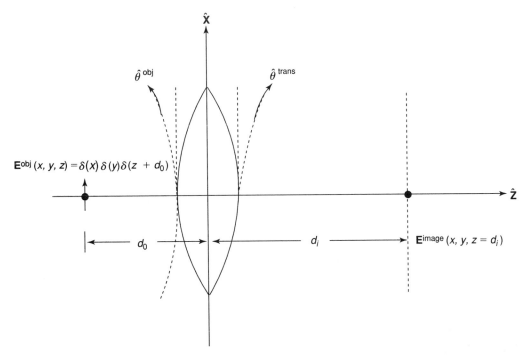

Figure 5.6 Converging spherical wave produced by a lens.

or, using the lens law to eliminate the quadratic phase terms in the integrand,

$$E(x, y) = jk \frac{e^{-jkd_i}}{2\pi d_i} e^{-j(k/2d_i)(x^2 + y^2)} \hat{x} \iint_{P_1} e^{j(k/d_i)(xx' + yy')} \, dx' \, dy' \qquad (5.13)$$

where we've assumed a paraxial ray problem wherein

$$\hat{\theta}^{\text{trans}} = \hat{x}$$

In other words, the field in the focal plane of a converging spherical wave is given as the *Fourier transform of the aperture function,* along with a diverging spherical wave phase function. When the converging wave isn't perfectly spherical, the field in the image plane will be the Fourier transform of the aberration function (given as the difference between the actual wavefront and a true spherical wavefront). The function in (5.13) that is the Fourier transform of the aperture function, is referred to as the *point spread function,* since it is a measure of the extent to which a point source in the object plane gets "spread out" or blurred in the image plane.

In our discussion of Maxwell's theorem in Chapter 4, we noted that a point source in the object plane will be imaged onto another point source in the image plane if and only if all of the Fourier (plane wave) components making up the delta function in the object plane are propagated to the image plane in such a way that they have the exact same amplitude and phase relationship in the image plane that they had in the object plane. In other words, the plane wave spectrum must have been passed through the equivalent of a perfect "all-pass" filter with flat amplitude and phase transmission properties. The extent to which the image point departs from the perfect object point is a measure of the "bandlimitedness" of the lens as a plane wave filter.

If the thin, quadratic lens in the previous discussion is circular, with radius a, then the point spread function can readily be shown (as in Chapter 1) to be the Airy pattern,

$$PSF(\rho) = 2\pi a^2 \frac{J_1\left(\frac{a}{d_i} k\rho\right)}{\frac{a}{d_i} k\rho} \qquad (5.14)$$

The spatial frequency variables are given by

$$\alpha = \frac{k}{d_i} x \quad \text{and} \quad \beta = \frac{k}{d_i} y \qquad (5.15)$$

Note that the Airy pattern is the point spread function for a perfect quadratic lens; it is the best point spread function you can get for a diffraction-limited aperture. Aberrations will cause the point spread function to be degraded from this ideal pattern.

Note from the equation above that as the radius a of the lens tends to infinity, the point spread function tends to a point. We'll talk more about this later.

We've mentioned the fact that the blurring in the image plane implies that a certain amount of plane wave filtering has taken place between the object plane and the image plane. From the discussion of Heisenberg's principle in Chapter 2, it's evident that the delta function spreads into the Airy pattern because the spatial frequency "passband" of the lens is limited; it is not infinite. It is of interest to calculate and quantify the actual plane wave filter response of the lens in Fig. 5.6.

One way to calculate the plane wave transfer function of the circular lens is to assume a Fourier transform expansion for the Airy function and then simply calculate the expansion coefficients. This method would yield the spatial frequency content of the Airy pattern. However, this is a mathematically difficult way of solving the problem since integrating the Airy function is a difficult problem. The mathematically simpler way of solving for the transfer function is to go back to the Fourier transform pairs

$$f(x, y) = \iint_{-\infty}^{\infty} F(\alpha, \beta) \, e^{j(\alpha x + \beta y)} \, d\alpha \, d\beta \qquad (5.16a)$$

$$F(\alpha, \beta) = \frac{1}{4\pi^2} \iint_{-\infty}^{\infty} f(x, y) \, e^{-j(\alpha x + \beta y)} \, dx \, dy \qquad (5.16b)$$

We know from the methods of Chapter 1 that if $F(\alpha, \beta)$ is the "circular pulse function" shown in Fig. 5.7, then $f(x, y)$ will be the circularly symmetric Airy function given in (5.14) above. Hence, these two functions are Fourier transform pairs. Thus, the ideal quadratic lens acts as a perfect low-pass filter for plane wave fields.

There is a very simple way to see why the ideal quadratic lens should act as a perfect low-pass filter for plane waves. Consider Fig. 5.8, which shows an axial point source in the object plane that is imaged onto the axial point source in the image plane. Let the lens be in the far field of both the object and image point sources; therefore, by the discussions on the Fraunhofer region from Chapters 1 and 2, the object and image *angular ranges correspond* to the object and image point *plane wave spectral bandwidths*. And the bandwidth of the image point is as shown. Only those spatial frequency (plane wave) compo-

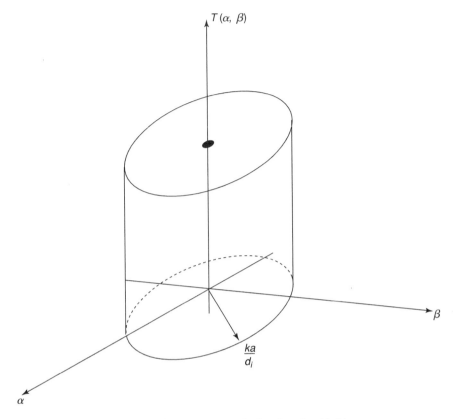

Figure 5.7 Perfect low-pass transfer function of an ideal lens.

nents lying within the conical regions shown contribute to the image of the point source in the object plane. And these spatial frequency (plane wave) components are the low spatial frequency components.

The preceding discussion gave us the image plane field due to a single axial point source in the object plane. When there is a distribution of sources in the object plane, we may use the superposition principle (from Chapter 1) to write the total field in the image plane in the form of an integral of the distribution function, convolved with the point spread function. With reference to Fig. 5.9, the point spread function for a laterally defocused point source is given as

$$E^{\text{image}}(x_i, y_i) = j\lambda \, \frac{e^{-jkd_0}}{d_0} \iint_{P_1} E^{\text{trans}}(x, y) \, e^{-j(k/2d_i)[(x_i-x)^2 + (y_i-y)^2]} \, dx \, dy$$

(5.17)

where

$$E^{\text{trans}}(x, y) = \hat{x} e^{-j(k/2d_0)[(x-x_0)^2 + (y-y_0)^2]} \, e^{j(k/2f)(x^2+y^2)}$$

(5.18)

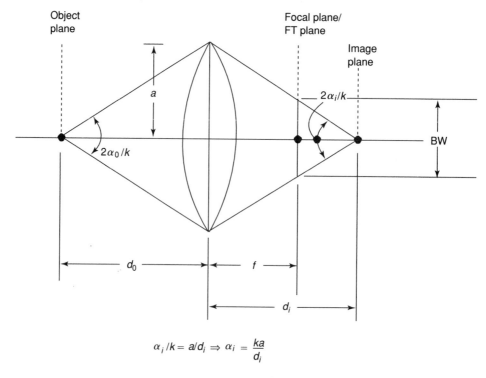

$$\alpha_i/k = a/d_i \Rightarrow \alpha_i = \frac{ka}{d_i}$$

Figure 5.8 On the low-pass property of an ideal quadratic lens. $2\alpha_0/k$ = range of spatial frequencies available from the object spectrum for the image spectrum. $2\alpha_i/k$ = range of spatial frequencies in the image. BW = spatial bandwidth of the image.

or

$$E^{\text{image}}(x_i, y_i) = j\lambda \frac{e^{-jkd_0}}{d_0} e^{-j(k/2d_0)(x_0^2 + y_0^2)} e^{-j(k/2d_i)(x_i^2 + y_i^2)}$$
$$\cdot \iint_{P_1} e^{jk[(x_i/d_i + x_0/d_0)x + (y_i/d_i + y_0/d_0)y]} \, dx \, dy \qquad (5.19)$$

whence

$$PSF(x_i, y_i) = j\lambda \frac{e^{-jkd_0}}{d_0} e^{-j(k/2d_0)(x_0^2 + y_0^2)} e^{-j(k/2d_i)(x_i^2 + y_i^2)} (2\pi a^2) \frac{J_1\left(\frac{ka\rho}{d_i}\right)}{\left(\frac{ka\rho}{d_i}\right)}$$

$$(5.20)$$

where now

$$\rho = \sqrt{(x_i + Mx_0)^2 + (y_i + My_0)^2} \qquad (5.21)$$

and

$$M = \frac{d_i}{d_0} \qquad (5.22)$$

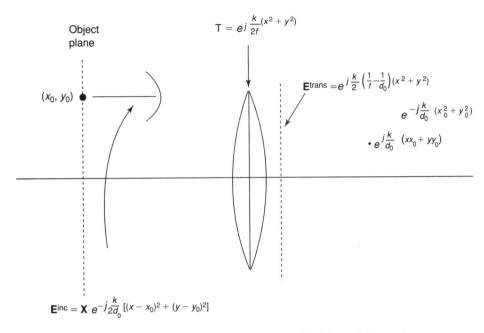

Figure 5.9 On the point spread function for a laterally defocused object point.

With this, the image of an extended object distribution may be written in convolution form as

$$E(x_i, y_i) = j2\pi a^2\lambda \frac{e^{-jkd_0}}{d_0} e^{-j(k/2d_0)(x_i^2 + y_i^2)} \iint e^{-j(k/2d_i)(x_0^2 + y_0^2)} \frac{J_1\left(\dfrac{ka\rho}{d_i}\right)}{\left(\dfrac{ka\rho}{d_i}\right)} dx_0 dy_0$$

(5.23)

This integral assumes that the point spread function is only a function of the differences: $x - x'$ and $y - y'$, not on the absolute location of x', y' in the object plane. A point spread function displaying this property is known as *shift invariant*. As stated by Born and Wolf [1], point spread functions are shift invariant only over limited regions of the object plane, known as *isoplanatic regions.*

At the beginning of this chapter, we saw that an *unfocused* aperture acts in a multiplicative way in the spatial domain and in a convolutional way in the spectral domain. Now we see that a *focused* aperture acts in exactly the opposite way; that is, it acts in a multiplicative way in the spectral domain and in a convolutional way in the spatial domain. This is because the lens—by adding a quadratic phase distribution to the aperture distribution—converts the aperture distribution into a *plane wave spectrum,* which converges in the image plane behind the lens. This plane wave spectrum has the same functional form as the original aperture distribution; hence, its image will be the Fourier transform of that aperture distribution. This is shown in Fig. 5.10. In the figure, we see that the lens brings the Fraunhofer region field from the sphere at infinity to the image plane behind the lens. So the spectrum of the image is the same as the original aperture distribution just in front of the lens (neglecting lens aberrations).

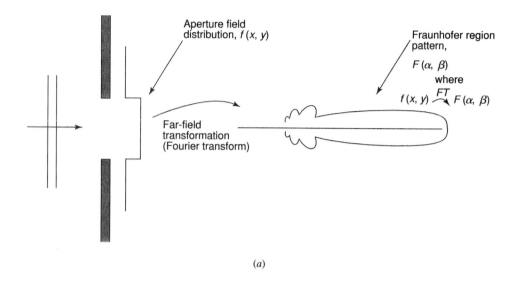

(a)

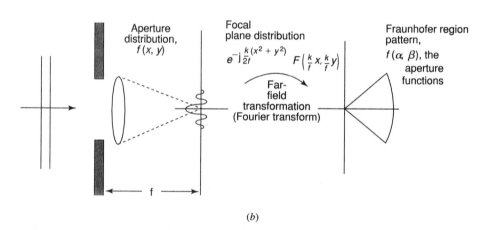

(b)

Figure 5.10 The spectrum of a focused aperture is the same as the aperture function: (a) Unfocused aperture and (b) focused aperture.

5.4 SUMMARY OF LENS ACTION, UNDER PLANE AND SPHERICAL WAVE INCIDENCE

In this section, we will summarize the action of thin convex and concave lenses on planar and spherical wavefronts. All combinations of planar, converging spherical, and diverging spherical wavefronts will be covered, for both types of lenses. This summary is important for understanding the operation of practical lens systems of the type presented in Chapter 7.

The general understanding of lens action presented in this section is important in modern optical system design. Even though sophisticated ray-tracing software now exists for designing and optimizing optical systems, it is still necessary for the engineer to have

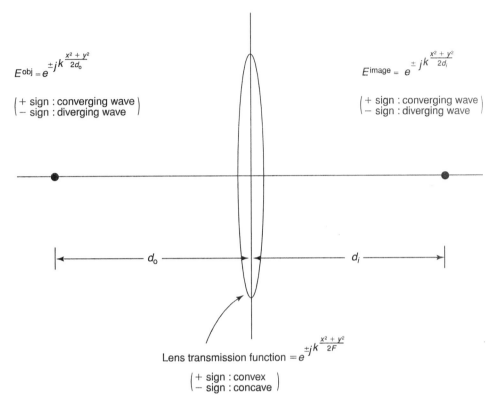

Figure 5.11 Generic lens and illumination field.

a baseline design layout intended for further optimization. Being able to lay out an optical system, determining when an image should lie within the focal length of a lens and when it should not, knowing when and how to invert an image—these are the skills necessary today in optical system design. The software will do the optimization, but the engineer must originate the design concept and configuration. The summary information in this section will permit quick analysis and configuration design for many types of optical systems.

First, consider Fig. 5.11, which shows the generic problem under consideration. The sign convention is given for incident and transmitted fields and for the lens transmission function. The reader should go through all sign conventions to make sure they are all understood.

For the first specific case, consider Fig. 5.12a, which shows a convex lens under plane wave illumination.

The incident field is

$$E^{\text{inc}} = e^{j\alpha x}$$

The lens transmission is

$$T(x, y) = e^{j(k/2F)x^2}$$

where we've only included the x-dependence of the lens transmission function, since the

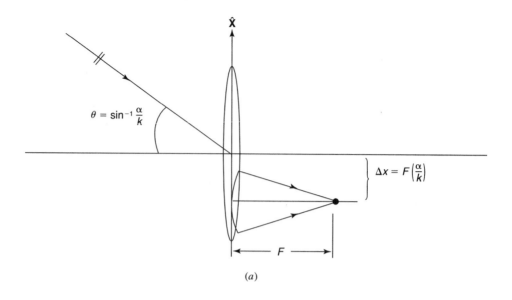

(a)

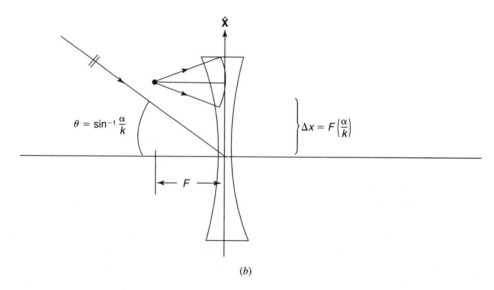

(b)

Figure 5.12 Plane wave incidence on thin lenses: (a) Plane wave incident on a convex lens and (b) Plane wave incident on a concave lens.

y-dependence is unaffected (to a first order) by a plane wave field incident in the $x - z$ plane.

Multiplying the incident field by the lens transfer function (and rearranging terms) gives (neglecting a constant-phase factor):

$$E^{\text{trans}} \;=\; e^{j(k/2F)(x + F\alpha/k)^2}$$

for the field along the back plane of the thin lens. This is a converging spherical wave of radius F, centered at the laterally defocused point, $x = -F\alpha/k$.

The second specific case is shown in Fig. 5.12b. This is a plane wave incident on a diverging lens. The incident field is the same as in the previous case, whereas the lens transmission function is the conjugate of the previous case. Therefore, the transmitted field is given by

$$E^{\text{trans}} = e^{-j(k/2F)(x - F\alpha/k)^2}$$

This is a diverging spherical wave of radius F, centered at the laterally displaced point, $x = F\alpha/k$. This wave has its center of curvature on the front side of the lens, the same side of the optic axis as the direction of arrival of the incident field.

The third case involves a convex lens illuminated by a diverging spherical wave. In this case, there are two possibilities for the transmitted field. The transmitted field can be either convergent or divergent, depending on the curvature of the incident field.

The problem is shown in Fig. 5.13. The incident field is

$$E^{\text{inc}} = e^{-jk(x - x_1)^2/2d_0}$$

and the lens transmission function is

$$T = e^{jk(x^2/2F)}$$

Multiplying the incident field by the lens transmission function gives the transmitted field over the back plane of the lens as

$$E^{\text{trans}} = e^{j(k/2d_i)(x + x_1 d_i/d_0)^2}$$

where

$$\frac{1}{d_i} = \frac{1}{F} - \frac{1}{d_0}$$

When the object distance is greater than the focal length, the image distance is positive (the image is behind the lens) and the image is the normal real (convergent transmitted wave) and inverted (phase center is on the opposite side of the optic axis from the source point) image of a convex lens. When the object distance is less than the focal length, the image distance is negative (the image lies in front of the lens). In the latter case, when the object distance is nearly equal to the focal length, the image distance is very large, as is the lateral magnification. This is the configuration used in an ordinary magnifying glass.

The fourth case involves a convex lens illuminated by a converging spherical wave, as illustrated in Fig. 5.14. The image in this case is always real and negative, and will be formed somewhere in front of the back focal plane of the lens. The incident field is

$$E^{\text{inc}} = e^{jk(x - x_1)^2/2d_0}$$

and the convex lens transmission function is as given previously. The transmitted field over the back plane of the lens is

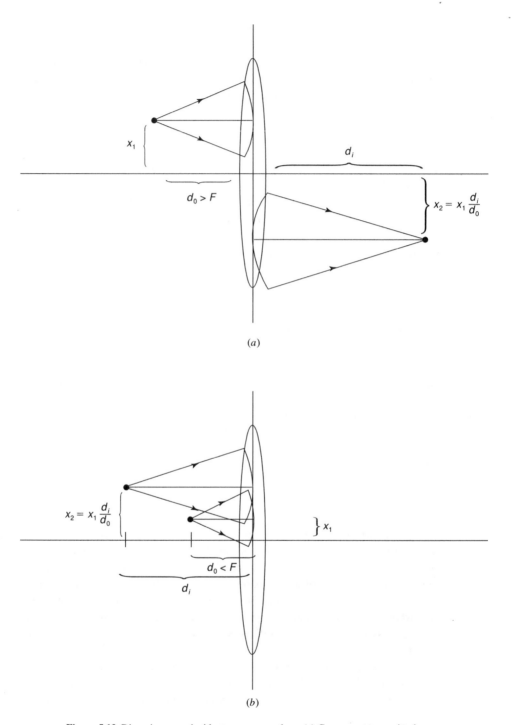

Figure 5.13 Diverging wave incident on a convex lens: (a) Convergent transmitted wave and (b) Divergent transmitted wave.

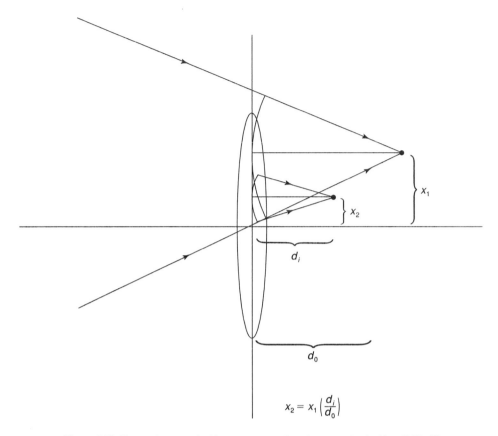

$$x_2 = x_1 \left(\frac{d_i}{d_0} \right)$$

Figure 5.14 Converging wave incident on a convex lens (a converging incident field will always converge more strongly after transmission through a convex lens).

$$E^{\text{trans}} = e^{j(k/2d_i)(x - x_1 d_i/d_0)^2}$$

where

$$\frac{1}{d_i} = \frac{1}{F} + \frac{1}{d_0}$$

which represents a converging spherical wave. Since the image distance is always less than the incident field radius of curvature, the magnification of this system will always be less than unity.

The fifth case involves a diverging field incident on a concave lens (see Fig. 5.15). The transmitted field in this case will always be diverging. We may reverse the sign of F in case 3 to obtain the transmitted field along the back plane of the lens as

$$E^{\text{trans}} = e^{-j(k/2d_i)(x + x_1 d_i/d_0)}$$

where

$$\frac{1}{d_i} = \frac{1}{F} + \frac{1}{d_0}$$

In this case, the image distance is always less than the object distance, and the magnification is always less than unity.

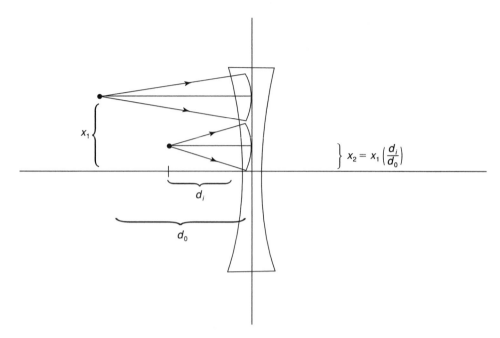

Figure 5.15 Diverging wave incident on a concave lens (a divergent incident field will always diverge more after transmission through a concave lens).

The sixth and last case is that of a convergent wave incident on a concave lens (see Fig. 5.16). In this case, we merely reverse the sign on F in case 4 to obtain

$$E^{\text{trans}} = e^{j(k/2d_i)(x - x_1 d_i/d_0)^2}$$

where

$$\frac{1}{d_i} = \frac{1}{d_0} - \frac{1}{F}$$

for the field in the back plane of the lens. Clearly, there will be two different cases, depending on whether the incident field curvature is greater or less than the focal length. These are indicated in Fig. 5.16a, b. When the radius of curvature of the incident field is greater than the focal length, the transmitted field will be divergent; when it is less than the focal length, the transmitted field will be convergent.

5.5 INCOHERENT IMAGING WITH LENSES

Despite the advances in modern laser technology, there will always be certain optical systems that use incoherent light, either from the sun or ordinary light sources. All optical instruments that process light for viewing by the human eye will fall into this category. Therefore, it is useful to look at the modifications that are made in the preceding theory in order to accommodate incoherent light waves.

Incoherent light by definition has no well-defined phase, so incoherent optical fields

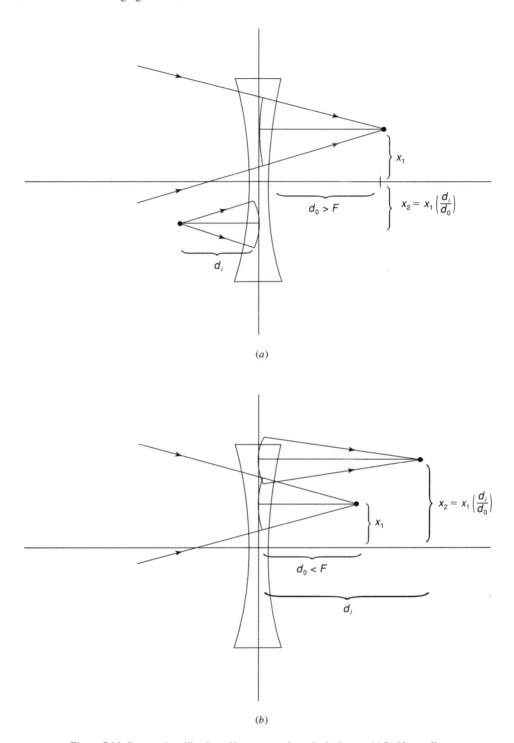

(a)

(b)

Figure 5.16 Concave lens illuminated by a converging spherical wave: (a) Incident radius
of curvature is greater than the focal length and (b) Incident radius of curvature is less than the focal length.

combine in a power sense rather than in a complex field sense. Therefore, we'll always deal with the energy associated with an incoherent wave, rather than the actual phasor field of the wave. The energy (or intensity) of an optical field is easily obtained from a consideration of the Poynting vector introduced in Chapter 1. In that chapter, we noted that the flow of electromagnetic power across a given surface is determined by the integral of the Poynting vector,

$$P = E \times H*$$

across the surface. Thus,

$$\text{Power} = \iint_s (E \times H*) \cdot ds$$

We can use Gauss's law to get an expression for the electromagnetic energy produced inside the closed surface, S, due to the flow of electromagnetic power across the surface. By Gauss's law,

$$\iint_s (E \times H*) \cdot ds = \iiint_v \nabla \cdot (E \times H*) \, dv$$

So, using Maxwell's equations, we can expand the term on the RHS of the equation above to get a type of continuity equation for electromagnetic power. Thus,

$$\nabla \cdot (E \times H*) = j\omega (\epsilon E \cdot E* - \mu H \cdot H*) - E \cdot J*$$

This is a continuity equation for electromagnetic power, similar to the continuity equation relating current and charge. It states that the divergence of the Poynting vector is equal to the time rate of change of electric minus magnetic field energy density, minus the power dissipated through loss. (The loss may be due to lossy dielectric materials or to lossy conductors.)

What's important for us is the first term inside the parentheses on the RHS of the equation above. This term represents the electric field energy density (usually called the field *intensity* in optics). This is the quantity we'll be looking at in this section in conjunction with incoherent optical fields.

From the form of the coherent point spread function given in the previous section, we may now obtain the incoherent point spread function by taking the magnitude and squaring it. Thus, for incoherent optical fields, the point spread function is given by

$$PSF(x_i, y_i) = \left(\frac{\lambda}{d_0}\right)^2 (2\pi a^2)^2 \frac{J_1^2\left(\frac{ka\rho}{d_i}\right)}{\left(\frac{ka\rho}{d_i}\right)^2} \tag{5.24}$$

We'll come back to this expression again in Chapter 6, when we look at spectral domain imaging properties of lenses.

5.6 FOURIER TRANSFORMING PROPERTY OF LENSES IN THE FRESNEL REGION

Returning now to optical systems using coherent laser light, we'll derive the Fourier transforming property of lenses, using the diffraction integral method. Each point source of light in the front focal plane gives rise to a plane wave in the back focal plane. Thus, the field in the back focal plane due to the point source in the front focal plane is (neglecting constants):

$$E(x, y) = E(x', y')\, e^{-j(2\pi f/\cos\theta)}\, e^{-jk(x+x')\sin\theta\,\cos\phi}\, e^{-jk(y+y')\sin\theta\,\sin\phi}$$

where

$$\theta = \frac{\sqrt{x'^2 + y'^2}}{f} = \frac{r'}{f}$$

$$\tan\phi = -\frac{y'}{x'}$$

By the superposition principle, the total field in the rear focal plane due to the entire electric field distribution in the front focal plane is given by the addition of all point source fields in the front focal plane. Thus, the total field is

$$E(x, y) = \iint_{\text{lens aper}} E(x', y')\, e^{-j(2\pi f/\cos\theta)}\, e^{-jk(x+x')\sin\theta\,\cos\phi}\, e^{-jk(y+y')\sin\theta\,\sin\phi}\, dx'\, dy'$$

$$(5.25)$$

Now, since

$$\cos\theta = \sqrt{1 - \sin^2\theta} \cong \sqrt{1 - (r'/f)^2} \cong 1 - \frac{1}{2}\left(\frac{r'}{f}\right)^2,$$

then

$$\frac{1}{\cos\theta} \cong 1 + \frac{1}{2}\left(\frac{r'}{f}\right)^2$$

and

$$\sin\theta = \frac{r'}{f}; \qquad \cos\phi = -\frac{x'}{r'}; \qquad \sin\phi = -\frac{y'}{r'}$$

the superposition integral expression for the field in the back focal plane becomes

$$E(x, y) = \iint_{\text{lens aper}} E(x', y')\, e^{j(k/f)(xx' + yy')}\, dx'\, dy' \qquad (5.26)$$

and we see that the field in the back focal plane is the Fourier transform of the field in the front focal plane.

5.7 GAUSSIAN BEAMS IN HOMOGENEOUS MEDIA: EVALUATION OF THE DIFFRACTION INTEGRAL IN THE FRESNEL REGION

We've looked at Fresnel region image and Fourier transform fields from the diffraction integral perspective. Now we look at a specific field distribution, which is unique in that both its image and its Fourier transform have the same functional form. This function is the Gaussian function. The Gaussian function is not just a mathematical curiosity either; many lasers produce output beams having Gaussian field distributions.

In Chapter 4, we studied paraxial optical fields having quadratic phase fronts. However, that study was based on the assumption of the Fraunhofer region/zero-wavelength limit in which we were able to represent the optical fields via a simple ray model. In this section, we're going to remove these restrictions to examine a rather unique form of field propagation that occurs in the near-field (Fresnel) region between two transverse planes of an optical system. Our field model for this study will be the Gaussian beam field discussed in the preceding paragraph.

We'll postulate the following transverse plane field (in practice, such a field may readily be produced by many types of lasers):

$$E(x', y') = \hat{u}\, e^{-(r'/w_0)^2}\, e^{jkr'^2/2r_0}$$

where the unit vector lies in the transverse plane. A sketch of the amplitude and phase of such a Gaussian wavefront is shown in Fig. 5.17. The quadratic phase part of this field is nothing new; we've been dealing with quadratic phase fields throughout this book in our study of Gaussian optics. The quadratic phase term merely indicates a converging or diverging spherical wave in the paraxial approximation.

The interesting aspect of this converging Gaussian wave is that it doesn't converge to a point, as our ray optic model would indicate that it should. Instead it "converges" to a plane phase front field, having a minimum-diameter Gaussian amplitude distribution. (This point is known as the "waist" of the beam and is nothing more than the quantity we've been heretofore referring to as the "image" of the convergent field.) We may easily use the diffraction integral equation [with the Fresnel region formulation, (5.6)] to obtain the transverse plane fields due to this distribution.

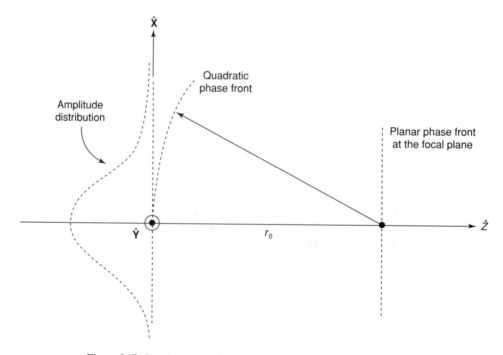

Figure 5.17 Gaussian amplitude and phase distribution in a transverse plane.

Thus,

$$E(x, y, z) = jk \frac{e^{-jkz}}{2\pi z} e^{-jkr^2/2z} \hat{u} \iint e^{-(r'/w_0)^2} e^{jkr'^2/2r_0} e^{-jkr'^2/2z} e^{jkrr'\cos(\phi - \phi')/z} r' \, dr' \, d\phi'$$

(5.27)

where

$$r^2 = \sqrt{x^2 + y^2} \quad \text{and} \quad r'^2 = \sqrt{x'^2 + y'^2}$$

The phi-integral is evaluated using the expression for the Bessel function of order zero, from Chapter 2. This yields

$$E(x, y, z) = jk \frac{e^{-jkz}}{z} e^{-jkr^2/2z} \hat{u} \iint J_0(krr'/z) \, e^{-[1/w_0^2 - j(k/2)(1/r_0 - 1/z)]r'^2} \, r' \, dr'$$

(5.28)

The remaining radial integral may be integrated in closed form over the infinite range, using Eq. 11.4.29 from Abramowitz and Stegun [2]. Thus, we obtain the following expression for the Fresnel zone field of the Gaussian field distribution:

$$E(x, y, z) \propto \frac{e^{-jkz}}{z} e^{-jkr^2/2z} e^{-(kr/2z)^2/[(1/w_0^2) - j(k/2)(1/r_0 - 1/z)]}$$

(5.29)

This equation describes the propagation of the Gaussian beam as a function of the distance, z, from the plane of the original Gaussian distribution. Let's now set $r_0 = \infty$ (this places the waist at $z = 0$) and see how the Gaussian amplitude beam propagates as it leaves the constant-phase plane. (It's a lot easier mathematically to start out at this plane and then look at how the field expands from this point.)

So, the magnitude of the Gaussian beam has the following form as a function of distance z from the waist

$$|E(x, y)| = e^{-(kr)^2/[(2z/w_0)^2 + (kw_0)^2]}$$

(5.30)

and the phase of the Gaussian beam has the following form as a function of distance from the waist

$$\angle E(x, y) = -\frac{kr^2}{2z} \frac{1}{1 + \left(\dfrac{kw_0^2}{2z}\right)^2}$$

(5.31)

By the first of these equations, we see that the effective width of the Gaussian beam varies as

$$\text{width}^2 = \left(\frac{2z}{kw_0}\right)^2 + w_0^2$$

(5.32)

where w_0 is the effective width of the Gaussian beam at its waist. And by the second equation, we see that the curvature of the Gaussian beam phase front varies as

$$R = z \left[1 + \left(\frac{kw_0^2}{2z}\right)^2\right]$$

(5.33)

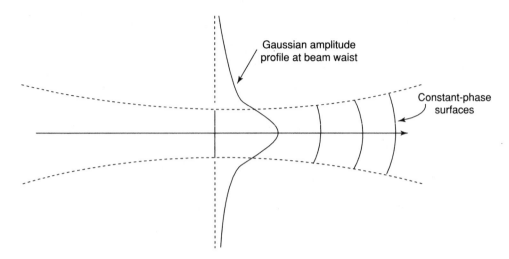

Figure 5.18 Sketch of a Gaussian beam near its beam waist.

At $z = 0$ (the assumed position of the beam waist), the curvature of the field is infinite (as we postulated), and as z tends to ∞, the curvature tends to z, as we'd expect from a Fraunhofer region analysis. In the middle region between the beam waist and the far field, things get interesting, however. There is a ''competition'' between two opposing terms for the curvature, one that varies linearly with distance z and one that varies inversely with distance z. Gaussian beams are often drawn in the form shown in Fig. 5.18. The figure shows how the beam narrows down to a waist and then diverges.

This ''beam waist'' phenomenon is not just a property of Gaussian beams, although the term is used primarily in connection with this type of beam. To reiterate, the Gaussian beam does not focus at a point, but rather, it narrows down to a waist, where the phase is uniform in the transverse plane. Now, as we saw in Example 2.2 from Chapter 2, the same sort of phenomenon occurs even for a uniformly illuminated aperture. In that context, we were looking at the Fraunhofer region field radiated by the aperture, but by reciprocity, we could equally well have applied the analysis to the problem of a spherical wave converging on its focal point. In any event, that analysis indicated that according to the diffraction integral formulation, a converging spherical wave does not converge to a point. Instead, it converges to a planar field distribution that is given by the inverse Fourier transform of the angular field distribution of the converging spherical wave. It just so happens that the Gaussian distribution is transformed into itself under both Fourier and Fresnel transformations, so it's most natural to think of this distribution as having a ''waist.'' However, all converging spherical waves exhibit this phenomenon.

As we've seen, in the near field of a planar aperture distribution, the ''rays'' (we're not really justified in using the term in the near field, since it is strictly a far-field term) do not propagate in straight lines—they ''curve.'' This is important to keep in mind when trying to analyze the operation of optical instruments, because it is easy to get into a sloppy thinking pattern of viewing rays as converging to a point. It's important to understand when this ray model is valid (at least for the purpose of understanding the operation of an optical device) and when it is totally inaccurate.

5.8 APPLICATIONS OF GAUSSIAN BEAMS: LASER CAVITY RESONATOR

Gaussian beams have practical importance, being the typical output beam produced by a number of different kinds of lasers. The reason for this is that many lasers use spherical mirror resonators to confine the optical fields in the active laser medium. When the two mirrors are close enough to be in each other's near field, the fields between the two mirrors will be represented by a Fresnel region form of the radiation integral.

Before we look at the action of spherical mirrors on Gaussian beams, it will be worthwhile to invert relations (5.32) and (5.33). To do this, we first divide (5.32) by (5.33) to obtain

$$\frac{2z}{kw_0^2} = \frac{kw^2}{2R} \tag{5.34}$$

Substituting this equation into (5.32) and (5.33) gives

$$w_0^2 = \frac{w^2}{1 + \left(\dfrac{kw^2}{2R}\right)^2} \tag{5.35}$$

$$z = \frac{R}{1 + \left(\dfrac{2R}{kw^2}\right)^2} \tag{5.36}$$

These inverse relations are very important because they give the beam waist width and beam waist location as a function of distance z from the transverse plane where the beam has a width w and radius of curvature R. (The wave is assumed converging; otherwise there will be no waist for positive z.)

By the last two formulas, it is clear that the Gaussian beam will be reflected back onto itself when the spherical mirrors have exactly the same curvature as the diverging Gaussian beam, as shown in Fig. 5.19. In this way, the Gaussian beam will be folded back onto itself on reflection, producing a resonant field.

The resonant properties of fields between mirrors has been likened to the propagation of optical fields in periodic lens waveguides [2,3]. This analogy holds for both ray fields and Gaussian beam fields. Consider Fig. 5.20, which shows the situation schematically for Gaussian beams. Often in thin lens analysis, we assume the lens to be in the Fraunhofer region of both the object and image distributions. That assumption, coupled with the zero-wavelength assumption, gives us the simple ray analysis that we derived in Chapter 2. If we now assumed the lens to be in the Fresnel region of both the object and image planes, lens analysis would proceed as follows. The lens law remains the same, except it now relates the radii of curvature of the object and image plane fields at the front and back faces of the lens, respectively. Thus, by the lens law, the radii of curvature R_0, R_i of the object and image wavefronts are related, as before, by the equation

$$\frac{1}{R_0} + \frac{1}{R_i} = \frac{1}{f} \tag{5.37}$$

where now

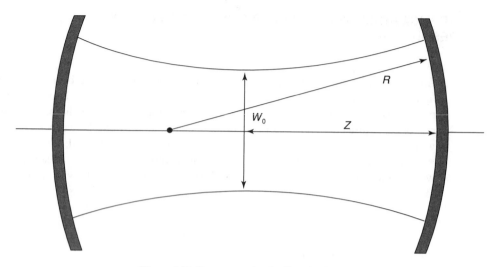

Figure 5.19 Resonant cavity for Gaussian beams.

$$R_0 = d_0 \left[1 + \left(\frac{kw_0^2}{2d_0} \right)^2 \right] \qquad (5.38a)$$

$$R_i = d_i \left[1 + \left(\frac{kw_0^2}{2d_i} \right)^2 \right] \qquad (5.38b)$$

and d_0, d_i are the object and image distances (along the z-axis) from the thin lens.

So, knowing how a thin lens transforms the radii of curvature of a Gaussian beam (the lens does not affect the width of the beam), along with the equations for calculating beam waist and waist position as a function of beam width and curvature, we may calculate the propagation of a Gaussian beam through a system of lenses. Note that the Gaussian beam tends to an ordinary ray field as

$$\frac{2z}{kw_0^2} \quad \text{tends to infinity}$$

In this limit, $R(z) = z$ and

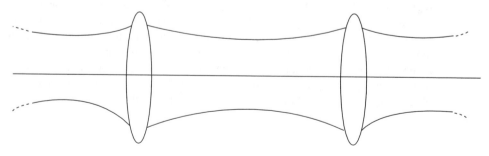

Figure 5.20 Lens waveguide for Gaussian beams.

$$w = \frac{2z}{kw_0}$$

5.9 GAUSSIAN BEAMS AND PULSE COMPRESSION

From (5.35), the equation for the waist width in terms of the beam width and radius of curvature, it's clear that the waist w_0 can be made arbitrarily narrow by making R small. Let's look at the implications of this property of Gaussian waves.

Consider Fig. 5.21, which shows a Gaussian beam at various points in an optical system. At plane P_1, the phase is planar and the beam has width $w_{0,1}$. The beam propagates to plane P_2, according to relations (5.32) and (5.33), and the beam width is now $w \gg w_{0,1}$. The lens L placed at P_2 alters the curvature R of the wavefront, making it highly convergent (i.e., R is made small). By (5.35), this wavefront now converges to a waist smaller than the original waist $w_{0,1}$. Pulse compression in this case is achieved via the greater angular bandwidth of the field on RHS of the lens versus that of the field on the LHS of the lens. This process, wherein the beam is expanded, multiplied by a quadratic phase, and then re-formed as a smaller spot, is called *pulse compression*. Pulse compression is also used in the time domain to generate pulses having faster switching times than the lasers that created them.

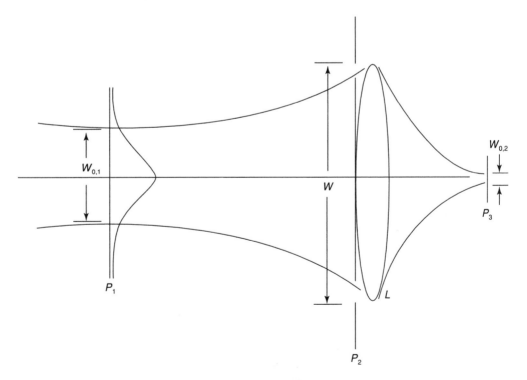

Figure 5.21 On pulse compression in the spatial domain. (Field in P_1 is a magnified version of field in P_3, and lens L alters the spectrum in P_2 according to the sine condition, to produce the reduced image in plane P_3.)

5.10 EDGE DIFFRACTION

In this section, we'll look at the form of the Fresnel region fields when the aperture is unbounded. The particular unbounded aperture we'll consider will have infinite extent in one direction and semi-infinite extent in the other. This "aperture" is the semi-infinite plane or *knife edge*.

The diffracted fields we'll calculate using this method are not exact. Our analysis will involve the so-called *physical optics* assumption that we've used all along for diffracted field calculations. Under this assumption, the aperture fields are simply taken as the incident fields themselves. In truth, the real aperture fields are not equal to the incident fields; this is only a convenient assumption made in order to obtain a tractable solution. As we know, at high frequencies, this assumption is generally a pretty good one, but it is still not exact.

It turns out that the half-plane problem has an exact solution [4], but it is extremely involved mathematically. Therefore, we'll look at edge diffraction using the approximate physical optics theory. Amazingly enough, both theories lead to the same type of functional description of the edge-diffracted fields, namely, in terms of Fresnel integrals.

We'll analyze the problem shown in Fig. 5.22, namely, that of calculating the fields on plane P_2 due to the edge diffraction caused by the half-plane in plane P_1. In the first section of this chapter, we obtained the equation for the Fresnel region fields as

$$E(x, y) = jk \frac{e^{-jkz}}{2\pi z} \iint_{P_1} E^{\text{tan}} (x', y') e^{-j(k/2z)[(x-x')^2 + (y-y')^2]} dx' \, dy' \qquad (5.39)$$

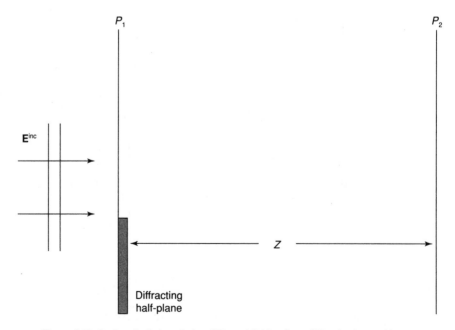

Figure 5.22 On the calculation of edge-diffracted fields using a diffraction integral formulation.

We'll assume a normally incident, unit-amplitude uniform plane wave field impinging on the "aperture" plane. (The incident field amplitude is unity.) The edge will lie along the y-axis, so that the integrals in the y-direction have infinite limits and those in the x-direction extend from 0 to infinity. The infinite integral in the y-direction is evaluated using the relation

$$\int_{-\infty}^{\infty} e^{-j(k/2z)(y-y')^2}\, dy' = \sqrt{\frac{2\pi z}{k}}\, e^{-j\pi/4} \tag{5.40}$$

This means the integral for the Fresnel region field is

$$E(x, y) = jk\frac{e^{-jkz}}{2\pi z}\sqrt{\frac{2\pi z}{jk}}\int_{0}^{\infty} e^{-j(k/2z)(x-x')^2}\, dx' \tag{5.41}$$

The change of variables defined by

$$u = x' - x$$

gives the edge-diffracted field as

$$E(x, y) = \sqrt{\frac{jk}{2\pi z}}\, e^{-jkz}\int_{-x}^{\infty} e^{-j(k/2z)u^2}\, du \tag{5.42}$$

Making the additional change of variables

$$v = \sqrt{\frac{k}{2z}}\, u$$

yields finally

$$E(x) = \sqrt{\frac{j}{\pi}}\, e^{-jkz}\int_{-\sqrt{kx/(2z)}}^{\infty} e^{-jv^2}\, dv \tag{5.43}$$

This is the expression we sought for the edge-diffracted field. This expression is readily expressed in terms of Fresnel integrals. The relation (5.40) is used to obtain

$$E(x) = \sqrt{\frac{j}{\pi}}\, e^{-jkz}\left\{\sqrt{\frac{\pi}{4j}} + \int_{-\sqrt{kx/(2z)}}^{0} e^{-jv^2}\, dv\right\}$$

$$= \sqrt{\frac{j}{\pi}}\, e^{-jkz}\left\{\sqrt{\frac{\pi}{4j}} + \int_{0}^{\sqrt{kx/(2z)}} e^{-jv^2}\, dv\right\} \tag{5.44}$$

Equation (5.44) is valid when $x > 0$ (i.e., when x lies in the illuminated portion of plane P_2).

When $x < 0$, that is, when x lies in the shadow portion of the plane P_2, then

$$E(x) = \sqrt{\frac{j}{\pi}}\, e^{-jkz}\int_{\sqrt{kx/(2z)}}^{\infty} e^{-jv^2}\, dv \tag{5.45}$$

Sommerfeld [5] gives a very good description of the Fresnel integrals in (5.44) and (5.45), as well as a description of their properties and the properties of the diffracted fields. If we define the Fresnel integral in the traditional form,

$$F(w) = \int_{0}^{w} e^{j(\pi/2)\tau^2}\, d\tau$$

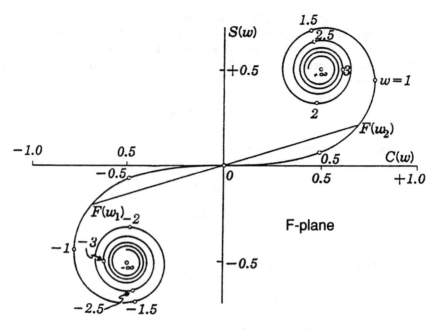

Figure 5.23 Cornu's spiral (after Sommerfeld).

then a graphical representation for the integral is given by Cornu's spiral, shown in Fig. 5.23. We see that the magnitude of the integral from 0 to some point w is oscillatory, whereas the integral from a point w to infinity is monotonically decreasing with increasing w. Therefore, it's not hard to see that the amplitude of the Fresnel region field on plane P_2 is given by the graph shown in Fig. 5.24.

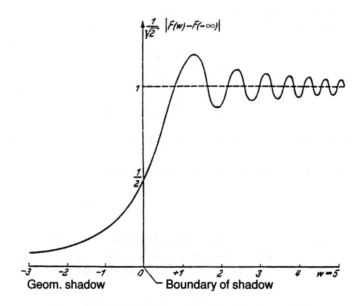

Figure 5.24 Field behind a straight edge (after Sommerfeld).

5.11 APERTURE STOPS AND DEPTH OF FIELD: AN INTRODUCTION TO SAMPLING IN THE SPATIAL DOMAIN

An aperture stop is nothing more than a transmissive window in an opaque screen, designed to pass light only through the aperture and to block it everywhere else. The best known example of an aperture stop to most of us is the pupil of the human eye. If the ambient light is too intense, the pupil automatically contracts, protecting the retina from photonic damage. If the ambient light is too dim, the pupil automatically expands, allowing the maximum amount of light onto the retina for improved visual detection. In this case, the pupil aperture stop acts as a power regulator.

Many types of aperture stops can be used in an optical system. The type of aperture stop discussed here (located at the entrance pupil of the system) serves mainly to restrict the amount of light energy entering into the system. This type of stop performs an additional function. This second function involves an important property of the imaging system known as "depth of focus." The depth of focus of an optical system is related to the far-field criterion discussed in Chapter 1. In that discussion, we determined the far-field distance as

$$R_{f.f.} = \frac{2d^2}{\lambda} \tag{5.46}$$

where d is the maximum lateral extent of the aperture. Clearly, the smaller the aperture, the closer the far-field range. All objects lying past the far-field range will be focused at the back focal plane of the system. When the image plane is located at the back focal plane of the system, the depth of focus will extend from the far-field range out to an infinite range. All objects located within these range limits will be in sharp focus. Objects located closer than this range will be out of focus at the back focal plane of the system.

When an optical system is focused on objects located a finite distance from the entrance pupil of the system, the entrance aperture stop still produces the same depth of field phenomenon. That is, the smaller the entrance pupil, the broader the longitudinal range of focus of the optical system. We can perform a quick calculation to show the effect of aperture extent on depth of focus. This may be done with the aid of Fig. 5.25. We are going to calculate the field at the axial image point I due to the axial object point source O located at an object distance $d_0 - \epsilon$.

The phase of the field due to the point source is

$$e^{j(k\rho^2)/[2(d_0 - \epsilon)]}$$

The phase of the field behind the lens is

$$e^{j(k\rho^2/2)[1/(d_0 - \epsilon) - 1/f]}$$

In the case of a circular aperture of radius a, a simple scalar Fresnel zone expression for the field at I is

$$E_I = \int_0^{2\pi} \int_0^a e^{(k\rho^2/2)[1/(d_0 - \epsilon) - 1/f + 1/d_i]} \, \rho \, d\rho \, d\phi \tag{5.47}$$

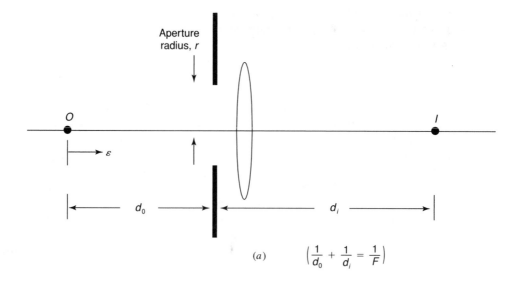

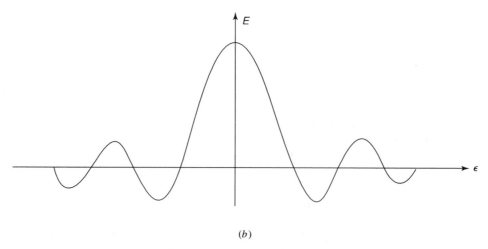

Figure 5.25 Field at the axial image point, I, due to a point source at axial object point located $d_0 - \epsilon$ from the lens: (a) Optical system and (b) Plot of relative field strength E at the image point (as a function of axial defocusing from object point).

which is easily integrated to yield a magnitude proportional to

$$\text{sinc}\,\frac{ka^2\eta}{4}$$

where

$$\eta = \frac{1}{d_0 - \epsilon} - \frac{1}{f} + \frac{1}{d_i}$$

Adding and subtracting $1/d_0$ to the RHS of the equation above, and enforcing the lens law,

$$\frac{1}{d_0} - \frac{1}{f} + \frac{1}{d_i} = 0$$

yields finally the electric field at the axial image point I due to the axially displaced object point as

$$E_I \cong \text{sinc } k\epsilon \left(\frac{a}{2d_0}\right)^2 \tag{5.48}$$

We see that as the ratio a/d_0 becomes less, the sinc function spreads out, making more and more of the object region in-focus (where focusing here is defined in terms of power level).

REFERENCES

[1] Born, M., and Wolf, E., *Principles of Optics,* 6th ed. New York: Pergamon Press, 1984, pp. 480–483.

[2] Kogelnik, H., and Li, T. "Laser Beams and Resonators," *Applied Optics,* vol. 5, no. 10, October 1966.

[3] Yariv, A., *Quantum Electronics,* New York: McGraw-Hill, 1989.

[4] Born, M., and Wolf, E., *Principles of Optics,* 6th ed. Chapter 11.

[5] Sommerfeld, A., *Optics,* New York: Academic Press, 1967.

6

Focusing and Imaging Properties of Lenses: The Plane Wave Spectrum Viewpoint

So far, we've looked at lens action from the ray optical and diffraction integral viewpoints. Now we'll go back and revisit this subject a third time, from the plane wave spectrum viewpoint, and see how it allows the incorporation of Fourier analysis techniques into the analysis and design of lenses and other optical systems. We'll also see how the plane wave spectrum viewpoint leads to new lens figures-of-merit for optical systems.

6.1 PLANE WAVE SPECTRUM ANALYSIS OF DIFFRACTION, IMAGING, AND FOURIER TRANSFORMING BY LENSES AND APERTURES

The point spread function introduced in Chapter 5 determines the image plane response of an optical system to a point source of light in the object plane. In this section, we'll introduce a complementary function in the spectral domain. This is the coherent transfer function, which determines the response of an optical system to a point source of (coherent) light in the spectral domain (i.e., a single propagating plane wave component out of an entire plane wave spectrum).

We briefly alluded to the plane wave spectrum approach to imaging in Chapter 5, using the logic that a 2-D delta function in the object plane is "smeared" into an Airy function in the image plane via a bandlimited (i.e., angle limited) plane wave filtering device. To continue with this concept, we're going to look at the "image" produced by an optical system illuminated by an incident plane wave. To begin, we'll consider the

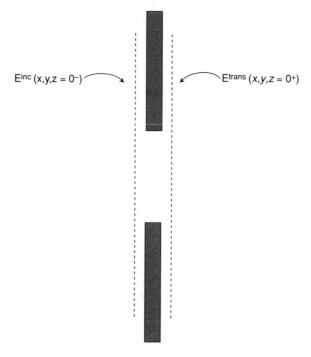

$E^{inc} (x,y,z = 0^-)$ $E^{trans} (x,y,z = 0^+)$

Figure 6.1 On the ''imaging'' of a plane wave by a diffracting aperture.

optical system consisting of the unfocused diffracting aperture shown in Fig. 6.1. (Later on, we'll consider the effects of focusing on the imaging of plane waves.) Since we're dealing with an unfocused, rather than focused, aperture, the ''image'' of the incident plane wave field will appear on the sphere at infinity. Imagine that a coherent field propagating in the $+z$-direction is incident on the aperture, and say that the transmitted field at $z = 0^+$ is given by a direct multiplication in the spatial domain of the incident field times the aperture function. Thus,

$$E^{trans}(x, y, z = 0^+) = A(x, y)E^{inc}(x, y, z = 0^-) \qquad (6.1)$$

where $A (x, y)$ is a windowing function defined by the relation

$$A (x, y) = 1 \qquad \text{inside the aperture}$$

$$A (x, y) = 0 \qquad \text{outside the aperture}$$

Note.   This kind of direct multiplication of the incident field by the aperture function produces accurate results when the aperture is many square wavelengths in lateral extent. When the aperture is on the order of a few wavelengths in extent, the transmitted field is not obtained so easily. In that case, it's necessary to go through a separate calculation (using a numerical integral equation formulation such as the method of moments) just to determine the field that is transmitted through the aperture.

To obtain a spectral domain expression for the transmitted field, we express the aperture function $A(x, y)$ and the incident field $E^{inc}(x, y)$ in spectral form using the Fourier transform pairs,

$$f(x, y) = \iint_{-\infty}^{\infty} F(\alpha, \beta) e^{j(\alpha x + \beta y)} \, d\alpha \, d\beta \tag{6.2}$$

$$F(\alpha, \beta) = \frac{1}{4\pi^2} \iint_{-\infty}^{\infty} f(x, y) e^{-j(\alpha x + \beta y)} \, dx \, dy \tag{6.3}$$

and then plug these spectral expressions into (6.1). Thus, by simply multiplying the spectral expansions for the incident field and the aperture function together, we can obtain the following expression for the (spatial domain) field at $z = 0^+$:

$$E^{\text{trans}}(x, y) = \iint_{-\infty}^{\infty} A(\alpha', \beta') e^{j(\alpha'x + \beta'y)} \, d\alpha' \, d\beta' \cdot \iint_{-\infty}^{\infty} E^{\text{inc}}(\alpha'', \beta'') e^{j(\alpha''x + \beta''y)} \, d\alpha'' \, d\beta'' \tag{6.4}$$

The Fourier transform of this spatial field is evaluated using (6.3) as

$$E^{\text{trans}}(\alpha, \beta) = \frac{1}{4\pi^2} \iint_{-\infty}^{\infty} \left\{ \iint_{-\infty}^{\infty} A(\alpha', \beta') e^{j(\alpha'x + \beta'y)} \, d\alpha' \, d\beta' \right\}$$
$$\cdot \left\{ \iint_{-\infty}^{\infty} E^{\text{inc}}(\alpha'', \beta'') \, e^{j(\alpha''x + \beta''y)} \, d\alpha'' \, d\beta'' \right\} e^{-j(\alpha x + \beta y)} \, dx \, dy \tag{6.5}$$

Using the definition of the delta function,

$$\begin{Bmatrix} \delta(x) \\ \delta(\alpha) \end{Bmatrix} = \frac{1}{2\pi} \int_{-\infty}^{\infty} e^{\pm j\alpha x} \begin{Bmatrix} d\alpha \\ dx \end{Bmatrix}$$

we can reduce this equation to the form

$$E^{\text{trans}}(\alpha, \beta) = \iint_{-\infty}^{\infty} E^{\text{inc}}(\alpha'', \beta'') A(\alpha - \alpha'', \beta - \beta'') \, d\alpha'' \, d\beta'' \tag{6.6}$$

It's evident then, that the action of an aperture in the spectral domain is not multiplicative, as it is in the spatial domain. In fact, if the incident field is a single plane wave function (a delta function in the spectral domain), that is, if

$$E^{\text{inc}}(\alpha'', \beta'') = \hat{u}\delta(\alpha'' - \alpha_0, \beta'' - \beta_0)$$

as we had indicated earlier at the beginning of the chapter, then

$$E^{\text{trans}}(\alpha, \beta) = \hat{u}A(\alpha - \alpha_0, \beta - \beta_0)$$

This says that a single plane wave spectral component incident on the aperture gives rise to an entire spectrum of plane waves, whose spectrum is determined by the Fourier transform of the aperture function. (For the case of a circular aperture, the Fourier transform of the aperture function is the Airy function described in Chapter 1.) And since the Fourier transform of the aperture function is the diffracted field that appears on the infinite sphere (this is where the "image" produced by the unfocused aperture will appear), the Fourier transform of the aperture function is then the "image" of the incident plane wave field.

As shown in Chapter 5, a point source illumination function in the object plane of a lens is imaged onto a pattern in the image plane, where the shape of that image plane pattern is the point spread function (given as the Fourier transform of the exit aperture function). Therefore, we see that a spatial point source in the object plane of a focused aperture is imaged in exactly the same way that a spectral point source is imaged by an unfocused aperture.

For the case of this unfocused aperture, we cannot use linear systems theory to analyze the diffraction effect of the aperture. In ordinary linear systems theory, a sinusoidal input function gives rise to a sinusoidal output function of the same frequency, changed only in magnitude or phase. In the case of this unfocused diffracting aperture, however, a (spatially) sinusoidal input gives rise to an entire *spectrum* of plane waves, containing many different spatial frequencies not present in the original incident field.

We will soon find out, however, that when a (spatially) sinusoidal field exists in the unit-magnification object plane (two focal lengths in front of a lens), the dominant field component in the image plane (two focal lengths behind the lens) will be a spatially sinusoidal field having the same spatial frequency as the object plane field. Other spatial frequency components will, of course, be present, due to the finite lens aperture, lens aberrations, and so on, but the dominant field will be an image of the object field. In this *special case* then, we can use linear systems theory to regard the lens as a device that multiplies the object-plane sinusoidal field by a complex "transfer function" to produce the image-plane field. The lens will be shown to be a low (spatial) frequency filter, passing paraxial plane wave fields but attenuating fields tilted with respect to the z-axis of the system (as shown in Fig. 5.7).

Inspection of (6.1) and (6.6) indicates that this particular optical system acts in the same fashion as an AM heterodyne system. The incident field

$$E^{\text{inc}}(x, y) = e^{j(\alpha_0 x + \beta_0 y)}$$

or

$$E^{\text{inc}}(\alpha, \beta) = \delta(\alpha - \alpha_0, \beta - \beta_0)$$

is a type of spatial frequency "carrier" and the function $A(x, y)$ is a type of "baseband waveform" that amplitude-modulates the sine-wave "carrier" according to (6.1). A graphical sketch of this phenomenon is shown in Figs. 6.2 and 6.3 for the case of two incident

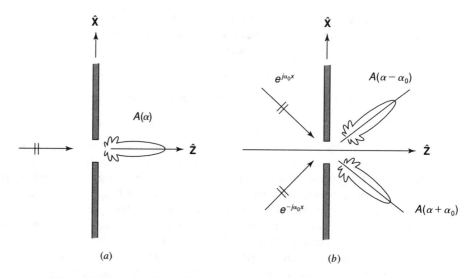

(a)   (b)

Figure 6.2 Illustration of the AM heterodyning phenomenon of an unfocused diffracting aperture (polar plot): (a) Far-field diffraction pattern of an unfocused aperture under plane wave illumination (normal incidence) and (b) Far-field diffraction patterns of unfocused aperture under plane wave illumination (oblique incidence).

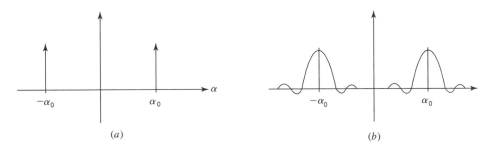

(a) (b)

Figure 6.3 AM heterodyning effect of unfocused diffracting aperture (cont.). $(x - y$ plot):
(a) Incident field spectrum and (b) Transmitted field spectrum.

plane waves. In the time/frequency domains, an amplitude-modulated carrier has the same spectrum as the baseband waveform, just centered about the carrier frequency. In the spatial/spatial frequency domains, an amplitude-modulated carrier has the same plane wave spectrum as the baseband waveform, only tilted by an amount equal to the tilt of the incident plane wave. Angular resolution of the two peaks shown in Fig. 6.3 is analogous to the radio problem of separating two closely spaced stations on the AM dial.

Let's now consider what happens when a plane wave is incident on a *focused* diffracting aperture. Consider the finite aperture located just in front of a lens, as shown in Fig. 6.4. The illumination is given as

$$E^{\text{inc}}(\alpha, \beta) = \hat{u}\delta(\alpha - \alpha_0, \beta - \beta_0)$$

and, as before, the diffracting aperture causes the incident plane wave field (a delta function in the spectral domain) to be spread out into a *spectrum* of waves. This plane wave spectrum is then incident on the lens. Each plane wave component of this spectrum is focused onto a spot in the focal plane, and each of these spots will have an intensity proportional to the intensity in the diffracted plane wave which converged on that spot. Thus, the intensity distribution in the focal plane is given by

$$I(x - \epsilon_x, y - \epsilon_y) = \left|E^{\text{trans}}(\alpha - \alpha_0, \beta - \beta_0)\right|^2$$

where

$$k\frac{\epsilon_x}{f} = \alpha_0, \qquad k\frac{\epsilon_y}{f} = \beta_0$$

If we now have a source distribution located a distance d_0 in front of the focused aperture (we'll now allow the object distance to be arbitrary to allow for magnification), then the object-plane distribution may be expressed in terms of a 2-D Fourier transform as

$$E^{\text{obj}}(x, y) = \frac{1}{4\pi^2} \iint_{-\infty}^{\infty} E^{\text{obj}}(\alpha, \beta)e^{-j(\alpha x + \beta y)} \, d\alpha \, d\beta$$

That is, the source distribution may be expressed in "spectral terms" as an integral of plane waves. Each propagating plane wave component in the source distribution will be focused by the diffracting aperture, as described above, to a spot in the back focal plane. Each of these spots in the back focal plane will have an intensity proportional to the intensity of the particular Fourier plane wave component in the object distribution

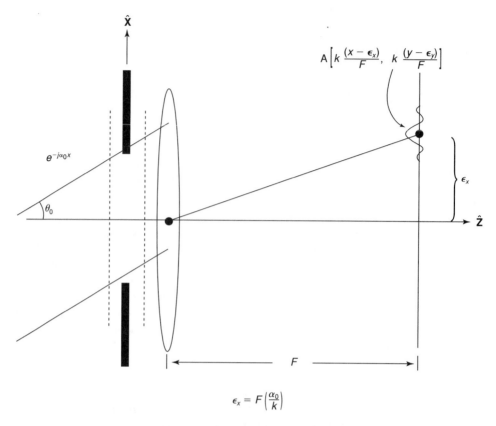

$$\epsilon_x = F\left(\frac{\alpha_0}{k}\right)$$

Figure 6.4 Diffracting aperture located just in front of a thin lens (oblique plane wave incidence).

which was focused to that spot (and it will have a pattern determined by the Fourier transform of the exit aperture function). *Note:* Even though each "spot" will have a finite pattern, we're going to ignore that fact for now and regard the spot as a perfect point.

As shown by the construction in Fig. 6.5, a quadratic phase taper will be associated with the Fourier-spectrum intensity distribution in the back focal plane. Therefore, the wave in the back focal plane of the lens will have a quadratic/spherical wave phase taper (and it will correspond to a *converging* spherical wave when $d_0 > f$). This means that (in a GO sense) the wave will converge to a point in the image plane as shown. In a physical optics (Fresnel region) sense, the wave will focus to a planar-phase image distribution equal to the Fourier transform of the angular plane wave spectrum distribution.

We may readily obtain the usual image magnification factor from Fig. 6.6. From the figure, we see that the object-plane plane wave component denoted by α is transformed into the image-plane plane wave component denoted by $-\alpha$. Therefore, the image-plane field distribution will be given by

$$E^{\text{image}}(x_i, y_i) = \iint E^{\text{obj}}(\alpha, \beta)e^{-j(-\bar{\alpha}x_i - \bar{\beta}y_i)}\, d\bar{\alpha}\, d\bar{\beta}$$

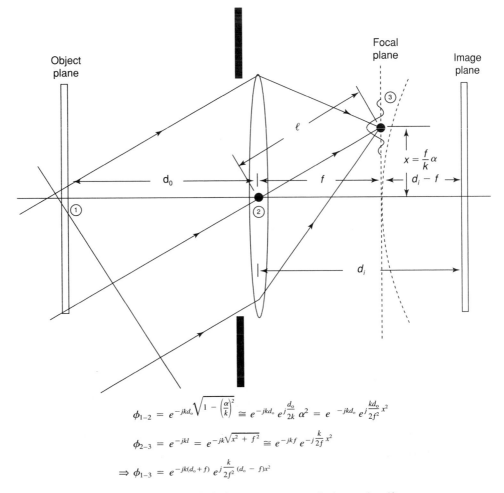

$$\phi_{1-2} = e^{-jkd_o}\sqrt{1 - \left(\frac{\alpha}{k}\right)^2} \cong e^{-jkd_o}\, e^{j\frac{d_o}{2k}\alpha^2} = e^{-jkd_o}\, e^{j\frac{kd_o}{2f^2}x^2}$$

$$\phi_{2-3} = e^{-jkl} = e^{-jk\sqrt{x^2 + f^2}} \cong e^{-jkf}\, e^{-j\frac{k}{2f}x^2}$$

$$\Rightarrow \phi_{1-3} = e^{-jk(d_o + f)}\, e^{j\frac{k}{2f^2}(d_o - f)x^2}$$

This represents a spherical wave convergent at the image plane if

$$e^{j\frac{k}{2f^2}(d_o - f)x^2} = e^{j\frac{kx^2}{2(d_i - f)}} \quad \text{or} \quad \frac{1}{f} = \frac{1}{d_o} + \frac{1}{d_i}.$$

Figure 6.5 Plane wave spectrum derivation of the lens law.

or

$$E^{\text{image}}(x_i, y_i) = \left(\frac{f}{d_i - f}\right)^2 \iint E^{\text{obj}}(\alpha, \beta)\, e^{j(\alpha x_i/M + \beta y_i/M)}\, d\alpha\, d\beta$$

$$= E^{\text{obj}}(-x_0, -y_0) \tag{6.7}$$

where

$$x_0 = \left(\frac{f}{d_i - f}\right) x_i = \frac{1}{M}\, x_i$$

$$y_0 = \left(\frac{f}{d_i - f}\right) y_i = \frac{1}{M}\, y_i$$

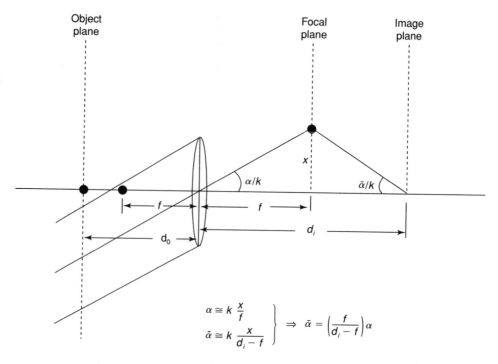

$$\left. \begin{array}{l} \alpha \cong k\, \dfrac{x}{f} \\[2mm] \bar{\alpha} \cong k\, \dfrac{x}{d_i - f} \end{array} \right\} \;\Rightarrow\; \bar{\alpha} = \left(\dfrac{f}{d_i - f} \right) \alpha$$

Figure 6.6 Image formation from a plane wave spectrum viewpoint.

which then gives the image-plane field as the appropriately magnified version of the object-plane field.

We may use a similar logical process to show how a lens performs a Fourier transform operation. Let $d_0 = f$ in Figs. 6.5 and 6.7. In this case, the field in the back focal plane is the Fourier transform of the object plane field, *with no quadratic phase*. Therefore,

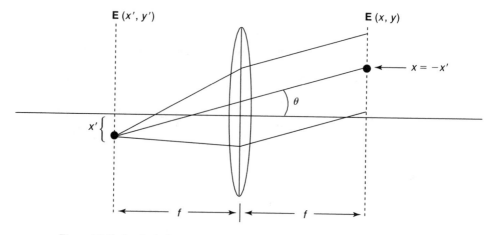

Figure 6.7 Each point in the source plane gives rise to a plane wave field in the observation plane.

the field in the back focal plane is now exactly equal to the Fourier transform of the object-plane field (neglecting the effect of diffraction by the unfocused aperture).

Another way of looking at imaging from a spectral domain point of view is to note that in the image plane, the plane wave spectral components of the object-plane distribution are reassembled in such a way as to have the same amplitude and (more importantly) the same phase relationships that they had in the object plane.

6.2 THE COHERENT TRANSFER FUNCTION: SHIFT INVARIANCE

We've seen how focused apertures can image fields, using the plane wave spectrum viewpoint. Now we'll look at the ways in which this plane wave spectrum viewpoint leads to concepts of lens action as well as to new measures of lens quality.

In previous chapters, we've seen how an object may be regarded as a superposition of point sources in the object plane (convolution in the spatial domain) or as a superposition of "plane wave sources" (plane wave spectrum representation). In the first case, the object-plane field is written as

$$O(x, y) = \iint_{-\infty}^{\infty} O(x_0, y_0)\delta(x - x_0)\ \delta(y - y_0)\ dx_0\ dy_0$$

where the subscripts indicate "object"-plane coordinates. In the second case, the object-plane field is written as

$$O(x, y) = \iint_{-\infty}^{\infty} O(\alpha, \beta)e^{j(\alpha x + \beta y)}\ d\alpha\ d\beta$$

Either representation may be used for describing the object-plane field. In the former "Huygens-type" representation, it's natural to think of the optical system as acting on each point source in the object plane, and then taking the total image plane field as the sum of all transformed point source fields in the image plane. That is, we would consider each point source to be transformed as

$$\delta(x - x_0)\delta(y - y_0) \rightarrow PSF(x - x_i, y - y_i)$$

where x_i, y_i is the GO image point of the object point x_0, y_0.

As we saw in a previous chapter,

$$PSF(x, y) = \frac{J_1(A\sqrt{x^2 + y^2})}{A\sqrt{x^2 + y^2}}$$

for a perfectly focused circular aperture. Where A is determined by the aperture radius and the half-angle of the ray cone from the aperture to the image point.

In the image plane, the previous Huygens-type convolution relation becomes

$$I(x, y) = \iint_{-\infty}^{\infty} O(x_i, y_i)PSF(x - x_i, y - y_i)\ dx_i\ dy_i \tag{6.8}$$

where the subscripts indicate image-plane coordinates.

In this case, we've implicitly assumed that the point spread function is independent of the image-plane coordinates x_i, y_i, that is, that the point spread function is *shift invariant*. In any actual imaging system, this will not be the case; at best, it is only approximately true over certain regions of the object plane.

So, if we know the point spread function of the imaging system, we may readily calculate the image-plane distribution from knowledge of the object-plane distribution.

In the case of the shift-invariant point spread function, (6.8) may be Fourier-transformed to produce

$$I(\alpha, \beta) = \frac{1}{4\pi^2} \iint_{-\infty}^{\infty} I(x, y) \, e^{-j(\alpha x + \beta y)} \, dx \, dy$$

$$= \frac{1}{4\pi^2} \iint_{-\infty}^{\infty} O(x_i, y_i) dx_i dy_i \iint_{-\infty}^{\infty} PSF(x - x_i, y - y_i) e^{-j(\alpha x + \beta y)} \, dx \, dy$$

$$= \frac{1}{4\pi^2} \iint_{-\infty}^{\infty} O(x_i, y_i) \, dx_i dy_i \, e^{-j(\alpha x_i + \beta y_i)} \iint_{-\infty}^{\infty} PSF(x - x_i, y - y_i)$$

$$\times \, e^{-j[\alpha(x - x_i) + \beta(y - y_i)]} \, dx \, dy$$

$$= 4\pi^2 O(\alpha, \beta) PSF(\alpha, \beta) \tag{6.9}$$

Thus, we see that for the shift-invariant lens system, the plane wave spectrum of the image-plane field is equal to the plane wave spectrum of the object-plane field modified by the plane wave response of the imaging system. (This relationship holds only for imaging optical systems, that is, optical systems that produce a likeness of the object plane distribution in the image plane).

As we've already seen in Chapter 5, the Fourier transform of the point spread function for a single lens is proportional to the complex (amplitude and phase) aperture illumination function, that is,

$$PSF(\alpha, \beta) \propto A\left[d_i\left(\frac{\alpha}{k}\right), d_i\left(\frac{\beta}{k}\right)\right]$$

where d_i is the image distance for the given lens and object distance.

For a perfect shift-invariant, diffraction-limited optical system, the image-plane electric field plane wave spectrum (PWS) is equal to the complex object-plane electric field PWS, multiplied by a pulse function in the shape of the exit aperture. As is well known in Fourier theory, this type of sudden truncation of a spectrum gives rise to a "ringing" phenomenon in the spatial domain, known as the Gibbs phenomenon [1]. The Gibbs phenomenon is manifested by the oscillatory nature of the function $J_1(x)/x$. If, instead of being suddenly truncated, the aperture is given a continuous amplitude taper, ranging from 100% transmittance at the center to some lower value of transmittance at the edge of the aperture, the ringing phenomenon in the point spread function can be reduced or eliminated, though at the expense of widening the point spread function. It is possible to go in the other direction as well, that is, to create a second discontinuity in the aperture field by obstructing the center of the aperture, causing additional ringing but also narrowing the width of the central spot. This latter technique is known in optics as *apodization* [2].

6.3 THE MODULATION TRANSFER FUNCTION (MTF) FOR INCOHERENT LIGHT

The coherent transfer function relating complex field amplitudes, though interesting and useful for introducing the transfer function concept in optics, has marginal utility in practice. This is because techniques of coherent detection have not yet reached the kind of broad usage in optics that they have in other areas of electrical engineering (e.g., in microwave technology). And the coherent transfer function relating actual field quantities is useful only in connection with coherent optical systems in which the magnitude and phase of the field can be recovered. The vast majority of optical systems use some form of incoherent detection based on received *power* level, rather than on actual electric field amplitude and phase, as in microwave systems.

So, now we'll introduce a second transfer function that has much broader usage in optics. This new transfer function relates power (field intensity) in the image plane to power in the object plane and is called the *optical transfer function* (OTF).

Whereas the coherent transfer function is the Fourier transform of the point spread function, the optical transfer function is the Fourier transform of the squared amplitude of the point spread function. (The squared amplitude is proportional to the electromagnetic power contained in the spot.) Thus, we now focus our attention on the power point spread function,

$$PPSF(x, y) = PSF(x, y)PSF^*(x, y)$$

(By the way, the notation PPSF is not a normal notation. It is used here merely to distinguish the power point spread function from the ordinary point spread function.)

The OTF is defined as the Fourier transform of the power point spread function. Thus,

$$OTF(\alpha, \beta) = \frac{1}{4\pi^2} \iint_{-\infty}^{\infty} PSF(x, y)PSF^*(x, y)e^{-j(\alpha x + \beta y)} \, dx \, dy \qquad (6.10)$$

where the point spread function is given as the Fourier transform of the aperture function via the relation

$$PSF(x, y) = \iint_{-\infty}^{\infty} A(\overline{\alpha}, \overline{\beta}) \, e^{j(\overline{\alpha}x + \overline{\beta}y)} \, d\overline{\alpha} \, d\overline{\beta}$$

For nonideal imaging systems, the "aperture function" is the complex illumination function of the exit aperture. Its conjugate is given similarly by

$$PSF^*(x, y) = \iint_{-\infty}^{\infty} A^*(\hat{\alpha}, \hat{\beta}) \, e^{-j(\hat{\alpha}x + \hat{\beta}y)} \, d\hat{\alpha} \, d\hat{\beta}$$

Substituting these last two expressions into the equation for the OTF yields the OTF in terms of the complex aperture illumination function as

$$OTF(\alpha, \beta) = \iint_{-\infty}^{\infty} A(\alpha + \hat{\alpha}, \beta + \hat{\beta}) \, A^*(\hat{\alpha}, \hat{\beta}) \, d\hat{\alpha} \, d\hat{\beta}$$

This equation is very important, and the reader who opts for a career in optics will see it again and again, especially in the field of optical testing. The equation, which is not terribly abstruse (being simply related to the convolution theorem for Fourier transforms), says that the OTF of a lens is equal to the *autocorrelation* (sometimes called the autocovariance [3]) of the complex aperture function of the lens. In reality, since the aperture distribution is in general complex, the OTF is not exactly the autocorrelation of the aperture distribution. It is actually the correlation of the aperture distribution with its conjugate [2].

The *magnitude* of the OTF, known as the *modulation transfer function* or *MTF,* is a purely real quantity and is readily measured. Note that the MTF, as with the aberration function and point spread function, is a function of the illumination (i.e., plane wave versus point source). In addition, note that the coherent transfer function is essentially the same as the aberration function added to a perfect spherical wave phase.

6.4 THE MTF OF A PERFECT LENS

Before going into too much detail about the relationship between MTF, the point spread function, and the lens aberration function, perhaps it's worthwhile to see what the MTF would be for a perfect lens. We know that the point spread function for a perfect lens is the Airy function, but what does the lens MTF look like? First, we'll define a perfect lens as a device that converts a uniform-amplitude incident plane wave field into a uniform-amplitude convergent spherical wave field. The converging spherical is shown in Fig. 6.8*a.* The normally incident plane wave, on passing through the perfect lens, will be converted into a uniform-amplitude aperture field distribution with spherical wave phase. This aperture distribution could, in principle, be Fourier transformed to obtain the true plane wave spectrum of the aperture distribution.

In the Gaussian regime (where spherical wavefronts are approximated by quadratic wavefronts), that spectrum would be very similar to the uniform-amplitude distribution shown by the solid line in Fig. 6.8*b,* and we can deduce this readily. We previously showed in Chapter 2 that the Fourier transform of an optical field could be identified with its Fraunhofer region radiation pattern. In the lens of Fig. 6.8*a,* the aperture distribution function has been transformed by the lens (by virtue of its adding a spherical wave phase to the aperture function) into the Fraunhofer region pattern of the image in plane *P.* It is in this way that the uniform-amplitude *aperture distribution function* gets converted into a uniform-amplitude *Fourier spectrum* of the image.

In reality, the Fourier spectrum is not exactly uniform-amplitude. Recall that the Fraunhofer region pattern is defined on a sphere of large radius, whereas the aperture distribution (soon to become the Fourier spectrum of the image distribution) is defined on a plane, namely, the plane of the exit aperture. Thus, rays propagating from aperture points near the rim of the lens to the image plane must propagate further than rays propagating from the lens vertex to the image plane. This additional propagation path induces a propagation path loss due to the $1/r$ spherical wave attenuation. For Gaussian optics, we may assume this additional propagation path loss to be negligible. We may see from Fig. 6.8 that the ratio of distances from the vertex to the focal point and from the rim to the focal point is equal to the cosine of the angle to the rim. Thus, the Fourier spectrum will be tapered by a cosine function. In the transmit case, the radiated field spectrum must be weighted by a secant distribution in order to compensate for this path length loss and produce a uniform aperture distribution [4].

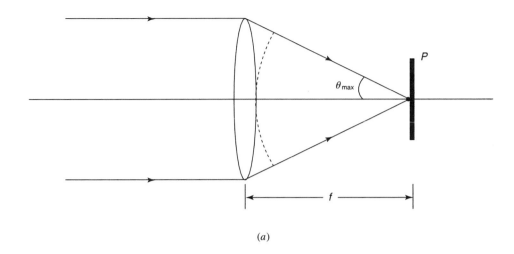

(a)

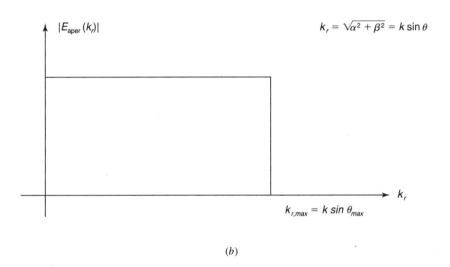

(b)

Figure 6.8 On the plane wave spectrum produced by an assumed perfect lens: (a) Spherical wave incident on the image plane, P, due to the normally incident plane wave field on the lens, and (b) Magnitude of spectrum of converging field.

Shown in three dimensions, the uniform aperture amplitude function is as shown in Fig. 6.9a, with the convolution function depicted in Fig. 6.9b. The MTF is calculated as shown in Fig. 6.10. The MTF is the area contained within the intersection of the two displaced circles and is given as indicated in the figure. A plot of the MTF of a circular lens under various amounts of axial defocus is shown in Fig. 6.11 [2]. In the figure, Born's variable f is the wavenumber (in cycles) in the x-direction and is equal to $\alpha/2\pi$. Also, R is the distance from the lens to the image plane, and a is the radius of the circular aperture.

For lenses, the MTF is in some ways a more useful figure-of-merit than the coherent transfer function. This is because image quality is more significantly degraded by phase errors than by amplitude errors in the aperture illumination function (equal to the coherent

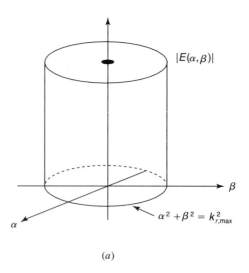

(a)

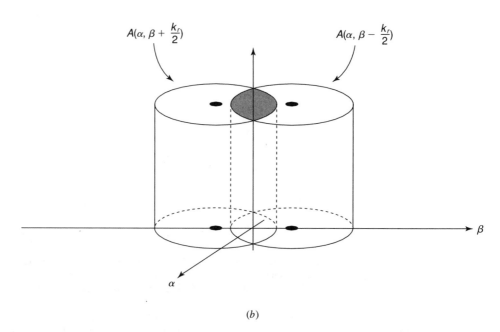

(b)

Figure 6.9 On the calculation of MTF for a perfect lens: (a) Spectrum of converging spherical wave in 3D, and (b) MTF is the area under the shaded portion common to the two shifted spectra.

transfer function). Thus, the amplitude of the complex aperture function/coherent transfer function will not yield significant insight to the quality of the imaging system. The MTF, however, is very sensitive to phase errors in the complex aperture illumination function. Hence, MTF magnitude as a function of spatial frequency yields an immediate graphical tool for visualizing lens quality, lumping all phase aberrations into a single (real-valued) curve.

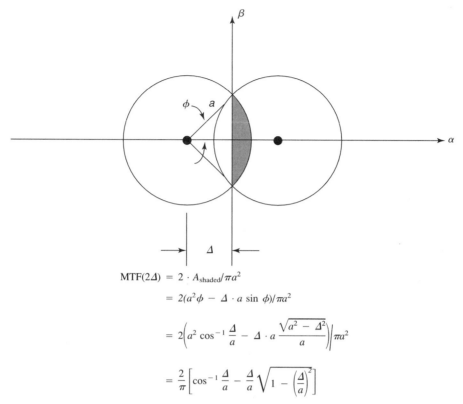

$$\text{MTF}(2\varDelta) = 2 \cdot A_{\text{shaded}}/\pi a^2$$

$$= 2(a^2\phi - \varDelta \cdot a \sin \phi)/\pi a^2$$

$$= 2\left(a^2 \cos^{-1} \frac{\varDelta}{a} - \varDelta \cdot a \frac{\sqrt{a^2 - \varDelta^2}}{a}\right)\bigg/\pi a^2$$

$$= \frac{2}{\pi}\left[\cos^{-1} \frac{\varDelta}{a} - \frac{\varDelta}{a}\sqrt{1 - \left(\frac{\varDelta}{a}\right)^2}\right]$$

Figure 6.10 On the MTF calculation for an ideal lens.

6.5 MEASURING THE MTF

It may at first seem virtually impossible to experimentally measure the MTF of a real lens since it involves a correlation between the aperture function and its complex conjugate. As far back as 1964, however, some novel techniques have been proposed for performing this measurement. One such technique involves the use of a shearing interferometer in which a pupil function is sheared with respect to itself.

Consider the shearing interferometer shown in Fig. 6.12 [5]. In this device, the aperture function of a lens is sheared with respect to itself by some fixed amount to produce twin beams of the type shown previously in Fig. 6.10. By moving the wedge prism P_2 up and down, we may introduce varying amounts of phase shift in one beam relative to the other, for a fixed amount of beam shear. We now show how this variable phase shift may be used to determine the MTF of the test lens.

With reference to Fig. 6.13, the total intensity of the combined field is

$$I^{\text{Tot}}(x, y) = \left| f\left(x + \frac{\varDelta}{2}, y\right) + e^{j\delta} f\left(x - \frac{\varDelta}{2}, y\right)\right|^2$$

$$= \left| f\left(x + \frac{\varDelta}{2}, y\right)\right|^2 + \left| f\left(x - \frac{\varDelta}{2}, y\right)\right|^2$$

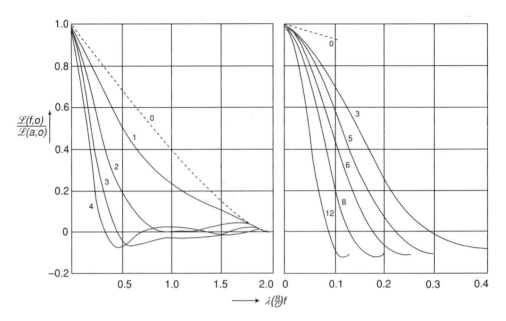

Figure 6.11 The normalized frequency response curves for incoherent illumination of a
system free of geometrical aberrations but suffering from defect of focus. Φ
$= (m\lambda/\pi)\rho^2$, $|G| = 1$. The number on each curve is the value of the parameter. $m = (\pi/2\lambda)\,(a/R)^2\,z$, z being the distance between the plane of observation
and the Gaussian focal plane. (After H. H. HOPKINS, *Proc. Roy. Soc.*, A, 231
(1955), 98.) (After Born and Wolf.)

$$+ f\left(x + \frac{\Delta}{2}, y\right) f^*\left(x - \frac{\Delta}{2}, y\right) e^{-j\delta} + f^*\left(x + \frac{\delta}{2}, y\right) f\left(x - \frac{\Delta}{2}, y\right) e^{j\delta}$$

If this field is focused onto a photodetector, the total power received by the detector
is proportional to the intensity integrated over the cross section of the collimated beam.
Thus,

$$P_{\text{RCV}} = \iint_{-\infty}^{\infty} I^{\text{Tot}}(x, y)\, dx\, dy = 2 \iint_{-\infty}^{\infty} |f(x, y)|^2\, dx\, dy$$

$$+ e^{-j\delta} \iint_{-\infty}^{\infty} f\left(x + \frac{\Delta}{2}, y\right) f^*\left(x - \frac{\Delta}{2}, y\right) dx\, dy$$

$$+ e^{j\delta} \iint_{-\infty}^{\infty} f^*\left(x + \frac{\Delta}{2}, y\right) f\left(x - \frac{\Delta}{2}, y\right) dx\, dy$$

or

$$P_{\text{RCV}} = 2P_0 + e^{-j\delta}P_0\, \text{OTF}(\Delta) + e^{j\delta}P_0\, \text{OTF*}\,(\Delta)$$

If we now divide the OTF up into its magnitude and phase parts as

$$\text{OTF}(\Delta) = \text{MTF}(\Delta)\, e^{j\phi(\Delta)}$$

then the total power received by the photodetector is

$$P_{\text{RCV}} = 2P_0\, \{1 + \text{MTF}(\Delta)\, \cos[\,\phi(\Delta) - \delta]\}$$

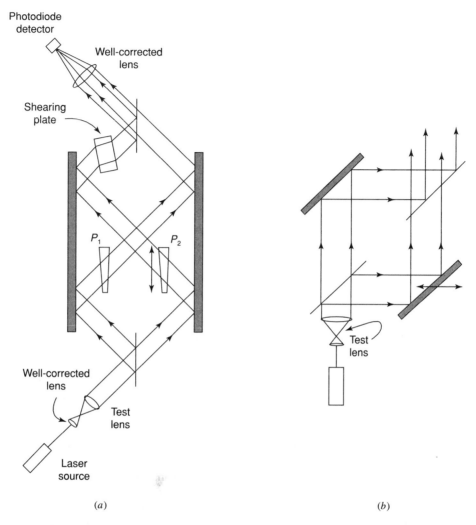

Figure 6.12 Shearing interferometers for measurement of lens MTF: (a) After Montgomery and (b) Mach-Zehnder, with mirror on carriage.

Thus, the received power consists of a constant term plus a "modulation term," which depends on the value of the relative phase shift δ between the two interfering beams. If this phase shift is varied (by sliding the wedge prism back and forth in the test setup shown) and the detector output is recorded, the value of the MTF can be obtained for various values of shear Δ.

Other techniques are also possible for measuring lens MTF [6,7].

6.6 SUMMARY OF LENS FIGURES-OF-MERIT

So far in this book, we've looked at a number of lens figures-of-merit including the aberration function, the point spread function, and the modulation transfer function. These all derive from the complex aperture function in fairly simple and direct ways. The aberra-

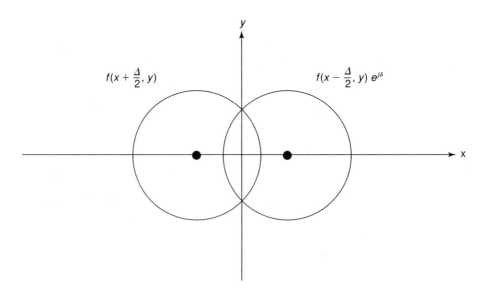

Figure 6.13 Two sheared pupil functions having variable phase shift, δ.

tion function is the phase of the complex aperture function; the point spread function is the Fourier transform of the complex aperture function; and the MTF is the magnitude of the autocorrelation of the complex aperture function. Therefore, in principle, if one of these quantities is known, the others may be deduced, with varying degrees of accuracy. In practice, it is common to measure each quantity independently and to specify each independently. However, the reader should be aware that each of these lens figures conveys roughly the same information, just in different forms.

One criterion of lens quality can be expressed in terms of two different figures-of-merit described above. This criterion is known as the *Strehl ratio* and is defined in either of the two equivalent terms. In the spectral domain, it may be defined as the ratio of the area under the MTF curve of the lens under test to that of an ideal lens. In the spatial domain, it may be defined as the ratio of the on-axis power point spread function of the test lens to the on-axis power point spread function of an ideal lens (an ideal lens being defined in both cases as one that converts an on-axis, perfectly spherical wave into another on-axis, perfectly spherical wave.

This equivalence between the two definitions of Strehl ratio is shown almost immediately. Letting $A(x, y)$ represent the complex aperture function, we have the following defining relations for MTF and PSF:

$$MTF(u, v) = \iint_{-\infty}^{\infty} A(\alpha, \beta) A^*(\alpha + u, \beta + v) \, d\alpha \, d\beta$$

and

$$|PSF(x = 0, y = 0)|^2 = \iint_{-\infty}^{\infty} A(\alpha, \beta) \, d\alpha \, d\beta \cdot \iint_{-\infty}^{\infty} A^*(\alpha, \beta) \, d\alpha \, d\beta$$

where in the last relation, the Fourier transform kernel is unity since $x = y = 0$. (The point spread function is assumed to be evaluated on-axis.)

Rearranging the first equation yields the area under the MTF curve as

$$\iint_{-\infty}^{\infty} \text{MTF}\,(u,\,v)\,du\,dv = \iint_{-\infty}^{\infty} \left\{ \iint_{-\infty}^{\infty} A(\alpha,\,\beta)A^*(\alpha\,+\,u,\,\beta\,+\,v)\,d\alpha\,d\beta \right\} du\,dv$$

or

$$= \iint_{-\infty}^{\infty} A(\alpha,\,\beta)\,d\alpha\,d\beta \cdot \iint_{-\infty}^{\infty} A^*(\alpha\,+\,u,\,\beta\,+\,v)\,du\,dv$$

which, through a change of variables is readily shown to be equal to the power point spread function on-axis.

So, the Strehl ratio is useful in showing how two lens figures-of-merit are related. As the magnitude of the PSF is diminished, so too is the area under the MTF curve. This reduction in area under the MTF curve is due primarily to nonuniformities in the coherent transfer function phase versus α and β.

6.7 DISCRETE SAMPLING OF FIELDS AND TRANSFORMS: SHANNON'S THEOREM

With today's advanced charge-coupled device (CCD) detector imaging arrays now capturing and digitizing images previously recorded on analogue film, the question arises as to how well a digitized image can theoretically be manipulated in order to perfectly reconstruct the continuous image. This question can be answered using Shannon's theorem, a well-known result from Fourier theory, which has direct application to the plane wave spectrum viewpoint of optical systems.

Shannon's theorem states that a 2-D bandlimited function may be exactly reconstructed from its sample values (taken over a regular 2-D sampling grid) whenever the sample spacing is at least twice the period of the highest spatial frequency present in the bandlimited signal. This theorem, as stated, assumes that an infinite array of data must be collected in order to reconstruct the original function. However, if the function is moderately space-limited, then only the significant sample values of the function are necessary for reconstruction, especially for reconstruction *within* the net of sample points. Outside the net of sample points, the approximation to the true function degrades fairly rapidly. Of course, a perfectly bandlimited function will never be perfectly space-limited (by Heisenberg's principle). However, in practice, the space-limited condition is often met to a very high degree of accuracy.

Before looking at Shannon's theorem in detail, it is good to remember that a photodetector does not sample the value of an optical field at an exact *point* in space. In reality, it averages the field over a finite *area* in the focal plane (occupied by the front face of the photodetector) to produce a single output signal. The fields in this finite aperture are averaged and summed to produce a net electrical signal from each CCD pixel. Therefore, when using Shannon's theorem in connection with data received through a physical detector array, it's important to realize that the data are imperfect. However, whenever the individual photodiodes are small in relation to the wavelength of the highest spatial frequency component in the image spectrum (i.e., when the detectors are equally sensitive to all plane wave spectral components of the image field distribution), the average field values received by the photodiodes can, to a high degree of accuracy, be taken as the actual field values at distinct points in the array.

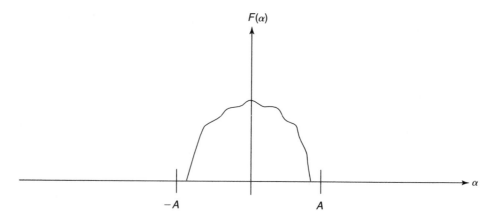

Figure 6.14 Bandlimited spectrum.

Shannon's theorem is obtained by considering the bandlimited spectrum shown in Fig. 6.14. We'll use the assumed transform pairs

$$f(x) = \int_{-A}^{A} F(\alpha) e^{j\alpha x} d\alpha$$

$$F(x) = \frac{1}{2\pi} \int_{-\infty}^{\infty} f(\alpha) e^{-j\alpha x} dx$$

where

$$\begin{Bmatrix} \delta(x) \\ \delta(\alpha) \end{Bmatrix} = \frac{1}{2\pi} \int_{-\infty}^{\infty} e^{\pm j\alpha x} \begin{Bmatrix} d\alpha \\ dx \end{Bmatrix}$$

and

$$F(\alpha) = 0 \quad \text{for } |\alpha| > A$$

A truncated (finite) spectrum of this type would be generated by the converging lens shown in Fig. 6.15. The finite spectrum may be laid out along the spatial frequency axis to produce the periodic function shown in Fig. 6.16. This periodic function may be represented in the mathematical form

$$F_p(\alpha) = \sum_{n=-\infty}^{\infty} F(\alpha - nA_p)$$

The original (aperiodic) function may be recovered from this periodic function by low-pass filtering the periodic spectrum with the unit-height pulse function shown in Fig. 6.16. This filtering process is given mathematically as

$$F(\alpha) = p_{2A_p}(\alpha) \sum_{n=-\infty}^{\infty} F(\alpha - nA_p)$$

The periodic extension of the finite spectrum may be represented in terms of a Fourier series of the form

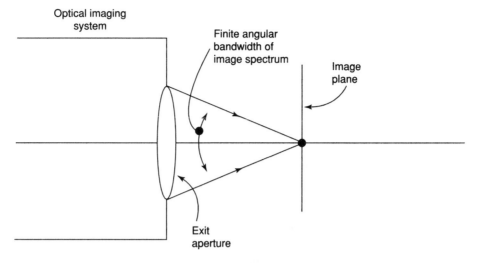

Figure 6.15 Image produced from a finite plane wave spectrum.

$$F_p(\alpha) = \sum_{n=-\infty}^{\infty} F(\alpha - nA_p) = \sum_{m=-\infty}^{\infty} c_m e^{-jx_m\alpha}$$

where

$$x_m = \frac{2m\pi}{A_p}$$

and A_p is the period of our "synthesized periodic function" in the spatial frequency domain. Using ordinary Fourier techniques, we may find the coefficients c_M of the Fourier series by multiplying both extreme sides of the equation above by the exponential function

$$e^{jx_M\alpha} \quad \text{for some fixed integer, } M$$

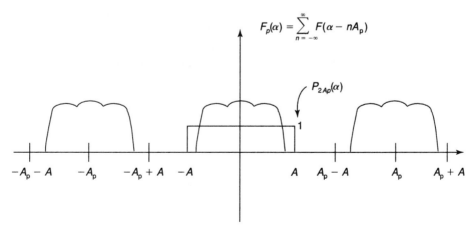

Figure 6.16 Periodic extension of $F(\alpha)$ with no overlap.

and integrating over one period (of length A_p) of the function. This yields the coefficients as

$$c_M = \frac{1}{A_p} \int_{-A_p/2}^{A_p/2} F(\alpha)\, e^{jx_M\alpha}\, d\alpha = \frac{1}{A_p} f(x_M)$$

So now we can go back and write the Fourier series expression for the periodically extended spectrum as

$$F_p(\alpha) = \frac{1}{A_p} \sum_{m=-\infty}^{\infty} f(x_m)\, e^{-jx_m\alpha}$$

Using the first of the two reciprocal Fourier transform relations shown above, which give the function $f(x)$ in terms of its Fourier transform $F(\alpha)$, we finally obtain the Shannon sampling theorem as

$$f(x) = \left(\frac{2A}{A_p}\right) \sum_{-\infty}^{\infty} f(x_m)\, \text{sinc}(x - x_m)A$$

where the sinc function is defined as

$$\text{sinc}\, x = \frac{\sin x}{x}$$

Therefore, by the equation above, it is possible to *exactly* reconstruct a bandlimited function anywhere, given only a knowledge of its values over some *discrete* net. Of course, the net is assumed infinite in extent, whereas any sum we'd evaluate in practice would naturally be finite. As it turns out though, the Shannon formula is very effective when the target point x lies within the sample net of grid points (interpolation), whereas the accuracy of the formula decreases rapidly for points outside the net of sample points (extrapolation).

In two dimensions, the Shannon sampling theorem becomes

$$f(x, y) = \frac{4AB}{A_p B_p} \sum_{m,n} f(x_m, y_n)\, \text{sinc}\,(x - x_m)A\, \text{sinc}(y - y_n)B$$

A real photodetector doesn't actually sample the electric field at an idealized *point* in space, but rather performs an averaging function of the field over the aperture area of the detector. Reconstruction via Shannon's theorem will be accurate when the spatial extent of the detector is much less than the equivalent wavelength of the highest spatial frequency present in the image. For example, say the x, y dimensions of the detector aperture are given as l_x, l_y. Then, if the maximum transverse wavenumbers are given as

$$\alpha_{\max} = k_0\, \sin\theta_{x,\max}$$

and

$$\beta_{\max} = k_0\, \sin\theta_{y,\max}$$

then,

$$\lambda_{\text{eq},x} = \frac{2\pi}{\alpha_{\max}} \gg l_x$$

and

$$\lambda_{\text{eq},y} = \frac{2\pi}{\beta_{\max}} \gg l_y$$

Equivalently, the pixel must be small enough so that its angular selectivity (when viewed as a type of receive antenna) is approximately constant over the range of spatial frequencies (plane wave components) present in the image. If the detector were a receive antenna, the criterion above would be that the antenna must be *omnidirectional* (i.e., equally receptive to plane waves incident from all directions). When the detectors in the array have *directivity* (nonuniform reception for different plane wave components), the received image spectrum is equal to the incident plane wave spectrum multiplied by the plane wave spectral response (the receive antenna pattern) of the individual detector elements. Similar considerations also hold in connection with near-field measurements of microwave antennas [9].

REFERENCES

[1] Hamming, R. W., *Digital Filters,* Englewood Cliffs, N.J.: Prentice-Hall, 1977.

[2] Born, M., and Wolf, E., *Principles of Optics,* 6th ed. New York: Pergamon Press, 1980.

[3] Blackman, R. B., and Tukey, J. W., *The Measurement of Power Spectra,* New York: Dover, 1958.

[4] Scott, C. R., *Modern Methods of Reflector Antenna Analysis and Design,* Norwood, MA: Artech, 1990.

[5] Montgomery, A. J., ''New Interferometer for the Measurement of Modulation Transfer Functions,'' *J. Opt. Soc. Am.,* vol. 54, no. 2, February 1964, pp. 191–198.

[6] Montgomery, A. J., ''Two Methods of Measuring Optical Transfer Functions with an Interferometer,'' *J. Opt. Soc. Am.,* vol. 56, no. 5, May 1966, pp. 624–629.

[7] Grimes, D. N., ''Optical Autocorrelator with Special Application to MTF Measurement,'' *Applied Optics,* vol. 11, no. 4, April 1972, pp. 914–918.

[8] Goodman, J. W., *Introduction to Fourier Optics,* New York: McGraw-Hill, 1968.

[9] Joy, E. B., and Paris, D. T., ''Spatial Sampling and Filtering in Near-Field Measurements,'' *IEEE Trans. Antennas Propagat.,* vol. AP-20, no. 3, May 1972, pp. 253–261.

PART III
REFLECTIVE AND REFRACTIVE OPTICS

7

Classical Optical Imaging Instruments

In this chapter, we'll look at some common optical imaging instruments. Some of these have historical value, and some are important in contemporary optical systems. For example, the classical astronomical telescope today continues to find widespread usage as a beam expander in laser-based optical systems. In addition, parabolic and Cassegrain reflecting telescopes find widespread use in the microwave region of the spectrum as antennas for satellite communications. Moreover, the devices presented herein illustrate a number of principles common to a wide variety of sophisticated optical imaging systems.

7.1 THE MAGNIFIER

The magnifier, shown in Fig. 7.1, can be understood from the summary discussion on lenses from Chapter 5. In that discussion, we obtained the relation

$$\frac{1}{d_1} = \frac{1}{d_0} - \frac{1}{F}$$

between object and image distances (valid when $d_0 < F$). When the object distance is made very small (the magnifier is placed against the object), the image distance is effectively equal to the object distance, and no magnification results. When the object distance is made larger (nearly equal to but still smaller than the focal length), the image distance increases, and by the equations from Chapter 4, lateral magnification results. Of course, the image is virtual and upright.

7.2 THE HUMAN EYE

The human eye is the first stage of a highly complex and—for the most part—still poorly understood optical system known as the human visual system (HVS). While the complex neural processes involved in the human perception of visual stimuli would be well beyond

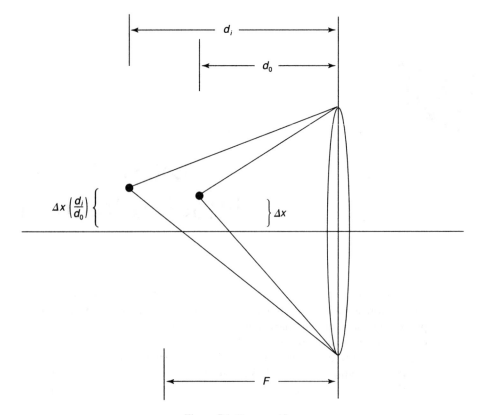

Figure 7.1 The magnifier.

the scope of this book, the basic optics of the eye are relatively easy to understand. It should be said, however, that the eye itself represents a surprisingly small part of the human visual response. For example, it is well known that many individuals whose eyes produce poor quality retinal images can perceive—and resolve—object detail much finer than the retinal image might at first indicate. In other words, their optical neural pathways have the ability to ''deconvolve'' the effects of ocular aberrations and imperfections to create a sharp, high-quality perceived image. Researchers in image reconstruction theory continue to search for techniques to achieve the remarkable levels of image improvement seen in the human visual system.

The eye is a single lens imaging system that produces an inverted object image on the retina. Since the distance from the lens to the retina is fixed (i.e., the image distance is fixed), whereas object distances are variable, the focal length of the lens must adapt in order to bring object points at various distances into sharp focus. We can quantify the change in focal length required to focus images at differing object distances. By the lens law, we have

$$\frac{1}{d_0} + \frac{1}{d_i} = \frac{1}{f}$$

or

$$f = \frac{d_i d_0}{d_i + d_0}$$

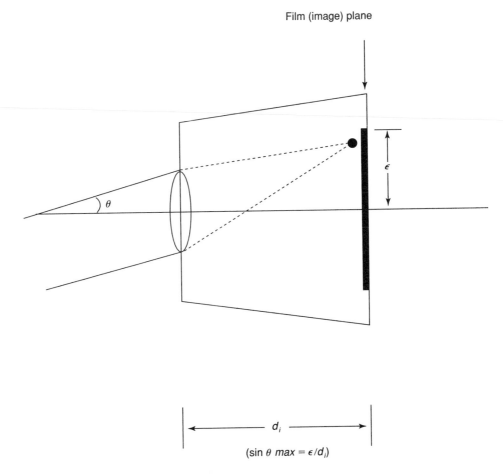

Film (image) plane

$$(\sin \theta \; max = \epsilon / d_i)$$

Figure 7.2 The single-lens camera.

Since d_i (the lens-to-retina distance) is fixed, focusing can only be achieved by varying the focal length, f, of the lens. As the object distance d_0 tends to infinity, the focal length of the lens must approach the (fixed) lens-to-retina distance, d_i. If the object distance changes to say, $9 \cdot d_i$, then the focal length of the eye must change to

$$f = .9d_i$$

that is, to 90% of its value for focusing objects at infinity. This gives an indication of the extent to which the eye lens must adapt in order to focus objects at a variety of distances.

The ordinary single-lens camera, shown in Fig. 7.2, is topologically identical to the eye. Unlike the eye, however, the camera achieves focusing by varying the image distance, d_i. Again, the image is real and inverted. For infinitely remote objects, the image distance is made equal to the focal length of the lens, and the field-of-view (FOV), θ, is given by $\sin^{-1} \epsilon / f$ where ϵ is one-half the lateral extent of the film and f is the focal length of the lens. A plane wave incident at an angle greater than this field angle will be imaged to a spot located off the film and will not be captured.

Simple telescopes are often designed using this single-lens topology, especially in the infrared (IR) regime. When high resolution and high magnification are required, the focal length, f, is made as large as practical. By the previous paragraph, a long focal length lens images an incident plane wave (at a given angle of incidence) farther away from the optic axis than would a short focal length lens. However, the diameter of the lens must also be made large so that the image contains enough plane wave content to minimize the point spread function.

7.3 THE GALILEAN TELESCOPE

The telescope is an image-forming instrument that acts on plane wave fields from infinitely distant objects and is defined by two principal conditions:

1. The telescope transforms incident plane wave fields into transmitted plane wave fields.
2. The magnification of the telescope is defined as the ratio of incident plane wave tilt angle to transmitted plane wave angle (as shown in Fig. 7.3).

The plane wave fields transmitted by the telescope are focused by the eye of the human observer to form a retinal image. Since telescopes are optical instruments whose purpose it is to "condition" an optical field for improved viewing by the human eye, it's useful to briefly review the three types of images that can be formed by an imaging instrument. These are shown in Fig. 7.3. All three instruments are assumed illuminated by a plane wave spectrum (angular spectrum) incident field, due to an infinitely remote source distribution.

The first type of instrument forms a *real image.* This means that each and every incident plane wave field is converted to a separate spherical wave field, which is then brought to a point in the focal plane of the lens. So, there are infinitely many spherical wave fields that are converged to spots in the focal plane.

The second type of instrument forms a *virtual* image. This means that each and every plane wave field incident on the instrument is converted into a diverging spherical wave, whose center is located at the front focal plane of the lens. Again, the field transmitted by the lens consists of infinitely many spherical wave fields whose centers lie in the focal plane of the diverging lens.

The third type of instrument doesn't form an image; rather, it forms a plane wave spectrum. It is very tempting to think of this spectrum as a diverging virtual image, but it is not. In this case, there is only *one* single *diverging* spherical wave, each of whose plane wave components corresponds to one of the plane wave components contained in the incident field. If this divergent spherical wave field is made to impinge on the eye, each plane wave component of the field will be focused by the lens of the eye into a single focal spot on the retina. The only difference between viewing the angular spectrum produced by the imaging instrument and viewing the original angular spectrum is that the original was a converging spectrum and the new one is a diverging spectrum. This distinction is not significant, however. What *is* important is that each plane wave component of the spectrum will be imaged on the retina as if it originated from infinitely far away. In this way, the lens of the eye can image the angular spectrum without strain.

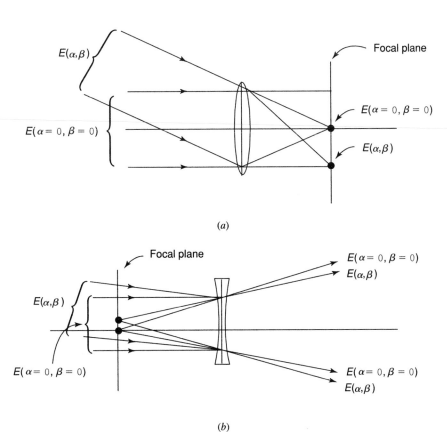

(a)

(b)

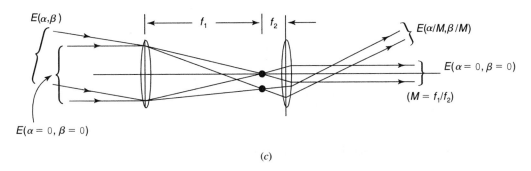

(c)

Figure 7.3 Three types of optical imaging instruments: (a) Real image; (b) Virtual image; and (c) Angular (plane wave) spectrum.

Sometimes a telescope may be adjusted to produce a slightly convergent real image or a slightly divergent virtual image, and the eye will adjust accordingly, just as it ordinarily adjusts for distance variations in ordinary scenes (with the viewer being largely unaware of the adjustments being made). Only when the field transmitted by the instrument is a

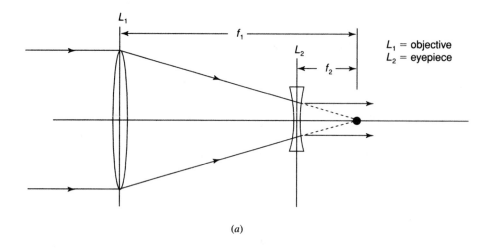

(a)

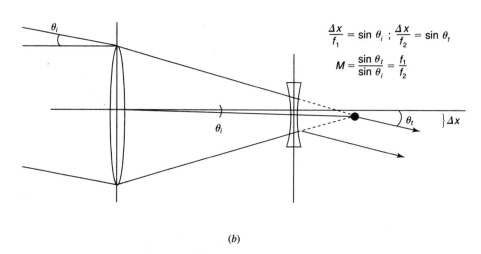

(b)

Figure 7.4 Galilean telescope: (a) Operation of the Galilean telescope for a normally incident plane wave and (b) Operation of the Galilean telescope for an obliquely incident plane wave.

divergent or convergent field that cannot be easily focused (or focused at all) without eye strain, will the focus error be noticeable.

A straightforward implementation of the telescope design is the Galilean telescope shown in Fig. 7.4. In optical parlance, the lens nearest the object is called the *objective*, and the lens nearest the eye is the *eyepiece*. In the case of the Galilean telescope, the objective lens is a convergent lens with a relatively long focal length and large diameter. The long focal length contributes to magnification, whereas the large diameter results in improved resolution by allowing increased spatial frequency content in the convergent field behind the objective.

The eyepiece of the Galilean telescope is a divergent lens that converts the convergent

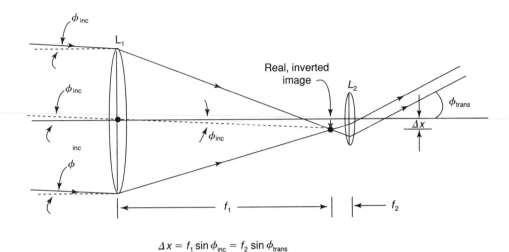

$$\Delta x = f_1 \sin \phi_{inc} = f_2 \sin \phi_{trans}$$

Figure 7.5 The astronomical telescope.

wavefront from the objective into a plane wave field suitable for viewing by the human eye. The image produced by the telescope is upright and virtual.

The astronomical telescope, shown in Fig. 7.5, is a two-lens imaging system similar in form to the Galilean telescope. The only difference between the two is that the eyepiece of the astronomical telescope is a convergent—rather than divergent—lens. The upshot of this difference is that the image produced by the astronomical telescope is real and inverted. For this reason, this configuration is far more useful for planetary observation than for use in ordinary terrestial applications.

7.4 THE TERRESTRIAL TELESCOPE

The terrestrial telescope is shown in Fig. 7.6. Its operation is easily understood from the summary discussion on lens operation from Chapter 5. Basically, it is a variation on the astronomical telescope theme, in which a second imaging section is incorporated in order to "invert the inverted image" and make it once again upright.

The objective lens L_1 first produces an inverted image, which is located within the focal length of the second lens L_2. Since Image 1 (the "object" for lens L_2) lies within the focal length of L_2, it will produce a virtual image, Image 2, as shown in the figure. Image 2 is the "object" for lens L_3, and this object lies outside the focal length of L_3. It produces Image 3 as shown in the figure. Image 3 is an upright image that lies in the focal plane of the eyepiece lens. The eyepiece collimates the rays as shown. This telescope is also discussed in Born and Wolf [1] where additional lenses are shown incorporated into the design for the purpose of shortening the total length of the telescope.

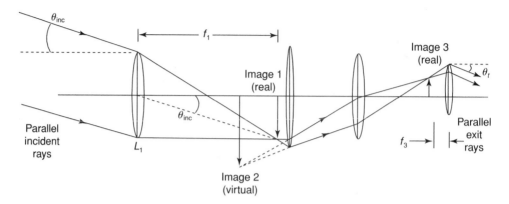

Figure 7.6 The terrestrial telescope.

7.5 THE REFLECTING (MIRROR) TELESCOPE

Telescopes can be made from reflecting (mirror) as well as refracting (lens) elements. The only disadvantage of parabolic mirrors, however, is a fairly large amount of coma aberration (see Chapter 9) arises for even very small off-axis incidence angles. Symmetric and offset parabolic mirror telescopes are shown in Fig. 7.7. In each case, the reflector surface can be generated in one of two ways. It can be formed either by intersecting a circular cylinder (parallel to the symmetry axis of the paraboloid) with the paraboloidal surface, or by intersecting a cone (centered at the focal point of the paraboloid) with the paraboloid [2].

The mirrors shown in Fig. 7.7 are designed to image infinitely remote scenes at the focal plane. This would be the configuration used when either film or a detector array is to be placed in the focal plane. When the image is to be viewed by the human eye, the converging spherical wave must be re-collimated using techniques discussed previously in connection with refracting telescopes. Alternatively, a convex parabolic submirror may be used to collimate the spherical waves.

An extensive body of work has been developed relative to the analysis, design, and general phenomenology of reflecting telescopes—in connection with microwave reflector antennas. For the interested reader, the IEEE Transactions on Antennas and Propagation contains a wealth of information on this subject, virtually all of which is readily applicable to optical mirror design. The analogy between optical mirrors and microwave reflector antennas even holds down to the focal plane array, which for the antenna is usually an array of waveguides or microstrip patches, and for the mirror is an optically active detector array. In both cases, however, the same design considerations apply for both devices.

7.6 BINOCULARS

Since binoculars are used for magnification of ordinary scenes on land, they must yield an upright image just as the terrestrial telescope. However, binoculars are distinguished from the telescope by the way in which they produce that upright image. Whereas the

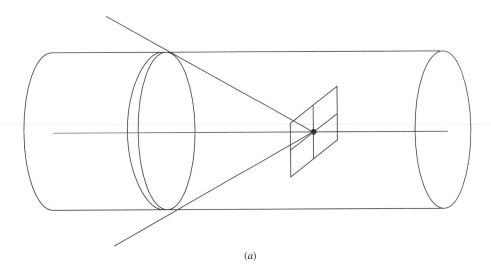

(a)

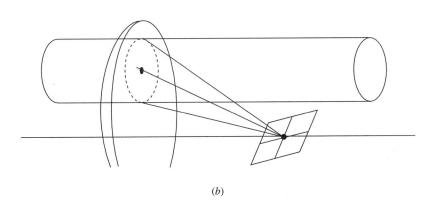

(b)

Figure 7.7 Parabolic mirrors and conic section constructions: (a) Symmetric mirror, with intersecting core and cylinder. (Focal plane array is shown orthogonal to symmetry axis.); (b) Offset reflector with intersecting zone and cylinder. (Focal plane array is shown orthogonal to cone symmetry axis.)

upright telescope image is produced using lenses, the upright image in the binocular is produced using prism reflectors. This use of prisms as reflecting surfaces is somewhat unique in that the reflection is produced using dielectric (rather than metallic) materials. Reflection is produced using the phenomenon of total internal reflection, discussed in Chapter 3 in connection with gradient index optical waveguides.

A typical binocular element is shown in Fig. 7.8. It is essentially a very compact—and lower magnification—telescope. Since binoculars are used for viewing remote scenes, as are telescopes, they are generally designed to convert an incident plane wave spectrum into an angularly magnified transmitted plane wave spectrum.

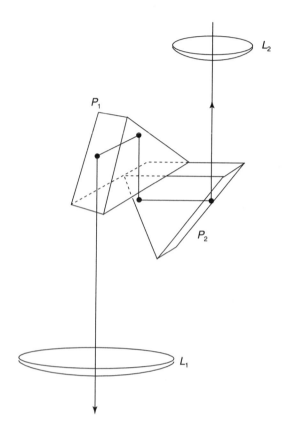

Figure 7.8 Basic binocular construction.

Figure 7.9 shows a portion of the optics in a binocular—the objective lens and one prism reflector, which is modeled as two 45 degree mirrors. The view presented is a side-view and shows the reversal of the image in the vertical plane due to the vertically oriented prism. Image reversal in the horizontal plane is due to the second (horizontal) prism. Thus, the combination of the two prisms achieves image reversal in two planes, creating an upright and unmirrored image suitable for viewing. The two orthogonal prisms only accomplish image inversion; they do not serve to magnify the image as the inverting lenses of the terrestrial telescope do.

The image is viewed using the eyepiece, which can consist of more than one lens for additional angular magnification.

7.7 THE MICROSCOPE

The microscope is essentially a sophisticated magnifier. Like the magnifier, it produces a virtual image of a very small (generally planar) object. Now, small objects have fine detail, which translates to wide-bandwidth plane wave spectra. Therefore, if a microscope is to produce a good image of a small object, it must be capable of transmitting a very

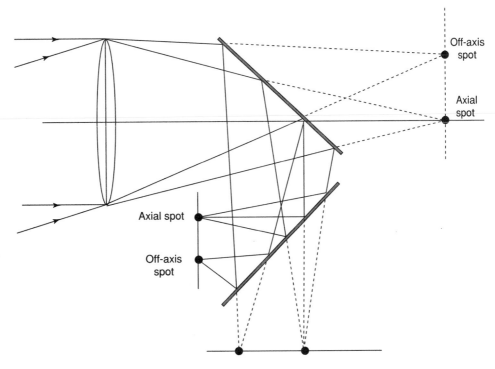

Figure 7.9 Image inversion in the vertical direction via two reflections through 45 degree mirrors.

wideband plane wave spectrum. The way to do this is to design the objective lens so that it intercepts a wide angular spectrum of plane waves from the object, as shown in Fig. 7.10.

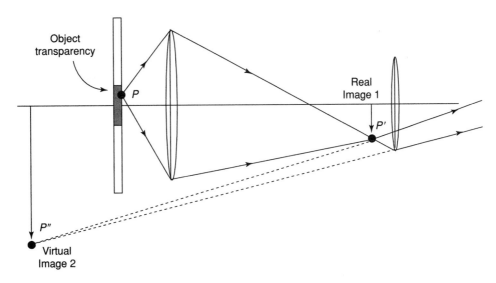

Figure 7.10 A simple, low-power microscope.

On the other hand, objects having broader spatial extent (with slowly varying spatial properties, as the magnified image of the small object is intended to have), have relatively narrow angular spectra. As shown in Fig. 7.10, the broad angular spectrum diverging from object point, *P,* is converted by the microscope into a narrow angular spectrum (diverging from *P″*), and the narrowing in the angular domain (the Fourier transform domain) is accompanied by a corresponding broadening in the spatial domain (i.e., the object appears enlarged). This is again due to the Abbe sine condition (Heisenberg's principle).

The principle of operation is readily evident from the figure. The object point *P* is imaged through the objective lens onto the real, inverted image point, *P′*. This image is spaced a distance slightly less than a focal length from the eyepiece lens. The image of *P′* in the eyepiece is the virtual image point *P″*. The point *P″* will be increasingly magnified as the point *P′* is brought nearer to the focal plane of the eyepiece. However, *P″* will also be further distant from the eyepiece, so that after a point no additional magnification is obtained. If greater magnification of the real image point *P′* is desired, additional lenses may be used in the eyepiece to create additional virtual images of *P″* having increased magnification.

After a point, additional magnification of the virtual images of *P″* will result in an image that is clearly blurred—even if the lens system were to have no aberrations whatsoever. The cause of the blur is the limited bandwidth of the object spectrum intercepted by the objective lens. This bandwidth can be increased in a number of ways. One of these is to increase the frequency of the light used to form the image. Another is to situate the object in a high refractive index liquid. And a third is to design an objective lens that intercepts a greater angular spectrum of fields. We'll examine the third of these techniques in greater detail.

High-power microscope objectives are constructed using spherical lenses having one cleaved face. The planar face is placed against the planar object, and the remaining spherical surface forms a virtual image of the object. Due to its high degree of curvature, this type of high-power objective cannot be regarded as a thin lens; therefore, another viewpoint must be used for understanding its operation. Prior to examining the high-power microscope objective in detail, it will be useful first to examine the imaging properties of spherical lenses. We'll follow the discussion in [1].

To this end, consider Fig. 7.11, which shows the problem of refraction of a light ray by a spherical dielectric ball. The index of the ball is given as *n,* and the index of the surrounding medium is unity. The radius of the ball is *a.* For graphical construction purposes, we draw two additional spheres, one interior to the original sphere, with radius *a/n,* and one exterior to the sphere, with radius *na.* With the aid of these two imaginary spheres, we may readily determine the direction of the refracted ray in the ball.

The incident ray intersects the dielectric sphere at point *P* as shown. Draw a construction line along the incident ray from *P* to point *Q,* where the imaginary extension of the incident ray intersects the exterior sphere. Now draw a second line connecting point *Q* with the center point *O* of the concentric spheres. The point *R,* where this latter radial line intersects the interior sphere, now determines the refracted ray. That is, the refracted ray lies along the line segment *PR* as shown. This may be proven using the triangle construction shown in Fig. 7.11*b.*

In the figure, triangle *OPQ* is similar to triangle *OPR.* This is easily seen since the ratio of the lengths of sides *OP* and *OR* of triangle *OPR* equals the refractive index, *n,*

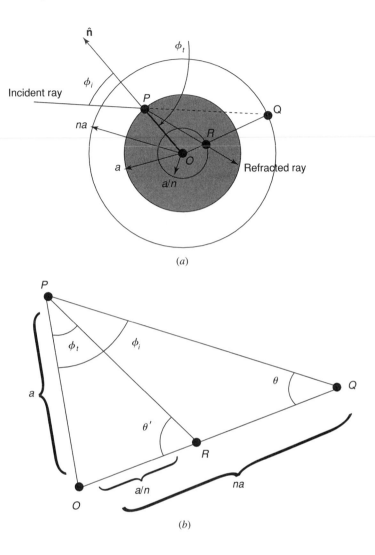

Figure 7.11 Refraction by a spherical dielectric ball: (a) Original problems and (b) Triangle construction.

the same ratio that exists between the lengths of sides OQ and OP of triangle OPQ. Therefore,

$$\theta = \phi_t \quad \text{and} \quad \theta' = \phi_i$$

With these relations, we may now apply the law of sines to triangle OPR to show that

$$\frac{a/n}{\sin\phi_t} = \frac{a}{\sin\theta'} = \frac{a}{\sin\phi_i}$$

and to triangle OPQ to show that

$$\frac{na}{\sin\phi_i} = \frac{a}{\sin\theta} = \frac{a}{\sin\phi_t}$$

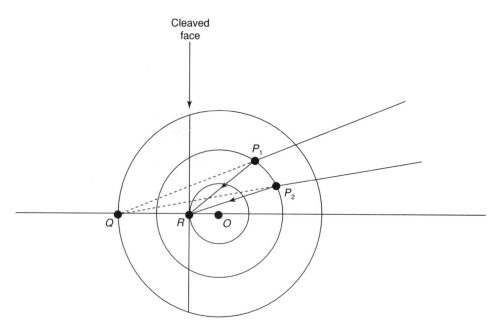

Figure 7.12 High-power microscope objective.

The first and last terms of both equations yield a Snell's law relation between the angles of incidence and refraction, demonstrating the validity of the graphical construction.

With this construction in hand, we may now analyze the high-power microscope objective, with the help of Fig. 7.12. Consider a family of radial rays that all would converge on point Q (if the dielectric ball had a unit refractive index). These rays intersect the sphere at points P_1, P_2, and so on. For the dielectric ball, however, the previous construction indicates that all the rays are focused at point R, shown residing on the planar face. Thus, R is a focal point for the cleaved spherical lens. The high curvature of the spherical lens means that rays having high angles of inclination will be captured by the objective, and this is precisely the requirement for good resolution of very fine object detail. As mentioned by Born and Wolf, this objective may be placed within a focal length of a converging lens, creating an even further magnified virtual image of the object.

7.8 THE CONFOCAL MICROSCOPE

The confocal microscope is a complicated-sounding device with a very simple principle of operation. Confocal microscopes are used to view thick, transparent objects. The device allows the viewer to observe consecutive lateral planes of the object for the purpose of generating a three-dimensional picture of the object. The basic confocal microscope is

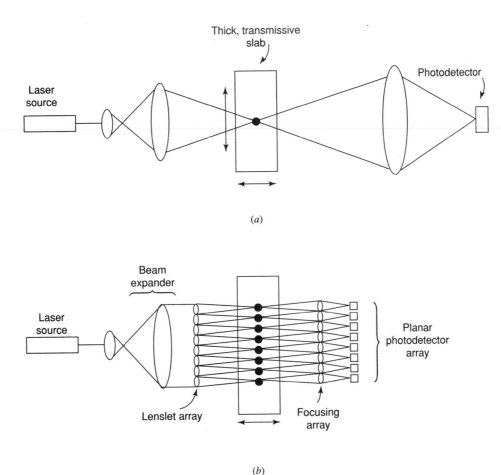

Figure 7.13 Confocal microscope topologies: (a) Original confocal miscroscope configuration and (b) Confocal miscroscope requiring only longitudinal movement.

shown in Fig. 7.13a. Light from a laser source impinges on a small lens (which can be either concave or convex) and then diverges spherically. The divergent beam is incident on the second lens, L_2, which focuses the beam onto a spot in the sample. Thus, only one spot in the entire sample is illuminated.

On the other side of the thick sample is a photodetector, which is focused onto the very *same* illuminated spot by the third lens (hence, the name *con*focal—both the source and detector are focused to the same spot in the sample). The primary problem with the arrangement shown is the need to mechanically move the sample both horizontally and vertically in order to obtain a complete image of the three-dimensional slab. One (of many) means for reducing the amount of mechanical motion is to have the light source illuminate a lenslet array as shown in Fig. 7.12b. Using this arrangement, in connection with a planar detector array, reduces the needed motion to one direction only (in the longitudinal direction).

Another solution approach would be to use the rotating mirror system (shown in

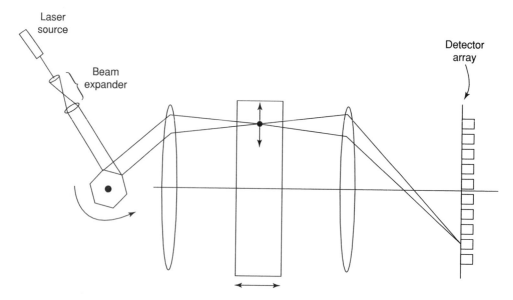

Figure 7.14 Confocal microscope using rotating mirror for lateral scanning.

Fig. 7.14) for lateral scanning. Numerous possible scanning configurations could be designed for the confocal microscope.

7.9 THE ZOOM LENS

The zoom lens is an optical imaging system in which the spacing between certain lenses can be adjusted in order to increase or decrease magnification of the system, while still maintaining focus. A natural consequence of the variability of the magnification is a change in field of view (FOV) (which decreases with increasing magnification, and vice versa, according to the sine condition and Heisenberg's principle). The combination of magnification and FOV changes produces the effect of "zooming" in on an object in a scene and isolating it for consideration.

A basic mechanical zoom lens system is shown in Fig. 7.15. The zoom effect is obtained by varying the positions of two lenses, L_2 and L_3. For magnification, L_2 is moved away from L_1. Before discussing the operation of the zoom lens in detail, we'll first look at some of the factors affecting the magnification of an optical system.

We've defined three types of imaging systems designed to view infinitely remote object scenes. Shown in Fig. 7.16, these are the telescope (plane wave converted to plane wave), the real-image producing system such as a camera (plane wave converted to converging spherical wave), and the virtual-image producing system such as a hologram (plane wave converted to a diverging spherical wave). By analogy with the single-lens system, we've defined *effective focal lengths* of these more general systems in terms of the ratio of the offset *angle* of an incident plane wave and the offset *distance* in the focal plane, where that plane wave component is focused. In Chapter 2, and earlier in this chapter in

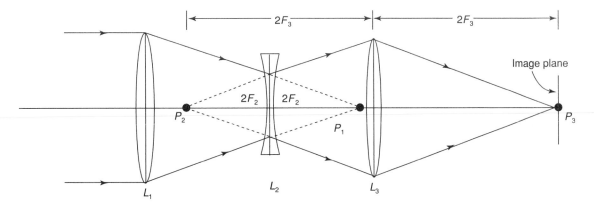

Figure 7.15 Basic zoom lens configuration (infinitely remote object scene).

the discussion on telescopes, the magnification of a telescopic system was defined as the ratio of tilt angles of the incident and transmitted plane wave fields as

$$M = \frac{\sin\theta_2}{\sin\theta_1}$$

as shown in Fig. 7.17. In the case of the real-image producing system, we cannot define a similar type of dimensionless magnification, since the incident far-field scene is an angular spectrum, whereas the resultant image is planar (lying on the image plane). What we *can* do however, is to define a *relative* magnification between two such real-image producing systems, relating the sizes of the images produced by the two different systems. By Fig. 7.16, the lateral magnification of a real-image producing system is related to the effective focal length of the system. Thus, a measure of the magnifying power of a real-image producing system is the effective focal length of the system.

Using the concept of effective focal length, we see that the real-image producing system shown in Fig. 7.15 may be equivalently unfolded (at least to a first order, for overall analysis purposes) as a single-lens system, as shown in Fig. 5.12a. This equivalence is used frequently in the design of front-fed and Cassegrain mirror design. Consider the two "equivalent" mirror telescopes shown in Fig. 7.18. The front-fed telescope shown in Fig. 7.18a is an "unfolded" version of the telescope shown in Fig. 7.18b. That is, if we draw lines from the focal point of the Cassegrain system out to the diameter of the objective lens, we will obtain the front-fed configuration.

An equivalent single-lens system may be found for any type of multiple-lens real-image producing system. The key parameter is the angle θ_2, shown in Fig. 7.19. This angle is related to the inverse of the effective focal length of the system. Thus, we may determine the relative magnification levels of two different imaging systems by taking ratios of sin θ_2 in the equivalent single-lens representations for the two systems. This is the key idea we'll need in the discussion on zoom lenses to follow, that is, that the smaller the half-angle of the equivalent system, the greater the magnification level, according to the sine condition. We really don't need to construct an equivalent single-lens system for every zoom lens. All that's necessary is to look at the sine of the angle extending from the very last lens of the system to the focal plane.

With this background in hand, we may now analyze the operation of the basic zoom

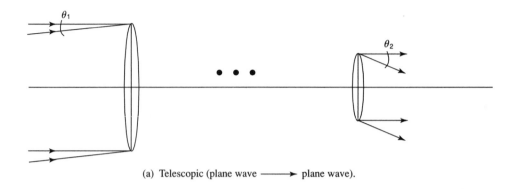

(a) Telescopic (plane wave ——→ plane wave).

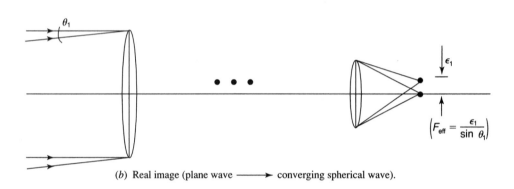

(b) Real image (plane wave ——→ converging spherical wave).

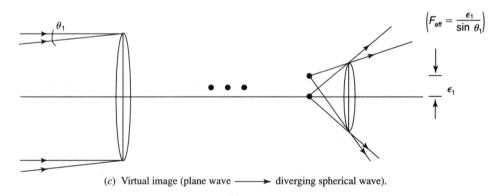

(c) Virtual image (plane wave ——→ diverging spherical wave).

Figure 7.16 Three types of systems that image far-field (angular spectrum) scenes.

lens. The lens in Fig. 7.15 is designed so that all lenses in the nominal, undisturbed configuration operate at unit magnification. That is, the image distance is designed to be equal to the object distance for all lenses after the first. This way, whenever the object distance increases/decreases by some small amount, the image distance will

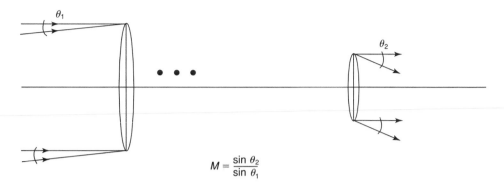

$$M = \frac{\sin \theta_2}{\sin \theta_1}$$

Figure 7.17 Magnification in a telescopic imaging system.

decrease/increase by the same small amount. This assumption is made for ease of analysis of the system.

Now, assume an axially incident plane wave field impinging on the objective lens L_1, as shown in Fig. 7.15. This plane wave field is converted to a converging spherical wave focused to the point P_1 in the focal plane of L_1. The divergent lens L_2 is placed so that the point P_1 is two focal lengths from L_2. Thus, the wave transmitted through L_2 is a divergent spherical wave, centered at point P_2, two focal lengths in front of L_2. Point P_2 is also located two focal lengths in front of lens L_3. Therefore, the divergent spherical wave incident on L_3 is focused at point P_3, two focal lengths behind L_3. This is the image plane of the zoom lens.

Now consider Fig. 7.19a, which shows the same nominal configuration in simplified form, with the lenses merely represented by vertical lines. A review of the geometry in the figure will show that the angle θ_2, the edge ray tilt angle in the image, equals angle θ_1, the tilt angle of the edge ray passed by L_1. So, no effective magnification has occurred due to L_2 and L_3.

Now consider Fig. 7.19b, in which L_2 has been moved a distance ϵ from L_1, and lens L_3 has been moved a distance ϵ toward L_1. Now, the effective object distance for L_2 (the distance from L_2 to P_1) has been reduced by ϵ. To a first order then, the image distance will be increased by ϵ. Therefore, the point P_2 from the nominal configuration still remains the imaginary focal point of L_2 in the magnified configuration. The field transmitted through L_2 is accordingly still shown as diverging from P_2. The wave diverging from P_2 is now incident on L_3, but now the distance from P_2 to L_3 has been decreased by ϵ (and at the same time, the tilt angle of the edge ray diverging from P_2 has been reduced to $\psi < \theta_1$). Since the "object distance" from P_2 to L_3 is now reduced by ϵ, the image distance will (to a first order) be increased by the same amount, ϵ, placing the point of focus again at point P_3, as in the nominal case. However, this time, the tilt angle of the edge ray in the image is reduced from the nominal case. Therefore, a magnified image is produced by the sine condition. This is the basic principle of operation of the zoom lens.

The zoom lens need not necessarily be designed in the way we've described. In fact, numerous variations on this theme are possible [3], and the interested reader is referred to the literature for details.

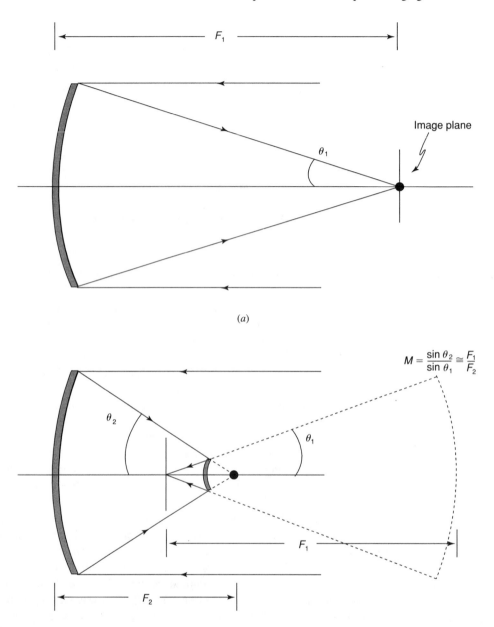

Figure 7.18 Front-fed and Cassegrain mirror configurations: (a) Front-fed parabolic telescope and (b) Cassegrain telescope (parabolic main reflector/hyperbolic sub-
reflector), showing unfolded, equivalent, single-mirror configuration.

7.10 ASPHERIC LENSES

Aspheric lenses, such as the Schmidt corrector, are used primarily to correct for the spherical aberration that can arise in ordinary lenses constructed from spherical surface segments.
The Schmidt corrector is a *shaped,* rotationally symmetric lens, whose generating arc is

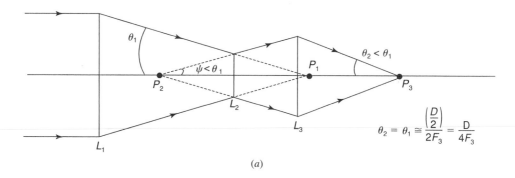

(a)

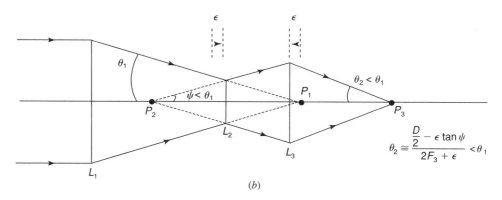

(b)

Figure 7.19 Nominal and magnifying configurations for the zoom lens: (a) Nominal configuration and (b) Image magnification.

determined via ray-tracing analyses. The outline shape is determined by enforcing a desired phase condition in the image plane and then working backwards to determine the lens shape that will produce the desired image phase. One of the most notable applications of a Schmidt corrector lens was in the Hubble space telescope, where one was used to correct large amounts of spherical aberration that had degraded the focusing properties of the instrument.

Even though modern aspheric lenses are now designed using sophisticated ray-tracing software, the general design approach can be demonstrated using a relatively straightforward analysis. For example, consider Fig. 7.20, which shows a shaped lens. If the incident illumination is regarded as a collimated plane wave, the goal of the design process is to bring all rays together at the focal point of the lens, as shown. The design algorithm can be expressed in terms of a difference equation.

At the vertex of the lens ($r = 0$), the lens curvature is prescribed so that an incident planar field (limited in lateral extent to points near the vertex) will be focused at the defined focal point. Thus, using (5.9) with $R_2 = 0$, we have

$$f''(0) = \frac{1}{R(0)} = \frac{1}{F(n-1)}$$

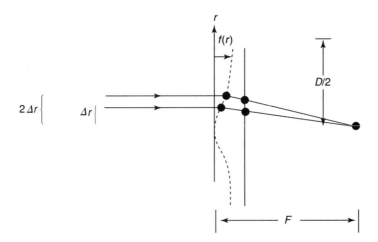

Figure 7.20 On the design of aspheric lenses.

where F = focal length of the lens
    $R(0)$ = lens radius of curvature at $r = 0$
    n = refractive index of the lens medium

From here, the construction proceeds as follows: $f''(0)$ is assumed given, and from this, $f(\Delta r)$ is obtained via the Taylor series as

$$f(\Delta r) = \frac{\Delta r^2}{2R(0)}$$

At this point, a ray-tracing analysis can be used to find the value of lens slope, that is, $f'(\Delta r)$, which refracts the ray directed toward the point $[\Delta r, f(\Delta r)]$ through the prescribed focal point. Once $f'(\Delta r)$ is known, the Taylor series expression

$$f(2\Delta r) = f(\Delta r) + \Delta r \, f'(\Delta r)$$

can be used to find $f(2\Delta r)$. Once again, ray tracing is used to find $f'(2\Delta r)$, and so on. The ray-tracing analysis depends on the assumed thickness of the lens at the lens vertex. Therefore, multiple values of thickness may have to be chosen so that the lens has the desired thickness at its maximum specified radius. Thus, some iterations on the design procedure may be required before the final lens design is obtained.

It turns out that generally aspherics deviate from ordinary spherical lenses only near the rim of the lens, where the lens becomes indented, as shown in the figure. Generally, aspherics are rather thick, although this is primarily due to the fact that thin lenses usually don't need aspherical shaping correction.

REFERENCES

[1] Born, M., and Wolf, E., *Principles of Optics,* 6th ed. Oxford: Pergamon Press, 1980.

[2] Scott, C. R., *Modern Methods of Reflector Antenna Analysis and Design,* Norwood, MA: Artech House, 1990.

[3] Mann, Allen, "Infrared Zoom Lenses in the 1980's and Beyond," *Optical Engineering,* vol. 31, no. 5, May 1992.

Other Common
Optical Components

In this chapter, we'll look at some practical optical components typically used in optical imaging systems.

8.1 NONPOLARIZING PRISMS

Prisms are generally used for one of two purposes: for beam deflection or reflection. The use of the prism as a beam deflector dates back to Newton, who used it to resolve white light into the color spectrum. In modern optical systems, prisms are used to deflect laser beams, as are the wedge prisms shown in this chapter. A single, mechanically rotating wedge prism is capable of steering a beam along the arc of a circle, while two such prisms can be used to steer a beam anywhere within a circular area.

As we saw in the previous chapter (in connection with binocular operation), prisms may also be used to reflect light. Prisms are often used in place of mirrors in systems where ruggedness is a concern. The phenomenon of total internal reflection produces exactly the same type of reflection property as a mirror. The following text and figures on prisms are taken from the Melles Griot Optics Guide 5.

> Prisms may be used in an optical system to deflect or deviate a beam of light. They can invert or rotate an image, they can disperse light into its component wavelengths, and they can be used to isolate separate states of polarization. The numerous applications for prisms make them an important component of any optical system.

> Prisms are, essentially, blocks of optical material with flat polished sides which are arranged to be at precisely controlled angles to each other. The angular specifications of prisms are of prime importance since small angular face errors can lead to incorrect function of the prism.

8.1.1 Explanation of Prism Drawings

> Prisms by themselves are incapable of forming real images. The meanings of the real images that we show in many of the prism drawings that follow require explanation. Using the right-angle prism as an example, we illustrate all of the detail necessary to visualize the effect of

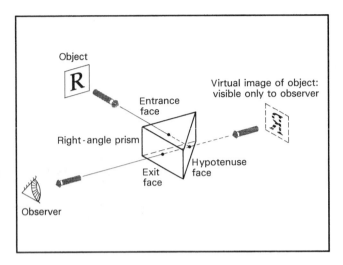

Figure 8.1 Left/right image inversion in right-angle prisms.

the prism on image orientation. In the drawing above, we show the situation as it actually occurs. A viewer looks through the prism at an object and sees a virtual image. This image may be displaced from the original object, such as in the example, or it may coincide with the object, as in the case of the dove prism. Furthermore, the orientation of the image may differ from that of the object; the right-angle prism shown in Fig. 8.1 has reversed left and right in this situation. Fig. 8.2 shows the same situation using the convention used in all subsequent prism drawings. A real image is shown; its orientation is a result of the action of the prism. It must be kept in mind, however, that a real image would only be formed if there were some sort of imaging optics present in the system. If no imaging optics are present, the image is virtual. The virtual image will have the same orientation as the real image we show, but it can only be viewed by an observer looking back through the prism system. Also, its location is not where the real image is shown in the drawing. Image orientation information

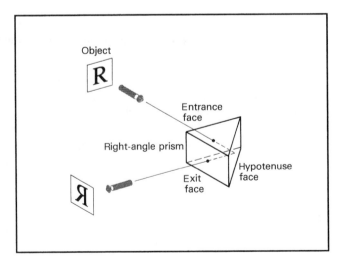

Figure 8.2 Shorthand convention equivalent to Fig. 8.1.

is of vital importance to the designer of optical systems using prisms. This illustration should be kept in mind while interpreting all subsequent prism drawings.

8.1.2 Aberrations Introduced by Prisms

The arrangement by means of which we have explained the prism drawings corresponds closely to the way in which prisms should actually be used in optical systems. If used in an optical system where the beam is either convergent or divergent, prisms will introduce aberrations that would not otherwise be present. These can be avoided or minimized by making sure that prisms are used only in collimated or nearly collimated beams. Conjugate distances that include prisms should be long. Alternatively, other elements of the system can be designed to compensate for aberrations introduced by the prism in the presence of convergence or divergence. The system as a whole (specifically including prisms) must be designed for an acceptable level of aberration.

8.1.3 Total Internal Reflection

Many prisms operate by total internal reflection (TIR). Rays internally incident upon an air/glass boundary, at angles greater than the critical angle, are reflected with 100% efficiency regardless of their initial polarization state. This is shown in Fig. 8.3. The critical angle is given by

$$\theta_c(\lambda) = \arcsin\left(\frac{1}{n_\lambda}\right),$$

and depends on the refractive index, which is a function of wavelength. If, at some wavelength, the refractive index should fall to less than $\sqrt{2} \approx 1.414$, the critical angle will

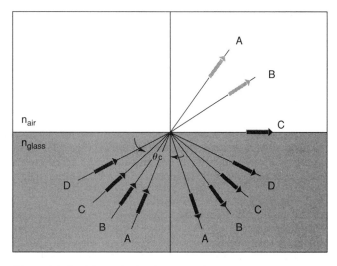

Figure 8.3 Total internal reflection occurs for all angles greater than θ_c.

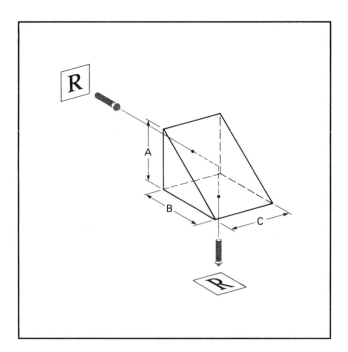

Figure 8.4 Right-angle prism used as a retroreflector.

exceed 45 degrees, and TIR will fail for a collimated beam internally incident at 45 degrees on the hypotenuse face of a right-angle prism. Reflectance decreases rapidly at angles of incidence less than the critical angle.

The right-angle prism is often preferable to an inclined mirror in applications involving severe acoustic or inertial loads, because it is easier to mount and deforms much less than a mirror in response to external mechanical stress. Very high transmission can be achieved by using the hypotenuse face in total internal reflection (TIR) and by antireflection coating the entrance and exit faces.

Alternatively, the hypotenuse face can be used in external reflection and coated with a metallic reflective coating.

If it is desired to use the hypotenuse face in internal reflection, but field-angle requirements exceed TIR acceptance limits, or the environment does not permit the hypotenuse face to be kept as clean as TIR requires, a right-angle prism with hypotenuse face aluminized will solve the problem.

As long as acceptance-angle limitations for TIR from the minor (roof) faces are not exceeded, the right-angle prism can serve as a retroreflector. This is illustrated in Fig. 8.4. If the hypotenuse face is antireflection coated, as shown in Fig. 8.5, reflective efficiency is very high and is insensitive to orientation about axes normal to the plane of the page. Reflective efficiency is extremely sensitive to orientation about an axis parallel to the hypotenuse face and in the plane of the page.

If the hypotenuse face is aluminized or silvered, there is also a beam path through the prism which guarantees precise right-angle deflection while allowing the prism a single degree of rotational freedom. (See Taylor & Hafner, ''Turning a Perfect Corner,'' *American Journal of Physics,* January 1979, pp. 113–114.)

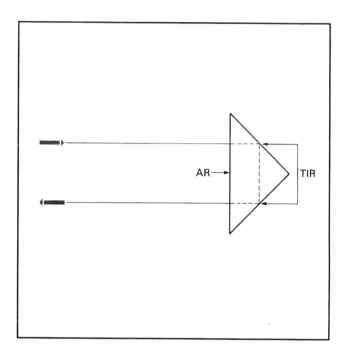

Figure 8.5 Antireflection coating applied to hypotenuse face.

The roof or Amici prism deviates or deflects the image through an angle of 90 degrees. Fig. 8.6 shows its effect on the image orientation, where the precise meanings of object and image were explained earlier. Basically, the roof prism is a right-angle prism whose hypotenuse has

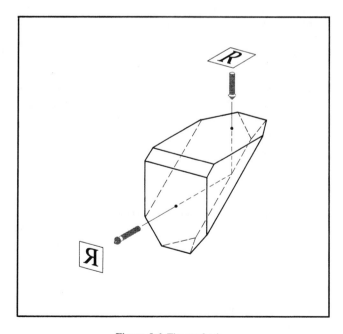

Figure 8.6 The roof prism.

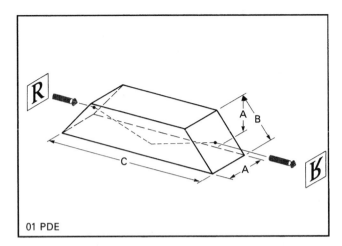

01 PDE

Figure 8.7 The dove prism.

been replaced by a 90-degree TIR roof. Its shape is further complicated by the fact that glass which does not contribute to useful aperture has been trimmed away to reduce size and weight. Cleanliness and quality of the TIR roof surfaces are important. The 90-degree deviation applies only to rays that are precisely normal to the entrance and exit faces, other rays being deflected by other amounts, exactly as in the case of the ordinary right-angle prism. The roof prism finds application in situations that demand both right-angle deflection and image uprightness (a combination of left-to-right reversion and top-to-bottom inversion, equivalent to a 180-degree rotation about the optical axis), as in some instrument viewfinders.

Dove prisms are used as image rotators in a variety of optomechanical systems. As the prism is rotated, the image passing through will rotate at twice the angular rate of the prism. Light entering the dove prism must be parallel or collimated, because the length of the prism is typically four or five times the dimension A. The hypotenuse face of a dove prism is the face of the largest area. Normally, the dove prism is used in the total internal reflection (TIR) mode, with the hypotenuse face unaluminized. Cleanliness of this face is important.

The useful aperture of a dove prism can be conveniently doubled by aluminizing or silvering the hypotenuse face of two identical prisms, and subsequently cementing these faces together (Fig. 8.7). The length of the combination is the same as for a single prism. Overall transmission is slightly reduced by the substitution of the metallic reflecting film for the usual TIR surface.

Penta-prisms have two important properties. The first is that the image is neither inverted nor reversed as it is deviated by 90 degrees. Second, the penta-prism is a constant deviation prism, meaning that the same exact 90-degree deviation angle applies to all rays transmitted by the useful aperture—regardless of the angles between these rays and the optical axis (or entrance and exit face normals). The deviation is thus independent of the orientation of the prism, making the penta-prism especially important in applications in which prism orientation cannot be precisely controlled (Fig. 8.8). This prism is commonly used in optical tooling, alignment, rangefinding, surveying, and other instrumental applications. For example, reflection normal to the surface of a calm pool of mercury, seen through a penta-prism, establishes an accurate horizontal reference.

The internal geometry of these penta-prisms is such that the reflecting surfaces must be coated with a metallic or dielectric reflective coating. TIR cannot be used. Standard penta-prism reflecting surfaces are coated with aluminum.

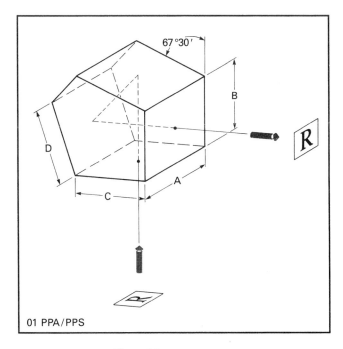

67°30'

B

D

A

C

R

01 PPA/PPS

Figure 8.8 The penta prism.

Wedge prisms are used as beamsteering elements in optical systems, playing a role in optics analogous to that of the wobble plate in mechanics. The minimum deviation or deflection experienced by a ray in passing through a thin wedge of apex angle θ_w is approximately given by

$$\theta_d = (n - 1)\theta_w$$

where n is the refractive index. The ''power'' (Δ) of a prism is measured in prism diopters, a prism diopter being defined as a deflection of 1 cm at a distance of 1 meter from the prism. Thus

$$\Delta = 100 \tan(\theta_d)$$

By combining two wedges of equal power (equal deviation) in near contact, and independently rotating them about an axis roughly parallel to the normals of their adjacent faces, a ray passing through the combination can be steered in any direction, within a narrow cone, about the path of the undeviated ray. The angular radius of this cone is approximately θ_d. Wedge prisms are most frequently used for laser beamsteering and Schlieren system calibration. Deviations for other wavelengths can be estimated from the above formula and appropriate nominal refractive index values. Apex angle is controlled to within very tight tolerances in the manufacturing process. As a result of the melt-to-melt index tolerance, deviation angles (functions of wavelength) are only nominally specified.

The deviation angles are specified with the assumption that the input beam is *normal* to the perpendicular face. At other input angles the deviation will, of course, be different. The equation to determine the deviation angle for the same input direction but other wavelengths is

$$\theta_d = \arcsin(n \sin\theta_w) - \theta_w$$

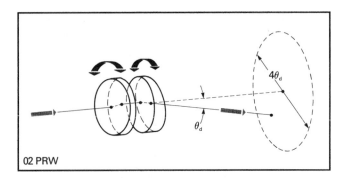

Figure 8.9 Laser beamsteering using wedge prisms.

where θ_d is the deviation angle, θ_w is the wedge angle, and n is the nominal index at the appropriate wavelength. This is shown in Fig. 8.9.

8.2 ANTIREFLECTION COATINGS

The antireflection (AR) coating is the optical analogue of a device from microwave theory known as the quarter-wavelength impedance transformer. The antireflection coating consists of several layers (the exact number depends on the bandwidth requirements) of thin dielectric films deposited on the surface of a lens, prism, or other transparent optical component. The purpose of the multilayer coating is to reduce the reflection coefficient (provide an impedance match) between the dielectric component and air. The derivation of the single-layer coating is relatively straightforward [1]. The analytical design of multilayer structures is somewhat more complex, though still well known, and is given in the microwave literature [2]. Modern design of AR coatings involves the use of sophisticated optimization software, which is now part of most optical CAD packages. AR coatings may be analyzed using the admittance tensor from Chapter 2 [4,5].

The following text and figures on AR coatings are taken from the Melles Griot Optics Guide 5.

In order to understand the operation of multilayer coatings, we will now examine the two-layer antireflection coating. It is useful to discuss them in some detail in order to understand the operation of multilayer coatings. It is beyond the scope of this text to cover all aspects of modern thin film design and operation. However, it is hoped that these notes will give the user some insight into thin films which will be of use when considering system designs and specifying cost-effective coatings.

8.2.1 Two-Layer AR Coatings

Two basic types of AR coating have been developed and are worth examining in detail: (1) Quarter/Quarter coating and (2) Broadband Antireflection coatings.

The Quarter/Quarter Coating

This type of coating was used as an alternative to the single-layer AR coating. The motivation for its development was the lack of a suitable material to improve the performance of single-layer coatings. The principle is very simple. The basic problem of a single-layer AR coating is that the refractive index of the coating material is too high, resulting in too strong a reflection from the first surface which cannot be completely canceled by interference of the weaker reflection from the surface of the substrate. In a two-layer coating, the first reflection is nullified by interference with two weaker reflections.

As shown in Fig. 8.10, a quarter/quarter coating consists of two layers, both of which have an optical thickness of 1/4 wave at the wavelength of interest. The outer layer is of a low

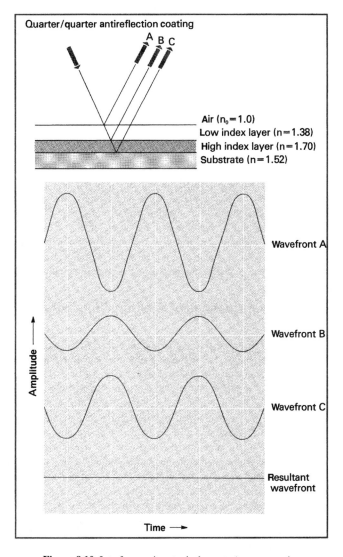

Figure 8.10 Interference in a typical quarter/quarter coating.

refractive index material, and the inner layer is of a high refractive index material (compared to the substrate). As the figure shows, the second and third reflections are both exactly 180 degrees out of phase with the first reflection.

As with any multilayer coating, the performance and design are calculated in terms of relative amplitudes and phases that are then summed to give the overall (net) amplitude of the reflected beam. The overall amplitude is then squared to give the intensity.

How does one calculate the required refractive index of the inner layer? Several methodologies have been developed over the last 40 or 50 years to calculate thin film coating properties and converge on optimum designs. The whole field has been revolutionized in recent years with the availability of powerful microcomputers.

With a two-layer quarter/quarter coating optimized for one wavelength at normal incidence, the required refractive indices can easily be calculated by hand. The formula for exact zero reflectance for such a coating is

$$\frac{n_1^2 n_3}{n_2^2} = n_0$$

where n_0 is the refractive index of air (approximated as 1.0), n_3 is the refractive index of the substrate material, and n_1 and n_2 are the refractive indices of the two film materials, as indicated in Fig. 8.10.

If the substrate is crown glass with a refractive index of 1.52, and the first layer is the lowest possible refractive index, 1.38 (MgF_2), then the refractive index of the high index layer needs to be 1.70. Either beryllium oxide or magnesium oxide could be used for the inner layer, but both are soft materials and will not produce very durable coatings. Although it allows some freedom in the choice of coating materials and can give very low reflectance, the quarter/quarter coating is very restrictive in its design. In principle, it is possible to codeposit two materials simultaneously to achieve layers of almost any required refractive index, but such coatings are not very practical in terms of production. As a consequence, thin film engineers have developed both multilayer antireflection coatings and two-layer coating designs that allow the refractive index of each layer to be chosen as that of a suitable material.

8.2.1.2 Two-Layer Coatings of Arbitrary Thickness

Interference is often simplistically thought of in terms of constructive or destructive interference, where the phase shift between interfering wavefronts is either 0 degrees or 180 degrees. For two wavefronts to completely cancel as in a one-layer AR coating, a phase shift of exactly 180 degrees is required. Where three or more reflecting surfaces are involved, complete cancellation can be achieved by carefully chosen arbitrary phase and relative intensities. This is the basis of a two-layer type of antireflection coating, where the thickness of the layers is adjusted to suit the refractive index of the available materials, instead of vice versa. For a given combination of materials, there are usually two combinations of layer thicknesses which will give zero reflectance at the design wavelength. These two combinations are of different overall thickness. For any type of thin film coating, the thinnest possible overall coating is used since it will have better mechanical properties (less stress). In this case, the thinner combination is also less wavelength sensitive. This is illustrated in Fig. 8.11.

Two-layer antireflection coatings are the simplest of the so-called V-coatings. The term *V-coating* arises from the shape of the reflectance curve as a function of wavelength, which is a skewed V shape with a reflectance minimum at the design wavelength. V-coatings are very

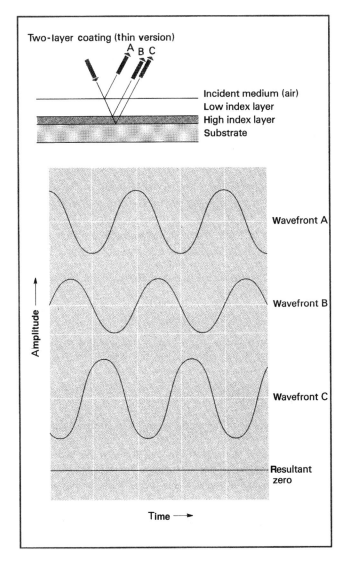

Figure 8.11 Two-layer coating (''thin'' version).

popular economical coatings for near monochromatic applications, such as optical systems using nontunable laser radiation (e.g., helium neon lasers at 632.8 nm). A typical curve of reflectance versus wavelength for a V-coating is shown in Fig. 8.12.

8.2.1.3 Broadband Antireflection Coatings

Many optical systems (particularly imaging systems) use polychromatic light; or light of more than one wavelength. In order for the system to have a flat spectral response, the transmitting optics are coated with a broadband or dichroic antireflection coating. The main technique used

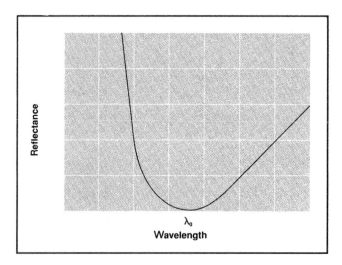

Figure 8.12 The V-coating derives its name from the characteristic shape of the performance curve.

in designing antireflection coatings which are highly efficient at more than one wavelength is to use absentee layers within the coating.

8.2.1.4 Absentee Layers

An absentee layer is a layer of dielectric material that does not change the performance of the overall coating at one particular wavelength, usually the wavelength for which the coating is being principally optimized. This results from the fact that the coating has an optical thickness of $\lambda/2$ wave at that wavelength. The effects of the extra ''reflections'' cancel out at the two interfaces since no additional phase shifts are introduced. In theory, the performance of the coating is the same at that wavelength whether or not the absentee layer is present.

At other wavelengths, the absentee layer starts to have an effect for two reasons. The ratio between the physical thickness of the layer and the wavelength of light changes with wavelength. Also, the dispersion of the coating material causes the optical thickness to change with wavelength.

8.3 SPATIAL LIGHT MODULATORS

The *spatial light modulator* (SLM) is not a classical optical component, but in recent years it has become the workhorse in many types of modern electro-optical systems. Numerous different types of SLMs are available, depending on their specific device physics and modes of operation. An in-depth presentation of their operation will not be attempted here, but rather simply a description of their input/output characteristics. These characteristics are quite easy to understand and are the important properties to know for optical system design.

We saw in Chapter 5 that a lens acts on an optical field by imparting a quadratic phase to it. We also saw in Chapter 1 that a Fresnel zone plate (FZP) creates a focusing effect by blocking the out-of-phase Fresnel zones and transmitting the in-phase Fresnel zones. Lenses and Fresnel zone plates are two types of spatial light modulators from classical optics. (The lens is a phase modulator, and the FZP is an amplitude modulator.) Both of these devices produce the same effect, however—that is, to transform incident planar or spherical wavefronts into spherical wavefronts with different radii of curvature. (The FZP produces two transmitted wavefronts, one converging and one diverging, as a result of the binary nature of the transmission function.) Other types of spatial light modulators used in optical system design include pinhole filters, diffraction gratings, holograms, and liquid crystal displays (LCDs). In this section, we'll emphasize the reconfigurable, pixelated spatial filters that are generally understood when the term *SLM* is used.

When a complex transmission or reflection function is implemented onto an SLM, we say the function has been written onto the device. SLMs exist which allow these patterns to be written electronically or optically and are (not surprisingly) referred to as *electronically addressed* and *optically addressed* SLMs, respectively. A sketch of an electronically addressed SLM is shown in Fig. 8.13 [3]. In this case, the SLM will alter the magnitude, phase, or polarization of the incident wavefront on a pixel-by-pixel basis in a binary fashion determined by the electrical inputs.

Some electronically addressed SLMs (such as liquid crystal SLMs) use an addressing scheme known as *matrix addressing* to minimize the number of electrical connections to the device. For an $N \times N$ pixel array, the number of electrical inputs required for

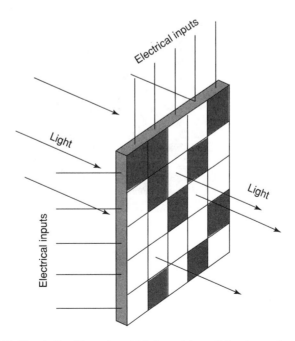

Figure 8.13 Electrically addressed spatial light modulator. (After Arsenault and Sheng.)

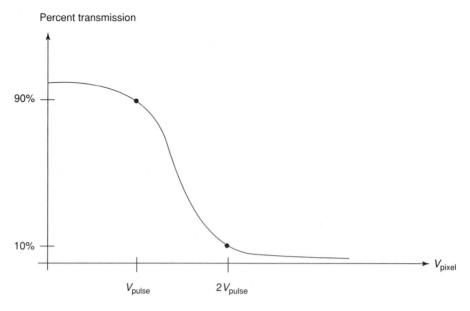

Figure 8.14 Typical nonlinear transmission characteristic of an electrically addressed SLM.

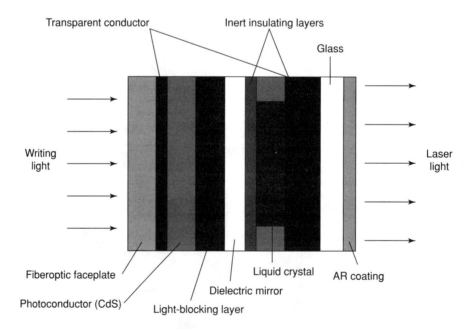

Figure 8.15 Hughes liquid crystal light valve construction (after Arsenault and Sheng).

a matrix-addressed array is only $2N$ rather than n^2, so the savings in pins is substantial. Matrix addressing works in the following way. Strobed pulse trains are sent down each of the wires in both the horizontal and vertical directions. These wires run the length of the SLM aperture and at each pixel are connected via the material within the pixel, just as two ordinary wires might be connected through a resistor (as in a typical read-only memory matrix). Depending on the timing of the pulses in the $2N$ horizontal and vertical wires, a pixel at any given moment may be energized by zero, one, or two electrical pulses.

The material in each pixel generally activates in a nonlinear fashion with respect to the voltage applied to it. For example, Fig. 8.14 shows a typical nonlinear response for a matrix-addressed pixel. If no pulses are incident on the pixel (zero voltage condition), then the pixel may be 100% transmissive. A single incident pulse (from either a horizontal or vertical wire) would produce a small decrease in transmissivity. That is, the voltage corresponding to a single incident voltage pulse would not be sufficient to cause a significant change in the pixel transmissivity from the zero voltage condition. Two coincident

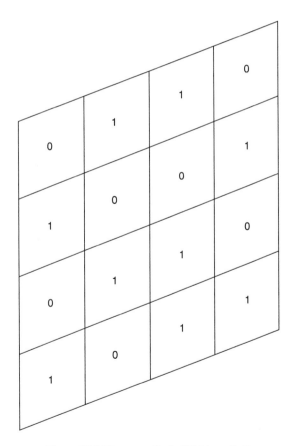

Figure 8.16 Binary amplitude SLM ($\tau = 0, 1$).

pulses would be sufficient to "switch" the pixel from a "mostly transmissive" state to a "mostly absorptive" state (say, 10% transmissive).

Depending on how the pixels are strobed, some pixels will never have two coincident pulses, whereas others will have them coincident at some times but never at all times. Therefore, the relaxation time of the pixel should be longer than the time between occurrences of the two coincident pulses. In this way, the pixel remains "set" in the opaque state at all times.

SLMs may also be optically addressable, as shown in Fig. 8.15. This figure shows a polarization-type SLM in which the polarization rotation is proportional to the intensity of the incident "writing" light.

Most SLMs are binary-state devices wherein the two states could be opaque/transmissive, transmissive with binary 0/180 degrees transmission phase, or transmissive with horizontal/vertical polarization. These three types of SLMs are illustrated in Figs. 8.16–8.18, respectively. Of increasing importance today are ternary phase and amplitude SLMs that have the following three transmission states:

$$\tau = 0, \pm 1$$

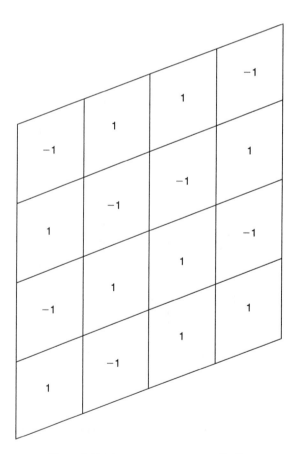

Figure 8.17 Binary phase SLM ($\tau = e^{j^0}, e^{j^\pi}$).

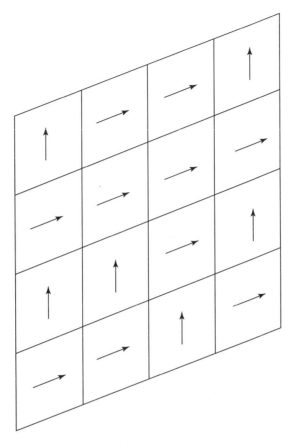

Figure 8.18 Binary polarization SLM (similar to the optically addressed Hughes LCLV).

8.4 INTERFERENCE FILTERS

An interference filter is structurally identical to an antireflection coating; that is, it is a multilayer stack of dielectric slabs. Unlike the AR coating, however, interference filters may be designed for bandpass or bandstop properties. An example of a single-layer bandpass interference filter is the Fabry-Perot interferometer from Chapter 2. Interference filters may be designed using transmission line analysis techniques and a trial-and-error process or specialized analytical techniques such as the Fourier transform method [6–8]. Modern software also exists for the design and optimization of interference filters. Examples of interference filters designed using the Fourier transform method are shown in Fig. 8.19 [8]. The figures compare the index profiles (versus optical path length through the stack) and transmittance of filters having various percentage bandwidths.

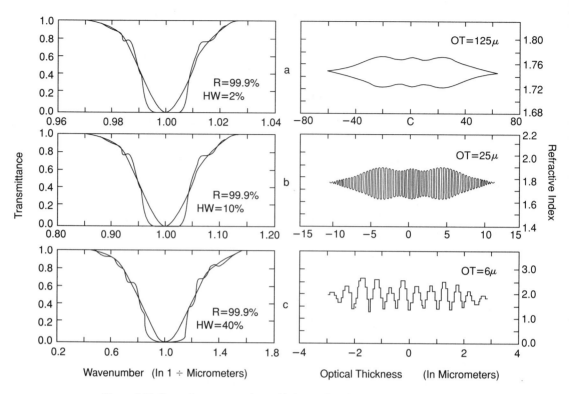

Figure 8.19 Transmittance properties and index profiles for three interference filters designed using the Fourier transform method (after Verly et al.).

REFERENCES

[1] Ramo, S., Whinnery, J. R., and Van Duzer, T., *Fields and Waves in Communication Electronics,* New York: Wiley, 1984.

[2] Matthaei, G. L., Young, L., and Jones, E. M. T., *Microwave Filters, Impedance-Matching Networks and Coupling Structures,* New York: McGraw-Hill, 1964.

[3] Arsenault, H. H., and Sheng, Y., *An Introduction to Optics in Computers,* Bellingham, WA: SPIE Press, 1992.

[4] Scott, C. R., *The Spectral Domain Method in Electromagnetics,* Norwood, MA: Artech House, 1989.

[5] Scott, C. R., *Field Theory of Acousto-Optic Devices,* Norwood, MA: Artech House, 1992.

[6] Dobrowolski, J. A., "Comparison of the Fourier Transform and Flip-Flop Thin-Film Synthesis Methods," *Applied Optics,* vol. 25, no. 12, June 15, 1986, pp. 1966–1972.

[7] Boivin, G., and St.-Germain, D., "Synthesis of Gradient-Index Profiles Corresponding to Spectral Reflectance Derived by Inverse Fourier Transform," *Applied Optics,* vol. 26, no. 19, October 1, 1987, pp. 4209–4213.

[8] Verly, P. G., Dobrowolski, J. A., Wild, W. J., and Burton, R. L. "Synthesis of High-Rejection Filters with the Fourier Transform Method," *Applied Optics,* vol. 28, no. 14, July 15, 1989, pp. 2864–2875.

9

Aberration Theory

The ideal optical system produces an image that has two important properties. First, each point in the image must be the focal point of a converging spherical wavefront that originally emanated from a corresponding point of the original object. Second, the two corresponding point locations must be related in a particular way. That is, if a point in the object space is taken as the origin for the object space, and its corresponding image point is taken as the origin of the image space, then for each pair of arbitrary object and image points, the vectors joining these points and their respective origins are related by an *affine mapping* of the form

$$r_i = MUr_0 + r_{\text{offset}}$$

where r_0, r_i are the object and image vectors, respectively, the scalar M is the magnification of the optical system, U is a unitary matrix (which represents either a rotation or reflection and whose rows and columns each form a set of orthonormal vectors), and r_{offset} is a translational term. This is the most general type of relation that can exist between the object and image spaces of a perfect imaging system. (Recall that by Maxwell's theorem, it is impossible to achieve perfect optical imaging unless M is unity. Thus, we're really engaging in "wishful thinking" when we speak as though M could ever be anything other than unity for a perfect imaging system.)

In this chapter we look at some of the ways in which an image becomes imperfect. Actually, there are only two ways, and both are related to the two properties of optical images listed here. Either the wave converging onto the image is nonspherical, or the image is not geometrically conformal with the original object. In the former case, the image will be poorly focused, but in proportion to the object. In the latter case, the image may be well focused, but it will be stretched or compressed in strange ways. (The distortion aberration is generally regarded as being less severe than the focusing aberration.) These image imperfections are the direct result of the imperfections in the converging wavefronts that produce the images. These latter imperfections are known as *aberrations*.

The aberration concept is but one of many possible ways to characterize the quality of an optical imaging system. We've already encountered two such "figures-of-merit," namely, the point spread function (PSF) and the modulation transfer function (MTF). The PSF and the aberrations function of an optical system are not independent descriptions of an optical system, as we've seen. (If the exit aperture is uniformly illuminated, the focal plane PSF is the Fourier transform of the complex exponential whose phase is the aberration function.)

9.1 THE ABERRATION FUNCTION

Consider the exit aperture shown in Fig. 9.1. Assuming a point source object (at either a finite or an infinite distance in front of the lens), we observe that the ideal optical imaging system will produce a perfectly spherical wave field in the exit aperture, centered on some point in the focal plane. The aberration function is equal to the deviation of the actual phase front (just behind the exit aperture) from a perfectly spherical phase front.

Our definition of the aberration function is deceptively simple. For example, a plane wave incident on a lens will produce a different aberration function from a converging or diverging spherical wave. Moreover, the aberration function of either type of wave (plane or spherical) will be a function of the direction of incidence of the wave. These same considerations are also true for the point spread function and MTF. Thus, we see that the point spread function, the MTF, and the aberration function depend not only on intrinsic properties of the lens itself, but also on the nature of the illumination. This extra complication sometimes makes it confusing to understand what is meant when the broad terms *point spread function, MTF*, and *aberration function* are used loosely.

In Chapters 5 and 6, we became somewhat lazy in our description of the fields transmitted by thin lenses. We replaced the more realistic *spherical waves* with the more mathematically convenient *quadratic waves* (in the Gaussian regime wherein spherical wavefronts may be approximated by quadratic wavefronts). That simplification made possible a relatively straightforward (closed-form) mathematical analysis of lens systems from a diffraction integral viewpoint. It should be noted that the diffraction integral technique is not dependent on the quadratic-phase front assumption; we merely made that assumption to permit closed-form mathematical solutions. In more exact diffraction integral calculations, the integral would be evaluated numerically, rather than in closed form, using a spherical wave kernel. The quadratic-phase simplification in analysis is justified for Gaussian optics, wherein the difference between a quadratic and spherical wavefront (i.e., the difference between a paraboloid and a sphere) is minimal. In Gaussian optics, terms larger than quadratic terms (e.g., quartic, etc.) are ignored. In aberration theory, these terms are not ignored. Thus, the reader should not be concerned over our newfound ''pickiness'' in dealing with wavefront terms higher than the quadratic. This is merely a degree of exactness that brings our analysis past the ''conceptual analysis'' level of the previous chapters to a more exact analysis of the type required for actual lens design. In practice, of course, this more exacting analysis is usually performed by modern ray-tracing software.

In discussing the aberration function in this book, we'll be referring to either plane wave illumination or point source illumination. As mentioned previously, the aberration function and point spread functions of a lens depend on the illumination. For example, an object scene may be regarded as the superposition of a number of point sources, all located at the individual points in an object. Each of these point sources will be located at unique positions in the object space, and in general, each of these point radiators will produce spot images having different point spread functions and aberration functions. In the language of optics, this means the point spread functions and aberration functions of these various point radiators are *space-variant;* that is, they vary from point to point in the object space. Space-variant lens systems give rise to images that vary in complicated ways in terms of the aberration and point spread functions.

For certain lenses, however, the aberration and point spread functions may be regarded as space-*invariant,* at least over certain regions of the object space known as

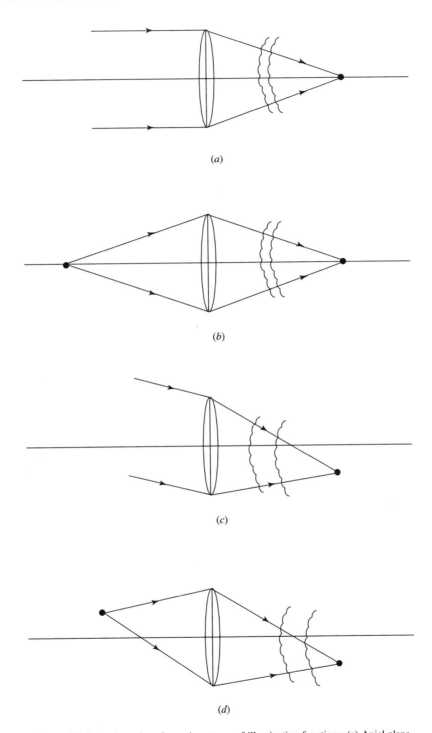

(a)

(b)

(c)

(d)

Figure 9.1 Lens aberrations for various types of illumination functions: (a) Axial plane
wave incidence; (b) axial spherical wave incidence; (c) off-axis plane wave
incidence; and (d) off-axis spherical wave incidence.

isoplanatic regions. For such lenses, we may merely evaluate the point spread function or aberration function for an on-axis point, and then use this function for all other points in the object space. Note that the point spread function contains somewhat less information than the aberration function, since it does not describe the way in which image points are related to their counterparts in the object space. That is, it does not reveal the distortion effects related to possible nonlinear mapping of object points to image points.

For shift-invariant systems, we may take the aberration function for any one point in the object space and use it to evaluate degradation effects for the entire object as it becomes transformed into an image.

9.2 PHASE VARIATIONS IN THE OPTICAL FIELDS

Phase aberrations are associated with the departure of the wavefront from a perfectly spherical wavefront. These aberrations degrade the point spread function of the system and cause the image to be blurred with respect to the original object. That is, fine detail is lost. Readers familiar with Fourier theory and filtering will recognize that this loss of fine detail is often the result of the loss of the high spatial frequency components of the object distribution, though in optical systems it is more often related to improper *phasing* between the Fourier (plane wave) components of the image, all of which may be present. These phase imbalances are critically important, and account entirely for the loss of image detail that occurs when an image is viewed in a plane axially displaced from the true focal plane.

In classical optics, before the advent of modern high-speed digital computers, wavefront aberrations were characterized in purely analytical terms. This placed severe restrictions on the types of aberrations that could be studied. Even today, with the availability of powerful computers, the primary aberrations of classical optics are widely used for describing optical system quality. One reason why is that in modern high-quality optical systems, aberrations of higher order generally are not present to a significant degree, and so the primary aberrations are still most significant.

The aberration function is defined as the difference in phase between the actual image wavefront and a perfectly spherical wavefront that converges on the intended image point. By this definition, aberration is measured by placing a point source in the object plane and then observing the resulting image of this point (and its associated converging wavefront) in the image region. Thus, aberration is a "local" phenomenon—it varies from point to point in the image region. Generally, aberrations are the least for object/image points located near the axis of the optical system and greatest for those at the outer rim.

We've mentioned that the concept of aberration relates to point source objects and their resultant images. If the aberrated image wavefront is as shown in Fig. 9.1, then the aberration function $\epsilon(x_0, y_0; x, y)$ is a function of the source point coordinates (x_0, y_0), as well as the coordinates (x, y) in the exit aperture. If we define

$$x_0 = r_0 \cos \phi_0 \qquad x = r \cos \phi$$

$$y_0 = r_0 \sin \phi_0 \qquad y = r \sin \phi$$

then the aberration function may also be defined in terms of the four variables

$$r_0, \, r, \, \phi_0, \, \phi$$

If the optical imaging system is rotationally symmetric, the aberration function only depends on the difference

$$\phi - \phi_0$$

and not on the absolute value of ϕ. The value of ϕ only fixes the ''origin'' of the aberration function. Therefore, the distance function depends only on the three variables

$$r_0, r, \phi_0 - \phi$$

in various combinations.

We may readily determine the types of combinations of these three variables that arise in rotationally symmetric optical systems. The phase from the object point to a point in the entrance pupil is given by the normal Euclidian distance function

$$\sqrt{r_{obj}^2 + r_{entr}^2 + r_{obj}r_{entr}\cos(\phi_{obj} - \phi_{entr}) + (z_{obj} - z_{entr})^2}$$

where the subscripts obj and entr refer to the object plane and entrance pupil, respectively. This may be thought of as the phase function of the incident field. If z_{obj} and z_{entr} are assumed constant (i.e., we limit our consideration to object and entrance pupil planes), this function may be expanded as a Taylor series in the three variables u, v, w where

$$u = r_{entr}^2$$

$$v = r_{obj}^2$$

$$w = r_{obj}r_{entr}\cos(\phi_{obj} - \phi_{entr})$$

An axially symmetric lens or lens system can alter the coefficients of the Taylor series, but cannot introduce new terms in the series. Therefore, the phase of the total field emerging from the system and the phase of the aberration field (the difference between the total field and a perfect spherical wave field phase) will both be expressible in terms of a power series in these three variables.

The aberrations of the form

$$f(x) = u^i v^j w^k \tag{9.1}$$

where

$$i + j + k = 2$$

are known variously as the *primary/Seidel/third-order aberrations*. Even today most lens design centers on minimizing these lowest order aberrations.

The assumption of rotational symmetry leads to a Zernike polynomial representation of the aberration function, as given in Born and Wolf [1]. While the validity of the axial-symmetry assumption for real lenses may be debated, this assumption has been in use in optics literally for centuries. In recent years, however, a more general functional representation for the aperture function has been put forward, in connection with microwave antennas. This functional representation, known as the *Jacobi-Bessel series* is general enough to describe an arbitrary (well-behaved) function on a circular aperture [2], not just a function arising from rotationally symmetric system. Nonetheless, the Zernike polynomial representation (of which the primary aberrations represent the lowest-order terms) remains a very good indicator of lens quality.

We'll not delve into diffraction theory at the level of the Zernike polynomials. This primarily analytical description is somewhat obsolete today anyway, in light of modern numerical diffraction-integral calculation techniques [3]. What we will do in this chapter is look at the primary aberrations and their effect on image quality.

9.3 THE SEIDEL ABERRATIONS

Generally, when the term *aberration* is used, it refers to the Seidel aberrations. We'll look at each of these individually and at their effect on image quality. The five nontrivial aberration functions satisfying (9.1) are shown in Fig. 9.2. These plots show the aberrations, which represent deviations from sphericity, in exit aperture coordinates. In Chapter 5, based on the Fresnel integral formulation, we showed how image formation can occur when the illumination function in the exit aperture contains a convergent quadratic phase. That is, the image in that case was given as the Fourier transform of the exit aperture illumination function. In the present case, the (assumed imperfect) lens will add an extra phase factor to this aperture field. That is to say, the field in the exit aperture will be given as

$$E(x_{ap}, y_{ap}) = E_0(x_{ap}, y_{ap})e^{jkf(x_{ap}, y_{ap})}$$

where E_0 = the exit aperture field just in front of the lens
         $f(x_{ap}, y_{ap})$ = the spherical wave plus the aberration function (expressed as a distance)

What the lens has done, then, is to introduce an extra phase term in the diffraction integral, and this phase term will affect the point spread function.

The extra phase terms due to aberrations, taken from Fig. 9.2, are

$$e^{jkA\rho^4}$$

where

$$\rho = \sqrt{x_{ap}^2 + y_{ap}^2}$$

for spherical aberration;

$$e^{jkA\rho_{obj}\rho^3\cos(\phi_{ap} - \phi_{obj})}$$

where

$$\phi_{ap} = \tan^{-1}(y_{ap}/x_{ap})$$

for coma;

$$e^{jkA\rho_{obj}^2\rho^2\cos^2(\phi_{ap} - \phi_{obj})}$$

for astigmatism;

$$e^{jkA\rho_{obj}^2\rho^2}$$

for a defect in focusing; and

$$e^{jkA\rho_{obj}^3\rho\cos(\phi_{ap} - \phi_{obj})}$$

for distortion.

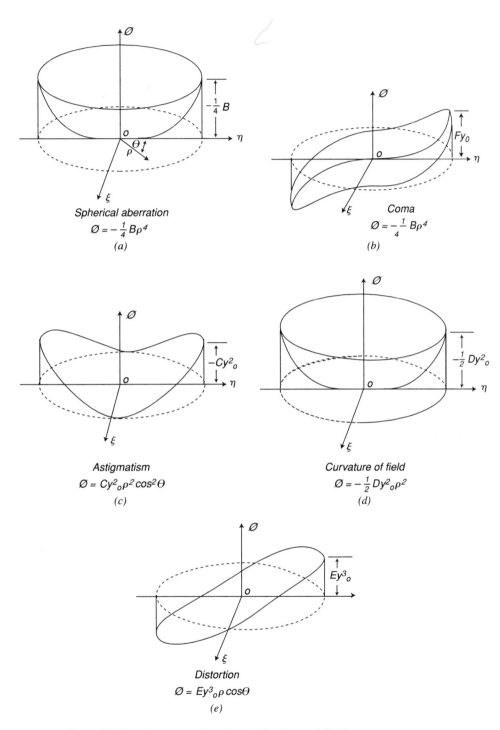

Spherical aberration
$$\varnothing = -\frac{1}{4} B\rho^4$$
(a)

Coma
$$\varnothing = -\frac{1}{4} B\rho^4$$
(b)

Astigmatism
$$\varnothing = Cy^2{}_0 \rho^2 \cos^2\Theta$$
(c)

Curvature of field
$$\varnothing = -\frac{1}{2} Dy^2{}_0 \rho^2$$
(d)

Distortion
$$\varnothing = Ey^3{}_0 \rho \cos\Theta$$
(e)

Figure 9.2 The primary wave aberrations. (After Born and Wolf.)

Each of these phase functions can be inserted into Eq. (5.6), and the result can be numerically integrated to determine the point spread function of an exit aperture containing the particular aberration function.

Astigmatism and the defect of focus aberration are basically the same thing, except an astigmatic aperture will have two different defects of focus in the two orthogonal planes. If the defect of focus aberration were shift invariant, it could be easily corrected by simply moving the focal plane to the point of focus dictated by the focusing defect. In that new plane, the point spread function will again be restored to its diffraction limited value. However, the defect of focus is a function of the object plane coordinates as well as the exit aperture plane coordinates. Therefore, the focusing defect will in general be different for different object plane points.

Since the astigmatism aberration results in two different focal point locations for the two orthogonal directions in the image plane, this aberration could only be corrected in one direction. The other direction will still have a defect of focus and a resultant broadening of the point spread function. In this case, the point spread function will be elliptical, with its longest axis in the direction of the defocus. Of course, the astigmatism aberration is also a function of the object plane coordinates; thus, any correction will only work for a limited set of points in the object plane.

The defect of focus aberration is quadratic in both the source and aperture coordinates; that is, the focal point varies in an axial direction, as a quadratic function of radial distance in the object plane. This gives rise to the quadratic focal surface shown in Fig. 9.3. This quadratic focal surface is known as the *Petzval surface* and typically has a focal length half that of the imaging system [3]. For a system that produces a real image (converging field), the Petzval surface curves toward the system; for a system that produces a virtual image (diverging field), the Petzval surface bends away from the system. One way to compensate for field curvature is to place one last (quadratic phase) lens just in front of the focal plane; this will flatten the phase of the image. Despite the fact that the image surface of best focus is the quadratic Petzval surface, virtually all imaging systems (save the human eye and a few others) incorporate a planar focal surface—the focal *plane* of the system.

The distortion aberration is unique among the primary aberrations in that it does not deform the shape of the point spread function—only its location. If the distortion

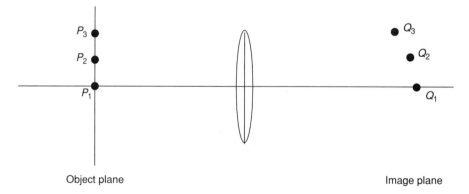

Object plane   Image plane

Figure 9.3 Variation of exit plane curvature (and image plane focus) as a function of object plane radius.

aberration were shift invariant, it would merely cause the entire image to be laterally displaced by some constant amount in the focal plane. Unfortunately, however, the distortion aberration is shift variant. Therefore, it causes the point-to-point mapping from the object space to the image space to be nonlinear.

9.4 CHROMATIC ABERRATIONS

Aberrations also exist in the frequency domain as well as in the spatial domain. These arise from the fact that the index of refraction of lens materials is a function of frequency at optical wavelengths. This frequency dependence is of no concern in optical systems illuminated by a single-frequency laser source. They arise primarily in (incoherent) color image processing systems where wide-bandwidth optical signals are required for good color reproduction.

By (5.9), the focal length of a real lens is dependent on the refractive index of the lens, and chromatic aberration occurs when the index is a function of wavelength, so that not all colors are in focus in the image plane. One solution is to design the imaging system to focus the color for which the eye is most sensitive (green). A more sophisticated solution is to design the system so that more than one color is focused at the same focal plane.

One way to do this is to replace all of the individual lenses in the system with *achromatic doublet lenses*. This type of lens is shown in Fig. 9.4. It simply consists of two lenses (having two different focal lengths and made from two different media) that are cemented together to form a single lens. Using the methods of Chapter 5, we may represent the transmission function of the total lens in the form

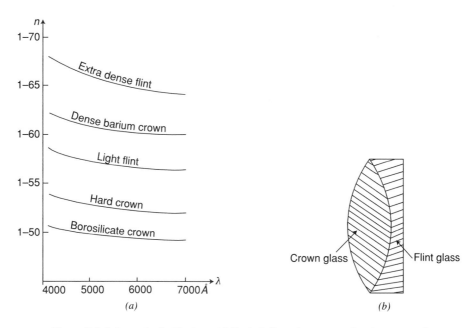

Figure 9.4 Achromatic doublet lens: (a) Typical dispersion curves of various types of glass and (b) an achromatic doublet. (After Born and Wolf.)

$$T(x, y) = e^{jk(x^2 + y^2)/2F_1(\lambda)} \cdot e^{-jk(x^2 + y^2)/2F_2(\lambda)} \qquad (9.2)$$

where the two focal lengths are explicitly written as functions of λ, and the second lens is assumed to be a diverging lens. If the refractive indices of these two lenses vary as shown in Fig. 9.4, then the negative sign attached to one of the indices in the exponent in (9.2) tends to cancel the frequency variation in the positive exponential. The goal of achromatic lens design is to set the focal length,

$$\frac{1}{F_{eq}} = \frac{1}{F_1} - \frac{1}{F_2}$$

equal to some specified value at two different frequencies. If the index functions are known in tabular form, an approximate polynomial may be formed for each of the focal lengths, and the effective focal length of the combination can be obtained in approximate analytic form. From that, the radii of curvature of the lenses may be found using numerical root-finding techniques.

REFERENCES

[1] Born, M., and Wolf, E., *Principles of Optics, 6th ed.* Oxford: Pergamon Press, 1980.

[2] Galindo-Israel, V., and Mittra, R., ''A New Series Representation for the Radiation Integral with Application to Reflector Antennas,'' *IEEE Trans. Antennas Propagat.,* Vol. AP-25, pp. 631–635, September 1977. (Correction, *IEEE Trans. Antennas Propagat.,* Vol. AP-26, July 1978, p. 628).

[3] Scott, C. R., *Modern Methods of Reflector Antenna Analysis and Design,* Norwood, MA: Artech House, 1990.

PART IV
OPTICAL INTERFERENCE PHENOMENA

10

Applications of the Plane Wave Spectrum Concept: Introduction to Diffraction Gratings

In this chapter, we begin to apply the plane wave spectrum concept first developed in Chapter 2. This concept is quite useful in many areas of physical optics and is an excellent vehicle for understanding the theory of optical diffraction gratings.

A diffraction grating is one of many types of masks for optical fields that are presented in this book. We've already encountered one type of mask in Chapter 1, the Fresnel zone plate. We'll encounter another type (the spatial filter) in Chapter 14 and yet another (the hologram) in Chapter 13. Amplitude and phase masks, especially modern adaptive masks made using liquid crystals, are useful in modern optical signal processing systems.

10.1 PERIODIC STRUCTURES AND DISCRETE PLANE WAVE SPECTRA

Since this book emphasizes electromagnetic phenomena in the optical regime, we'll concentrate on the typical one-dimensional (line) diffraction grating that is most often encountered in typical optical systems. (In the microwave regime, a two-dimensional diffraction grating, known as a *frequency selective surface* or *FSS*, is more frequently encountered.) An example of a typical one-dimensional grating is shown in Fig. 10.1. The reader interested in more general types of planar and volumetric diffraction gratings is referred to [1 and 2].

We'll start and by looking first at infinite gratings and then at finite gratings. Infinite diffraction gratings present a nice mathematical simplification when the illumination field is an infinite plane wave. Under this condition, the diffraction grating gives rise to the

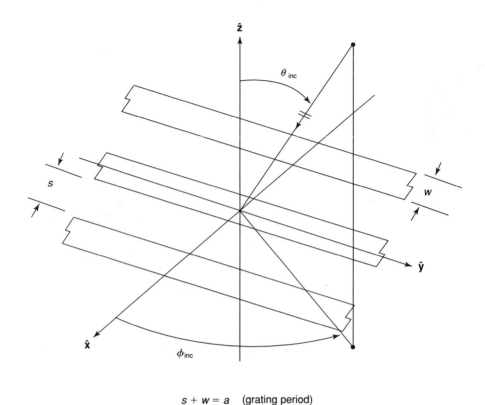

$$s + w = a \quad \text{(grating period)}$$

Figure 10.1 One-dimensional strip grating under plane wave incidence.

discrete plane wave spectrum. This can readily be seen with the aid of Floquet's ("Flo-KAY's") theorem.

Floquet's theorem may be stated as follows. Assume that a (single-frequency/coherent) plane electromagnetic wave is incident on a planar periodic surface of the type shown in Fig. 10.1. Under this type of illumination, the reflected and transmitted fields will be periodic in the tangent plane. Moreover, these fields will have exactly the same periodicity as the structure itself. That is, all fields or currents at any two points (x_1, y_1) and (x_2, y_2) on the surface will be related via the equation

$$E(x_2, y_2) = E(x_1, y_1) \, e^{j(\alpha_{inc}\Delta x + \beta_{inc}\Delta y)} \tag{10.1}$$

where

$$\alpha_{inc} = k \sin \theta_{inc} \cos \phi_{inc}$$

$$\beta_{inc} = k \sin \theta_{inc} \sin \phi_{inc}$$

$$\Delta x = x_2 - x_1 = ma; \qquad m = \dots -2, -1, 0, 1, 2 \dots$$

$$\Delta y = y_2 - y_1 = \text{arbitrary}$$

In (10.1), the tangential component of the incident plane wave field is assumed given as

$$E_{inc}(x, y) = E(\alpha_{inc}, \beta_{inc}) \, e^{j(\alpha_{inc}x + \beta_{inc}y)}$$

Floquet's theorem states that, up to the linear incident field phase, the fields and currents are periodic (with period, a, the same period as the grating itself) in the x-direction. Thus, all field quantities *repeat* every cell period, with the only difference from one cell to the next being the linear incident field phase. This fact is significant because it allows for tremendous simplification of problems involving periodic surfaces. All that's necessary to do is to find the fields and currents over one period of the grating and then they will be known everywhere else.

It is well known from the theory of Fourier series that a periodic function may be expressed in terms of a Fourier series, which physically represents a *discrete* plane wave spectrum. The following Fourier series for the scattered electric field can be shown to satisfy (10.1):

$$E(x, y) = \sum_{m = -\infty}^{\infty} E(\alpha_m, \beta)e^{j(\alpha_m x + \beta y)}$$

where

$$\alpha_m = k \sin \theta_{\text{inc}} \cos \phi_{\text{inc}} + 2m\pi/a$$

$$\beta = k \sin \theta_{\text{inc}} \sin \phi_{\text{inc}}$$

This result shows that the theory of optical diffraction gratings is closely related to the theory of Fourier series. We will find this Fourier representation of the scattered field to be very useful in understanding the operation of diffraction gratings in optical systems.

A hologram is a special type of diffraction grating that has no repetitive periodic structure; it is *aperiodic*. As a result, the fields scattered by it form a continuous, rather than discrete, plane wave spectrum, which may represent a real or virtual optical image.

10.2 PHYSICAL OPTICS DIFFRACTION BY AN INFINITE STEP AMPLITUDE GRATING

In this book, the term *physical optics* (PO) always refers to an analytical technique wherein the total field at any point on the surface of a lens or mirror is taken as the field that would exist at that point if the incident field was an infinite plane wave and the lens (or mirror) was an infinite plane tangent to the real (curved) surface. In other words, the physical optics approximation will permit a relatively simple mathematical solution of the problem, since the field at the surface of the object is, in effect, assumed to be known at the outset. All that's needed in order to find the diffracted field is to integrate these PO fields using the methods of Chapter 1. The physical optics approximation is generally quite accurate in the optical regime.

To develop the physical optics solution to diffraction by an infinite strip grating (of the form shown in Fig. 10.1), we may refer to Fig. 10.2. Under the physical optics approximation,

$$E^{\text{trans}}(x, y) = \tau(x, y)E^{\text{inc}}(x, y)$$

where $\tau(x, y)$ is the complex transmission function of the grating. The transmission function consists of a magnitude part, relating to the attenuation properties of the grating, and a phase part, relating to the propagation phase through the grating. A grating for which

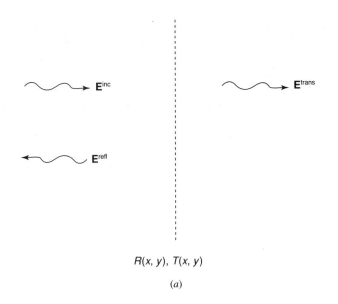

$R(x, y), T(x, y)$

(a)

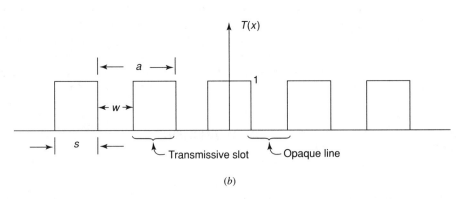

(b)

Figure 10.2 On the physical optics approximation in grating analysis: (a) Incident, reflected, and transmitted fields and (b) transmission function of a step grating.

the transmission function is purely real is called an amplitude grating, and a grating that
has unit-amplitude transmission accompanied by a phase delay is called a phase grating.

The transmission properties of an amplitude grating may be analyzed using ordinary
concepts from Fourier series. Say, for simplicity, that a normally incident plane wave
impinges on the step-amplitude grating in Fig. 10.2*b*. In this case, the magnitude of the
transmitted field is given as the transmission function itself. Thus, the transmitted field
is a periodic function expressible in terms of a Fourier series of the form

$$E(x, y) = \sum_{m=-\infty}^{\infty} E(\alpha_m)e^{j\alpha_m x}$$

where

$$\alpha_m = \frac{2m\pi}{a}$$

Using the Fourier methods from Chapter 2, we may invert the relation above to obtain

$$E(\alpha_m) = \frac{1}{a} \int_{-s/2}^{s/2} E(x)e^{-j\alpha_m x} \, dx$$

or

$$E^{\text{trans}}(\alpha_m) = \frac{s}{a} \operatorname{sinc} \alpha_m s/2 = \frac{s}{a} \operatorname{sinc} m\pi \frac{s}{a}$$

Thus, the transmitted electric field is of the form

$$E(x, z) = \frac{s}{a} \sum_{m=-\infty}^{\infty} \operatorname{sinc}\left(m\pi \frac{s}{a}\right) e^{j(\alpha_m x - k_z z)}$$

where

$$\alpha_m = 2m\pi/a$$

$$k_z = \sqrt{k^2 - \alpha_m^2}$$

The transmitted field is a (discrete, weighted) infinite spectrum of plane waves. The dominant plane wave component is the $m = 0$ component (referred to as the *zero order*), and the other components are weighted according to the sinc function. The order of each plane wave component is determined by its index in the Fourier series. If the slot width is much less than the periodicity, the first few diffracted orders will have nearly equal amplitudes. The diffraction grating in this case serves as a type of "power splitter" in the spectral domain, sending energy into the lower diffraction orders with roughly equal power distribution. If the various plane wave components are brought incident on a lens, then a detector array (or an array of optical fibers) can be placed in the focal plane to receive the energy from the focal spots of the various diffracted orders. Doing so will yield the Fourier (power) spectrum of the grating function.

A key parameter in the evaluation of diffraction gratings is the *diffraction efficiency*. This refers to the magnitude of a particular plane wave diffraction order in relation to the magnitude of the incident plane wave field. Diffraction efficiency is related to the contrast between the transmissive and the opaque parts of the grating. For example, we selected the highest contrast possible in assuming the opaque portion of the screen to have zero transmission and the slot part of the grating to have unit transmission. This contrast gives the highest diffraction efficiency. If, however, we had chosen the transmissive and opaque portions of the screen to have transmissivities equal to 0.8 and 0.7, respectively, then the diffraction efficiency would have been much less (i.e., most of the transmitted light energy would have remained in the zero order).

The plane waves comprising the various diffraction orders will propagate at various angles with respect to the grating normal. The angle from normal is given as

$$\theta = \tan^{-1} \frac{\alpha_m}{k_z}$$

Of course, when abs $(\alpha_m) > k$, the plane waves are evanescent.

One significant example of a square-wave grating is the Ronchi grating, wherein the widths of the opaque and transmissive strips are equal. We'll study applications of this grating in detail in the following chapter.

The step grating produces an infinite (discrete) spectrum of transmitted and reflected plane waves. Sometimes this is not desired; often only two or three orders are wanted. In this case, a sinusoidal grating can be useful. Sinusoidal gratings are especially useful for analysis purposes as well, as we'll see in the following chapter. (This is particularly true for coarse gratings with periods much longer than the optical wavelength.) The sinusoidal grating has roughly the same transmittance properties as the step grating, but only three plane wave terms (diffraction orders) are involved in the transmission function:

$$\tau(x, y) = \frac{1}{2} + \frac{1}{4j} [e^{j(2\pi/a) x} - e^{-j(2\pi/a)x}]$$

10.3 PHYSICAL OPTICS DIFFRACTION BY AN INFINITE SINUSOIDAL PHASE GRATING

The phase grating is completely analogous to the amplitude grating in that the transmission phase—rather than the transmission amplitude—is modulated by some periodic function. Thus, for a sinusoidal phase grating, the transmission function is given by

$$\tau(x, y) = e^{j\phi_{max}[1/2 + 1/2 \cos (2\pi/a)x]}$$

Once again, this can be represented in terms of a Fourier series of the form

$$E(x, y) = \sum_{m=-\infty}^{\infty} E(\alpha_m) e^{j\alpha_m x}$$

where

$$\alpha_m = \frac{2m\pi}{a}$$

As before, this may be inverted using the usual relation,

$$E(\alpha_m) = \frac{1}{a} \int_{-a/2}^{a/2} E(x) e^{-j\alpha_m x} dx$$

Plugging the transmission phase equation into this relation gives the Fourier coefficients as

$$E(\alpha_m) = \frac{1}{a} \int_{-a/2}^{a/2} e^{j\phi_{max}[1/2 + 1/2 \cos (2\pi/a)x]} e^{-j\alpha_m x} dx$$

or

$$E(\alpha_m) = \frac{1}{a} e^{j(K/2)} \int_{-a/2}^{a/2} e^{j[K/2 \cos (2\pi/a) x - \alpha_m x]} dx$$

If we assume $a_0 = 0$ (normal incidence), then

$$\alpha_m = \frac{2m\pi}{a}$$

and the equation above becomes

$$E(\alpha_m) \cong J_m(\phi_{\max}/2)$$

where J_m is the Bessel function of order m.

By adjusting the maximum phase excursion in the sinusoidal phase transmission function, certain diffraction orders may be emphasized. The order is emphasized that most closely satisfies the relation

$$m = \phi_{\max}/2$$

Thus, one clear difference betwen sinusoidal amplitude and phase gratings is that the first gives rise to only three diffraction orders, whereas the second has an infinitude of diffraction orders, all weighted according to the Bessel function. Clearly, the math involved in connection with phase gratings is much more complicated than that for amplitude gratings. However, phase gratings possess one very important advantage over amplitude gratings, and this is seen by realizing that the phase grating is no different from an ordinary lens. Whereas a lens has a quadratic phase function, a sinusoidal phase grating has a periodic phase function. And like a lens, the phase grating tends to negligibly attenuate an optical field passing through it. This is in contrast to the amplitude grating, which blocks a portion of the light it transmits. For this reason, phase gratings are often preferable to amplitude gratings whenever low-input power levels or signal-to-noise ratios are areas of concern.

10.4 DIFFRACTION GRATINGS AND COLOR SEPARATION

The classic practical application of diffraction gratings is in the area of spectrum analysis. Consider Fig. 10.3, which shows a plane wave incident on a planar line grating. The incident wave consists of a spectrum of frequencies (colors), each of which may be considered individually. By Floquet's theorem, the grating produces an infinite number of diffraction orders, whose propagation constants in the x-direction are given by

$$\alpha_m = \frac{2\pi}{\lambda} \sin \theta_0 + \frac{2m\pi}{L}$$

Here our primary interest is in the $m = -1$ order. Now, the incident field is assumed to be comprised of a (frequency) spectrum of plane waves, each having a unique frequency and wavelength, but all propagating in the same direction. This means that every frequency component has a unique value of α_0, given as

$$\alpha_0(\lambda) = \frac{2\pi}{\lambda} \sin \theta_0$$

And the x-direction propagation constant for the $m = -1$ order is given as a function of frequency/wavelength as

$$\alpha_{-1}(\lambda) = \alpha_0(\lambda) - \frac{2\pi}{a}$$

where a is the period of the grating. We'll pick one frequency near the center of the spectrum and denote its wavelength as λ_0. If we choose the period, a, of the grating such that

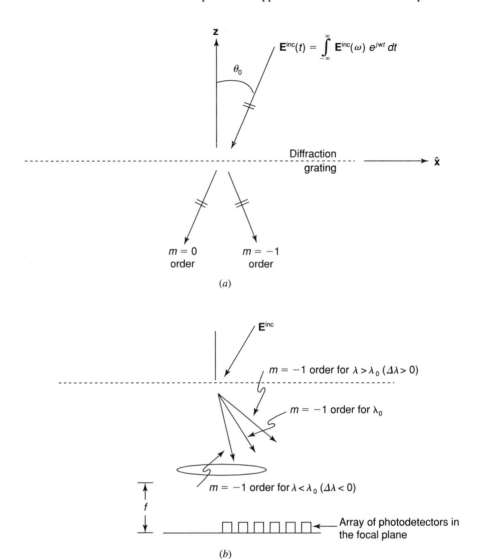

Figure 10.3 Diffraction grating as an optical spectrum analyzer: (a) An incident plane
wave consisting of an entire frequency spectrum and (b) focusing the $m =
-1$ orders on the focal plane.

$$a = \frac{\lambda_0}{2 \sin \theta_0}$$

then for λ_0, we'll stipulate that

$$\alpha_{-1} = -\alpha_0$$

When this condition is satisfied, we may show that

$$\alpha_{-1}(\lambda) \cong -\alpha_0(\lambda_0) \left(2 - \frac{\lambda_0}{\lambda} \right)$$

for the other wavelengths in the incident field spectrum.

From this equation, we can find the deflection angle of the $m = -1$ order for each frequency as

$$\sin \theta = \frac{\alpha_{-1}(\lambda)}{k}$$

$$= -\sin \theta_0 \left(2\frac{\lambda}{\lambda_0} - 1 \right)$$

In other words, the $m = -1$ order of each frequency component is deflected in a different direction by the grating. If these various $m = -1$ plane wave diffraction orders are focused by a lens onto a photodetector array (as shown in Fig. 10.3), then the frequency content of the incident optical field is deduced. Basically, when operated in this fashion, the diffraction grating is an optical demultiplexer, taking the various frequency components of an optical field and breaking them up into individual signal channels.

This ''spectrum analyzing'' property of diffraction gratings makes them very useful in filtering applications as well. Diffraction gratings are often used in this connection to filter the fields present in laser cavity resonators. Laser cavities are often formed using Fabry-Perot resonators, which are very long in terms of optical wavelengths. Typically, these resonators are many thousands of wavelengths in length, and many resonant modes can coexist simultaneously. For example, consider Fig. 10.4, which shows a typical laser cavity. (This cavity is a ''folded'' form of a periodic structure, and a condition similar to Floquet's theorem can be used to determine the resonant frequencies of the cavity.) If we assume a planar optical field trapped inside the slab and bouncing back and forth in the horizontal direction as shown, then, assuming positive, real-valued reflection coefficients at each of the dielectric/air boundaries, the resonance condition may be stated as

$$kL = 2N\pi$$

or

$$\frac{2\pi}{\lambda} L = 2N\pi$$

or

$$N = \frac{L}{\lambda}$$

where k is the wavenumber in the dielectric laser cavity medium. This equation leads to a discrete frequency spectrum, in the same way that Floquet's theorem leads to a discrete spatial frequency spectrum. When this periodicity condition is satisfied, the fields remain in phase after any number of round trips through the cavity. When the cavity length is many thousands of wavelengths, and when lasing can take place over a relatively broad band of optical frequencies, many closely spaced resonant frequencies will be excited by the lasing process. Generally, however, only one output frequency is desired from the laser. This is the whole purpose of the laser for most applications—that is, to produce a perfect monochromatic (single-frequency) optical field.

The unwanted frequency components are filtered out by placing a diffraction grating at an angle with respect to the cavity, as shown in Fig. 10.4b. The angle is chosen such that

$$\alpha_{-1} = -\alpha_0$$

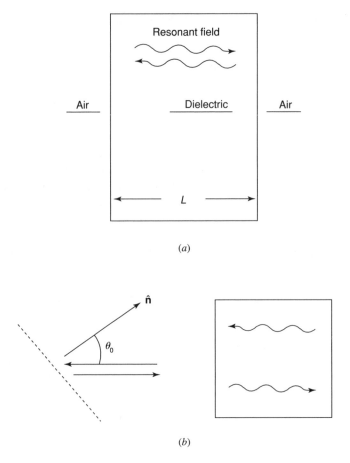

(a)

(b)

Figure 10.4 Laser cavity resonator with external line grating—for single-frequency operation: (a) A typical Fabry-Perot laser resonator and (b) laser activity with external diffraction grating.

for the frequency of interest. Thus, this frequency component is reflected directly back to the laser cavity, whereas the other resonant frequency components are deflected away in other directions. In this way, a laser can be designed to produce a single-frequency output signal.

10.5 FINITE DIFFRACTION GRATINGS

Finite diffraction gratings are analyzed in a straightforward way using familiar concepts from the theory of Fourier analysis. Consider Fig. 10.5, which shows a finite diffraction grating as the product of an infinite grating transmittance function and a finite aperture function, $A(x, y)$. By the results of Chapter 1, the far-zone electric field transmitted by the grating is given as the Fourier transform of the electric field just behind the grating. If the incident field is assumed to be normally incident in the positive-z-direction (from

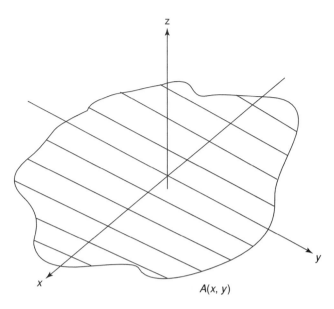

Figure 10.5 Finite diffraction grating as the product of an infinite grating function and an aperture function $A(x, y)$.

the negative-z side), the incident field is uniform in amplitude and phase across the grating aperture. In this case, the field just behind the grating aperture is simply the product of the grating transmittance function and the aperture function. Thus,

$$E^{\text{trans}}(x, y, z = 0^+) = \tau(x, y) \cdot A(x, y)$$

Using the convolution theorem, we see that the far-zone electric field—which is the Fourier transform of this transmitted field—is proportional to the convolution of the FT of the grating transmittance function and the FT of the aperture function. Thus,

$$E^{\text{far zone}}(\alpha, \beta) = \int_{-\infty}^{\infty} A(\alpha' - \alpha, \beta' - \beta) \, d\alpha' \, d\beta' \cdot \left\{ \sum_{m=-\infty}^{\infty} E_m \delta(\alpha' - \alpha_m) \right\}$$

or

$$E^{\text{far zone}}(\alpha, \beta) = \sum_{m=-\infty}^{\infty} E_m A(\alpha_m - \alpha, \beta_m - \beta)$$

where we've neglected any constants of proportionality in the previous expression. For the case of a rectangular aperture $A(x, y)$, the diffracted field given by the expression above is shown plotted in Fig. 10.6 as a train of sinc functions. From the equation above and the plot, it's evident that the finite aperture "smears" out the lines in the diffraction pattern for the infinite grating.

10.6 THREE-DIMENSIONAL DIFFRACTION GRATINGS

Situations arise in physics wherein three-dimensional diffraction gratings can be produced. Such situations include three-dimensional (volume) holograms, acousto-optic Bragg cells, and X-ray diffraction by atomic lattices. Three-dimensional diffraction gratings yield phe-

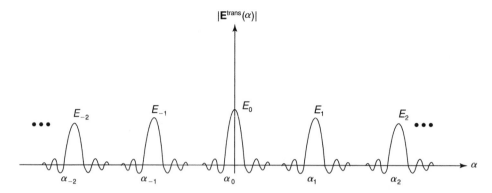

Figure 10.6 The plane wave spectrum produced by a finite line grating.

nomena that are significantly different from those found in connection with two-dimensional gratings. In particular three-dimensional gratings can be designed which allow no electromagnetic energy to propagate through them. These particular types of 3-D gratings are known as *photonic bandgap structures.*

In the case of a three-dimensional diffraction grating, periodicity constraints exist in all three Cartesian directions. The extra constraint in the third dimension places severe restrictions on the types of fields that may exist in the three-dimensional structure. With reference to Fig. 10.7, we will consider a volume grating in which Floquet's theorem

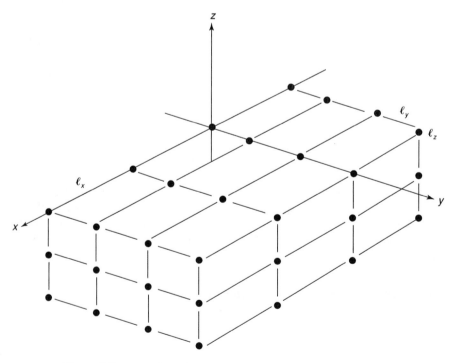

Figure 10.7 A three-dimensional periodic medium (lattice).

applies. When we apply Floquet's theorem to a 3-D volume, we no longer have a plane wave spectrum of fields, as in the case of 2-D gratings. That is, even though the fields satisfy a 3-D periodicity condition, the individual spectral components of the field may not be regarded as plane waves. We may write the discrete plane wave spectrum phase function in the volumetric medium as

$$E(x, y, z) = \sum_{m,n,p} E_{mnp} \, e^{j(\alpha_m x + \beta_n y + \gamma_p z)}$$

In this equation,

$$\alpha_m = \alpha_0 + \frac{2m\pi}{l_x}$$

$$\beta_n = \beta_0 + \frac{2n\pi}{l_y}$$

$$\gamma_p = \gamma_0 + \frac{2p\pi}{l_z}$$

where now

$$\alpha_m^2 + \beta_n^2 + \gamma_p^2 = k_{mnp}^2$$

for all possible combinations of m, n, p.

Note that k_{000} (which we'll just call k_0 for simplicity) is not in general equal to the parameter

$$k = \omega\sqrt{\mu\epsilon}$$

for the medium between the scatterers. This is the primary difference between two- and three-dimensional periodic media (periodic surfaces versus periodic volumes). In the instance of periodic surfaces, the Floquet spectrum is nothing more than a discretized version of the plane wave spectrum, wherein the propagation constants of all the plane waves have the same absolute magnitude. In the case of the periodic volume, the root of the sum of the squares of the propagation constants in the three Cartesian directions is not the same for all the terms in the triply-periodic sum above (though in some cases, as in Bragg scattering, some of the terms can be arranged to satisfy such a condition). The relation between the propagation constant k_0 and the parameter k (related directly to frequency) of the medium is in general nonlinear and is given by the ω-β diagram of the 3-D periodic medium.

Under certain special circumstances (i.e., the Bragg condition alluded to above), some spectral components may have the same propagation constant,

$$k_{mnp} = \sqrt{\alpha_m^2 + \beta_n^2 + \gamma_p^2}$$

Consider for simplicity, the case of a 3-D medium for which l_x, l_y are infinite—that is, a medium that is periodic in one dimension only. Note that this 1-D periodic structure is still a periodic *volume*, in contrast to the diffraction gratings studied in previous sections of this chapter, which are periodic *surfaces*. The distinction between the two is crucial to both the analysis and the effects produced by the two types of structures.

Two Floquet modes (propagating in the y - z plane) may have the same propagation constant magnitude (with opposing sign) in the z-direction when $n = 0$ and $p = 0, -1$. In order for

$$\gamma_{-1} \quad \text{to equal} \quad -\gamma_0$$

we must have

$$\gamma_{-1} = \gamma_0 - \frac{2\pi}{l_z} = -\gamma_0$$

where

$$\gamma_0 = k_0 \cos \theta_0$$

and

$$k_0 = \frac{2\pi}{\lambda_0} \quad (\lambda_0 = \text{cell-cell wavelength})$$

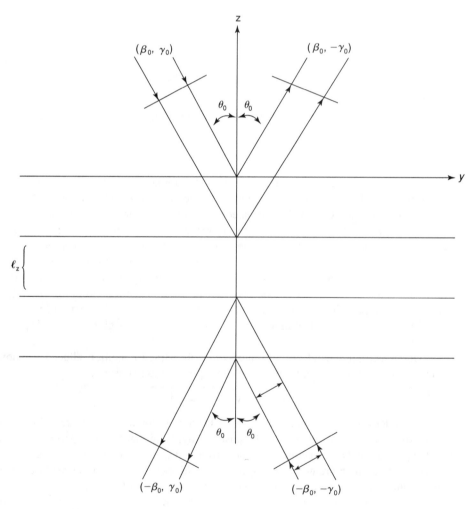

Figure 10.8 The Bragg condition in a one-dimensional periodic medium showing the four Bragg modes.

and θ_0 is the incident angle from normal. These two modes are characterized by the propagation constants

$$(\beta_0, \gamma_0) \quad \text{and} \quad (\beta_0, -\gamma_0)$$

The condition above implies that

$$\lambda_0 = 2\, l_x \cos\theta_0$$

which is known as the *Bragg condition,* shown in Fig. 10.8. Note that instead of using the $n = 0, p = 0, -1$ modes, we could have also used the $n = 0, p = 0, -2$ plane waves to obtain another Bragg condition as

$$2\lambda_0 = 2\, l_z \cos\theta_0$$

or in general,

$$n\lambda_0 = 2\, l_z \cos\theta_0$$

Two additional modes defined by

$$(-\beta_0, \gamma_0) \quad \text{and} \quad (-\beta_0, -\gamma_0)$$

make up a total of four Bragg modes in the structure, which all have the same propagation constant magnitude in the z-direction.

One well-known type of one-dimensional periodic structure in which Bragg scattering takes place is the acousto-optic Bragg cell [2]. When detailed mode-matching techniques are used to analyze the four Bragg modes of structure, as in [2], it can be shown that energy transfer (mode coupling) can take place between the Bragg modes.

REFERENCES

[1] Scott, C. R., *The Spectral Domain Method in Electromagnetics,* Norwood, MA: Artech House, 1989.

[2] Scott, C. R., *Field Theory of Acousto-Optic Signal Processing Devices,* Norwood, MA: Artech House, 1992.

11

Introduction to Optical Moiré Techniques

In this chapter, we examine the remarkable phenomenon known as *optical moiré*. Moiré patterns constitute a special type of interference pattern, generated when line gratings (typically Ronchi rulings) are used to illuminate and view certain types of objects. Frequently, two gratings are used for this purpose, one of which is placed in the illuminating beam and the other which is used for viewing the object (though a single grating can often serve both functions). The grating in the illuminating beam casts curvilinear shadows onto the illuminated object. When these curvilinear shadows are viewed through a second (usually identical) ruled grating, the overlaid linear and curvilinear lines produce contour maps that are superimposed on the object. These lines delineate height contours for diffusely reflective surfaces (slope contours for specularly reflective surfaces) and refractive index contours for transmissive media.

The name *moiré* comes from a French word meaning *mohair,* and this term is rather descriptive of the character of moiré interference patterns. The means by which the moiré contours are produced (i.e., by overlaying—or multiplying—two curvilinear contour patterns) causes the resultant product pattern to have a slightly "hairy" appearance. This stems from the fact that the two original patterns are always visible along with the moiré product pattern. In this way, a moiré pattern is often quite different visually from an ordinary type of interference pattern generated by adding together two coherent optical waveforms via phase addition and cancellation (although certain *recorded* interference patterns—such as holograms—also display this "hairy" appearance due to the linear relationship between film development and incident field *intensity*). Holograms are closely related to moiré patterns as we'll see, and the distinction between interferometry, moiré, and holography can often become quite blurred.

The primary difference between moiré on the one hand, and holography and ordinary interferometry, on the other, is that moiré is usually performed with incoherent light, with the interference taking place between the shadows (or images of the shadows) of the rulings. This is at least true when the period of the rulings is much greater than the optical wavelength. In this case, diffraction effects are minimized and GO effects dominate. Moiré can also be performed using fine amplitude rulings (and even phase diffraction gratings) having periodicities on the order of the optical wavelength. In this case, analysis of the moiré phenomenon is based on diffraction phenomena and is somewhat different from the G.O "shadow" case, although the principles are similar. One author [1] denotes the incoherent, GO-based technique as "moiré" and the coherent, diffraction-based technique as "moiré interferometry."

In the 1960s, the moiré phenomenon gained widespread attention through "op art," [2] in which moiré patterns create visually dazzling repetitive patterns. In modern optics, the phenomenon is used in scientific applications ranging from surface contouring (topography) to fluid flow measurement.

11.1 INTRODUCTION TO THE PRINCIPLE OF OPTICAL MOIRÉ: IN-PLANE MOIRÉ ANALYSIS

In its most general form, the moiré phenomenon can in principle occur whenever two contour maps are overlaid, one on top of the other. In order for the resultant moiré to appear, both contour maps must have light and dark contours of roughly equal width, and these light and dark contours must have nearly equal widths on both maps. When these conditions are met, superposing the two contour maps will generally produce some type of moiré interference pattern.

In this section, we look at a number of different types of moiré phenomena starting from the very simple and proceeding to the mildly complex, showing the mathematical foundations of moiré phenomena and describing some of the many practical applications of these phenomena. We can start with Fig. 11.1, which shows one simple and useful

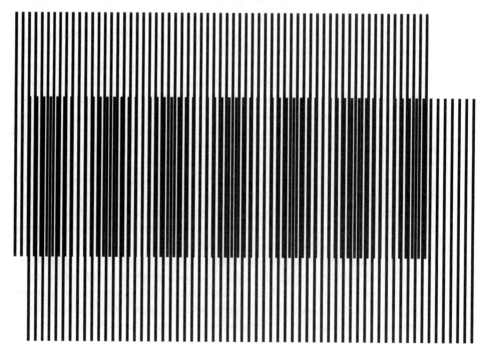

Figure 11.1 Superimposed Ronchi rulings.

application of optical moiré. This figure—created using a standard drawing program for personal computers—shows the superposition of two Ronchi rulings. The lower ruling has a period slightly longer than the upper grating; it was created by copying the upper grating, and then grabbing the "handles" of that grating and elongating it horizontally. The moiré appears in the region of overlap of the two gratings.

This type of moiré has two practical applications. First, it is used for quality control in the manufacture of Ronchi rulings. The patterns that appear in the region of overlap determine the precision of the manufactured grating with respect to a master. The period of the moiré lines determines the period of the test grating with respect to the master. A second application of this type of pattern is in deformation studies of bars under pure tension or compression. In this application, a grating is applied to a bar in the uncompressed state. The bar grating has exactly the same periodicity as the reference (viewing) grating, so that overlaying the two produces no moiré. When the bar is placed in tension or compression, the periodicity of the bar grating changes in response to the deformation of the bar. The period of the resultant moiré directly determines the deformation of the bar. This is one example of using moiré to determine *in-plane* displacements of a body under mechanical load.

This type of moiré pattern is readily analyzed. For simplicity, it's customary to assume the Ronchi rulings to have sinusoidal, rather than step function, transmittance functions. Since the light used is generally incoherent, the transmittance functions will be taken as the intensity transmittance functions, to save the mathematically cumbersome operations of squaring the transmitted field distributions. For most of our analyses (involving ordinary moiré), these simplifications will not affect the results.

So, the total intensity transmittance of the two overlaid gratings will be given as the product of the two individual grating transmittance functions:

$$I^{\text{Tot}}(x) = \left[1 + \sin \frac{2\pi}{L_1} x \right]\left[1 + \sin \frac{2\pi}{L_2} x \right]$$

which is expanded to yield

$$I^{\text{Tot}}(x) = 1 + \sin \frac{2\pi}{L_1} x + \sin \frac{2\pi}{L_2} x + \sin \frac{2\pi}{L_1} x \sin \frac{2\pi}{L_2} x$$

The first term is a dc term, indicating constant illumination. The second and third terms are the patterns of the two individual gratings—the "hair" of the "moiré." The fourth term defines the actual moiré. This is expanded using the trig identity

$$\sin a \sin b = \frac{1}{2} \left[\cos(a - b) - \cos(a + b) \right]$$

to yield the "sum" and "difference" moiré patterns. (Here, we're interested primarily in the "difference" pattern.) The difference pattern takes the form

$$I(x) = \cos 2\pi \left(\frac{1}{L_1} - \frac{1}{L_2} \right) x$$

The equivalent period of the difference moiré is then

$$L_{\text{eq}} = \frac{L_1 L_2}{\left| L_2 - L_1 \right|}$$

We can use this equation to see how the period of the moiré is used for quality control applications. Say the master grating has 20 lines/inch (period = .050″), and the manufactured grating has a period of .055″ (in error by 5 mils). Then, the equivalent length of the difference moiré is:

$$L_{eq} = \frac{5 \cdot 10^{-2} \times 5 \cdot 10^{-2}}{5 \cdot 10^{-3}} = 0.5 \text{ in.}$$

This type of tolerance is typical for many types of Ronchi rulings manufactured for educational use.

One annoying aspect of moiré patterns is the "hair" in the fringes due to the presence of the original gratings. This "hair" may be removed by moving the two gratings in a particular fashion. If we take the product pattern above and allow each grating to be translated with constant velocity, then we have

$$I^{\text{Tot}}(x) = \left[1 + \sin \frac{2\pi}{L_1} (x - c_1 t) \right]\left[1 + \sin \frac{2\pi}{L_2} (x - c_2 t) \right]$$

In this case, the difference moiré takes the form

$$I(x) = \cos\left[2\pi \left(\frac{1}{L_1} - \frac{1}{L_2} \right) x - 2\pi \left(\frac{c_1}{L_1} - \frac{c_2}{L_2} \right) t \right]$$

When the two gratings are translated in the same direction, with velocities in the ratio

$$\frac{c_1}{L_1} = \frac{c_2}{L_2}$$

the difference moiré remains stationary (i.e., all time dependence vanishes in the equation above), while all of the other patterns oscillate with time. If the pattern is recorded on film or an integrating array of photodetectors, all patterns except the difference fringes integrate to zero, leaving only that term in the final interference pattern. Note that this is *not* the same as simply translating both gratings horizontally; the two gratings move in the same direction, but (in general) at *different* speeds.

As the next step in complexity, we may look at the moiré produced by two identical line gratings, wherein one grating is inclined at an angle with respect to the other. This is shown in Fig. 11.2. In the area of overlap, rather wide bright and dark fringes are formed (which are seen more readily if the figure is viewed from the edge of the page, sighting along the x-direction). It turns out (and we'll prove shortly) that for small tilt angles between the two grids, the moiré fringes bisect the angle between the two line gratings.

To prove this assertion, let's represent the two line gratings mathematically as simple sinusoidal transmission gratings of the form:

$$\tau_1(x, y) = 1 + \sin \frac{2\pi}{L} x$$

$$\tau_2(x, y) = 1 + \sin \frac{2\pi}{L} (x \cos \delta + y \sin \delta)$$

where δ is the tilt angle of the second grating from horizontal.

Directly multiplying these two transmission functions together produces the resulting moiré given by

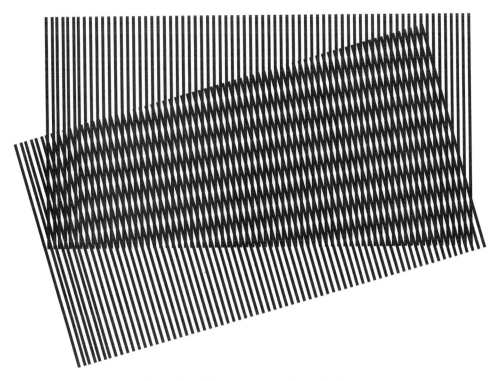

Figure 11.2 Oblique superimposed Ronchi rulings.

$$M(x, y) = 1 + \sin\frac{2\pi}{L}x + \sin\frac{2\pi}{L}(x\cos\delta + y\sin\delta)$$

$$+ \sin\frac{2\pi}{L}x\sin\frac{2\pi}{L}(x\cos\delta + y\sin\delta)$$

The first term is simply a dc bias term, so that the transmittance function is never negative. The second and third terms are the two original grating patterns. (These contribute to the "hair" of the moiré.) The fourth term is the term of interest here, which produces the actual moiré.

We'll ignore the first three terms on the RHS of the previous equation and focus on the fourth. Using the usual trig identity,

$$\sin a \sin b = \frac{1}{2}[\cos(a - b) - \cos(a + b)]$$

we expand the fourth term as

$$M(x, y) = \cos\frac{2\pi}{L}[(1 - \cos\delta)x - y\sin\delta]$$

$$= \cos\frac{2\pi}{L}\left(2\sin\frac{\delta}{2}\right)\left(\sin\frac{\delta}{2}x - \cos\frac{\delta}{2}y\right)$$

with,

$$L_{eq} = \frac{L}{2\sin\delta/2}$$

where we've only retained the "difference term" describing the wide fringes. The usual trigonometric "half-angle" formulas were used in the derivation of the equation above.

This equation is an equation for a sinusoidal moiré grating, the direction of which is roughly orthogonal to the two original gratings. By the equation above, the long lines of the grating bisect the angle between the two original gratings. The equivalent period L_{eq} of the moiré grating is significantly longer than the periods of the original two gratings, being increased by the factor $2 \sin \delta/2$. This type of moiré is useful in shear strain studies and is a second example of in-plane strain analysis [3,4].

As in the case of the two parallel gratings in Fig. 11.1, the "hair" can be removed from this moiré by moving the two gratings in a specific fashion relative to each other. If we allow for linear movement of the two individual gratings, we may represent the difference pattern above as

$$M(x, y) = \left[1 + \sin \frac{2\pi}{L}(x - c_1 t)\right]\left\{1 + \sin \frac{2\pi}{L}[(x - c_2 t)\cos \delta + y \sin \delta]\right\}$$

$$= \cos \frac{2\pi}{L}[(1 - \cos \delta)x - y \sin \delta - t(c_1 - c_2 \cos \delta)]$$

$$= \cos \frac{2\pi}{L}\left(2 \sin \frac{\delta}{2}\right)\left(\sin \frac{\delta}{2} x - \cos \frac{\delta}{2} y\right)$$

when

$$c_1 = c_2 \cos \delta$$

In this case, the difference term has no time dependence, whereas all of the other terms (except the dc term) have sinusoidal time variation. When the two grid velocities are related as above, the difference moiré may be separated from the other terms to produce clean interference fringes.

It is also possible to remove the high spatial-frequency moiré "hair" by passing the pattern through a low-pass spatial filter (basically, a circular aperture in the Fourier transform plane of a lens). The circular aperture is an almost perfect low-pass filter, which is effective in removing the annoying ripples typical of traditional moiré.

11.2 INTRODUCTION TO THE PRINCIPLE OF OPTICAL MOIRÉ: OUT-OF-PLANE MOIRÉ ANALYSIS

We've seen how the moiré method may help determine deformations of a body in the surface plane of the body. Optical moiré may also be used to determine many types of body characteristics normal to the surface of the body. For example, the moiré method may be used to determine constant-height contours on a three-dimensional object, as well as out-of-plane deformations of bodies under mechanical stress. In this section, we introduce the fundamentals of this type of moiré analysis.

The subject of out-of-plane contouring may be approached using a simple computer

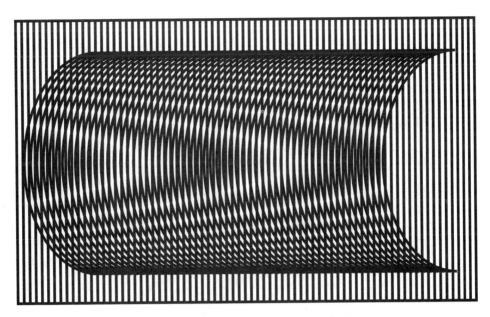

Figure 11.3 Overlap of linear and curvilinear gratings.

drawing package of the type used to produce Figs. 11.1 and 2. Consider Fig. 11.3, which shows the overlap of a line grating and an array of circular arcs. In the overlap region, a moiré is produced. This fringe pattern is significant because it corresponds to the pattern that would be seen if the side of a soda can were viewed through a Ronchi ruling as in Fig. 11.6a (i.e., it has real physical significance). The fringe patterns in this case correspond to level-height contours, where "height" in this case is measured in a direction perpendicular to the grating.

Certain dangers are inherent in "electronic moiré" which are also present in actual experiments using Ronchi rulings. If the array of circular arcs is stretched horizontally, as in Fig. 11.4, the character of the fringes changes; that is, they become nearly straight-line segments. This corresponds to viewing the soda can as shown in Fig. 11.6b. If this array is stretched further still, as in Fig. 11.5, the moiré fringes shift directions altogether and become completely nonphysical. This corresponds to viewing the soda can as depicted in Fig. 11.6c. This transformation of the character of the fringes is known as *aliasing,* and it occurs whenever a function (in this case, the "height function" of a soda can) is sampled too coarsely. In these days of digital analysis of interference patterns, aliasing is an important phenomenon to consider.

We may readily analyze the phenomenon of moiré contour fringes. With reference to Fig. 11.7, consider a plane wave impinging on the diffusely reflecting surface shown. A Ronchi ruling is placed just above the surface, so that it casts a sharp shadow onto the surface. The equation of the grating transmittance is

$$I(x') = 1 + \sin \frac{2\pi}{L} x'$$

where again, we'll assume incoherent illumination and the transmittance to be an intensity transmittance. Translating the equation above to the plane of the object, we see that this intensity distribution becomes

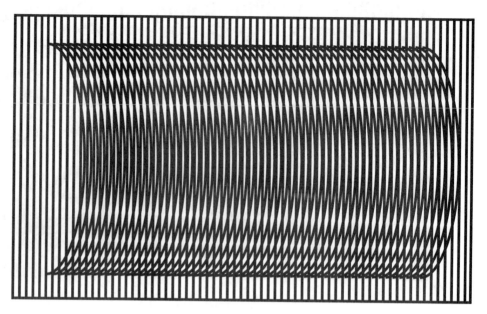

Figure 11.4 Overlap of linear and stretched curvilinear gratings.

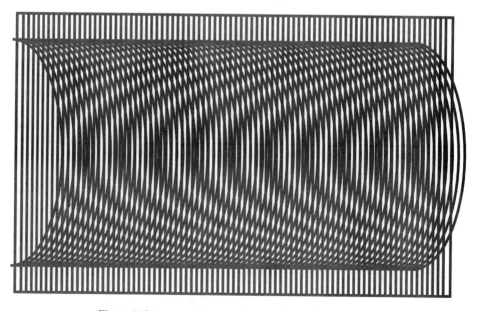

Figure 11.5 Overlap of linear and stretched curvilinear gratings.

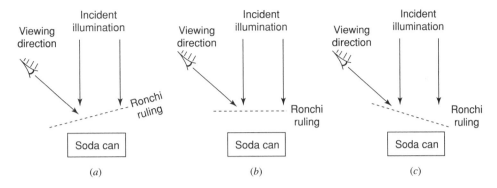

Figure 11.6 Different views of the soda can.

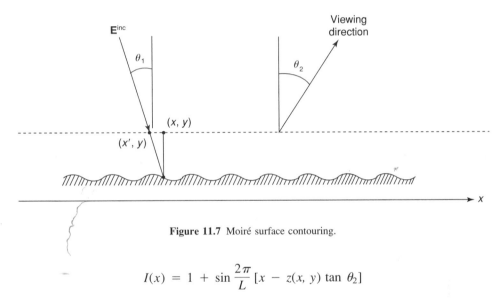

Figure 11.7 Moiré surface contouring.

$$I(x) = 1 + \sin \frac{2\pi}{L}[x - z(x, y) \tan \theta_2]$$

This distorted grating pattern is then viewed through the transmittance function of the reference grating (which, in this case, is simply the original grating):

$$I(x) = 1 + \sin \frac{2\pi}{L}[x + z(x, y) \tan \theta_2]$$

We multiply these two transmittance functions together (to model the combined shadowing effects of the illumination and viewing gratings) in order to obtain the sum and difference fringes in the usual way as in the previous section. The difference fringe system is proportional to

$$I(x, y) = \cos \frac{2\pi}{L} z(x, y)(\tan \theta_1 + \tan \theta_2)$$

Thus, up to a constant, the fringe pattern is directly related to the height function [5–10]. As in the case of in-plane moiré, the linear motion of the two gratings may be used to wash out the high-frequency fringes (due to the original gratings) and isolate the desired moiré fringes [11,12].

11.3 ALTERNATE TECHNIQUES IN MOIRÉ CONTOUR ANALYSIS

A number of laboratory variations exist on the basic moiré theme presented in the previous section for surface contouring. For instance, it is not necessary to use a Ronchi ruling to produce a grating shadow onto an object. The shadow may be synthesized artificially (albeit, using coherent light) by combining two laser light beams at different angles of inclination to the surface [5,13]. Of course, this removes one of the primary conveniences of conventional moiré, that is, that it allows the use of ordinary incoherent light. (For very high-resolution moiré, in which gratings having periodicities on the order of the optical wavelengths are used, diffraction phenomena dominate the GO shadow effects and coherent light must be used anyway.)

For ordinary surface contouring, a diffusely reflecting surface is necessary because light must be scattered in *all* directions from the surface, not simply in the specular direction, in order for the moiré patterns to form. This does not necessarily mean that specularly reflective surfaces cannot be used in moiré, however. When the surface is reflective, moiré patterns still form, but their interpretation now is in terms of slope contours rather than height contours. This type of moiré is sometimes termed *reflection moiré*.

Heretofore, we've looked at moiré patterns that are produced when a line ruling is placed in close enough proximity to a surface that the shadow cast onto the surface by the ruling is in sharp focus (i.e., GO effects predominate and diffraction effects are negligible). This is termed *shadow moiré*. Sometimes the ruling can be located remotely, with the grating imaged onto the surface being contoured. That is, the grating is placed in the object plane of a focusing lens, and the surface to be contoured is located in the image plane of the lens. This is termed *projection moiré*. One shortcoming of projection moiré is that the depth of the surface being contoured cannot exceed the depth of focus of the grating image. Otherwise, the grating will be in-focus at some locations on the surface and out-of-focus at others. The resulting moiré pattern will be of high quality only at those locations where the grating is in sharp focus.

A typical setup for projection moiré is shown in Fig. 11.8 [14].

11.4 USE OF HIGH SPATIAL FREQUENCY GRATINGS AND PHASE GRATINGS IN MOIRÉ ANALYSIS

The shadow and projection moiré techniques—depending as they do on the sharpness of the GO shadow (or its focused image)—are seriously limited by diffraction effects. In other words, if a plane wave is given a sinusoidal or step function amplitude modulation by a grating, that wavefront amplitude modulation will maintain its functional form over significant propagation distances as long as the plane wave spectral content of the modulation function is limited to a narrow range of spatial frequencies (tilt angles) about the propagation direction of the original unmodulated plane wave. Stated another way, the wavefront modulation will retain its form as long as the period of the modulation is much longer than the optical wavelength. This relates to the concept of the "distance-bandwidth product," which states that the narrower the plane wave spectrum bandwidth, the greater

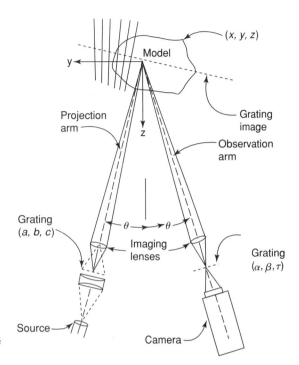

Figure 11.8 Optical system for projection moiré contouring. (After Doty [14].)

the propagation distance over which the waveform shape will be maintained, and conversely.

If the grating period is large in comparison to wavelength, the plane wave spectral bandwidth of the amplitude modulation will be limited in terms of tilt angles from the nominal propagation direction of the wave. Therefore, diffraction effects will be minimized. If the grating is very fine—with periodicities on the order of optical wavelengths—then numerous propagating diffraction orders will be produced by the grating and the waveform shape will be a function of the propagation distance.

We may readily analyze a moiré contouring system using high spatial frequency line gratings, by employing a plane wave spectrum approach similar to that used for diffraction gratings in the previous chapter. Consider Fig. 11.9, which shows the situation. A unit-amplitude plane wave is incident on the line grating, which is assumed to produce only two diffracted orders in addition to the zero order. Assuming that the grating imparts a sinusoidal amplitude modulation of the form,

$$A(x) = 1 + \cos \alpha x$$

we observe that each of the first two diffracted orders will have amplitudes equal to one-half the amplitude of the zero order. We may follow the methods of the previous chapter and write the field just behind the grating as

$$A(x) = 1 + \frac{1}{2} e^{j\alpha x} + \frac{1}{2} e^{-j\alpha x}$$

Propagating this field down to the surface to be contoured, we obtain the field at the surface as

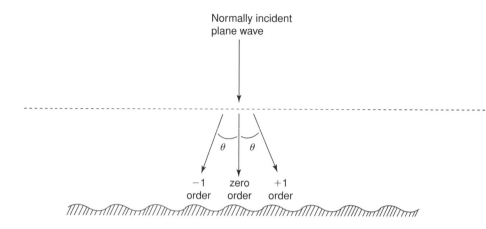

Figure 11.9 Contouring a surface using a high spatial-frequency diffraction grating.

$$A(x) = e^{-jkz(x,y)} + e^{-jk\cos\theta\, z(x,y)} \cos \alpha x$$

$$= e^{-jkz(x,y)} \left[1 + e^{jk(1-\cos\theta)\, z(x,y)} \cos \alpha\right]$$

where

$$\cos \theta = \frac{\alpha}{k}$$

This field is then reflected back from the surface under test, which is assumed to be diffusely reflecting (i.e., the surface converts the line spectrum created by the grating into the smeared line spectrum shown in Fig. 11.10). We'll consider only the three discrete diffracted plane wave orders:

$$\alpha = 0, \pm k \sin \theta$$

which are reflected back from the surface.

Propagating each of these three plane waves back from the surface to the plane of the diffraction grating, we find the field just below the grating to be

$$A(x) = 1 + e^{j2k(1-\cos\theta)z(x,y)} \cos \alpha$$

neglecting constants. We now multiply this field by the grating function,

$$A(x) = 1 + \frac{1}{2} e^{j\alpha x} + \frac{1}{2} e^{-j\alpha x}$$

and isolate terms propagating in the ''quasi-planar'' zero order, normal to the plane of the diffraction grating. These terms may be isolated since they propagate in different angular directions from the zero-order image; this technique has long been used in holography to isolate desired images from undesired ones. So, we retain only those terms that do not contain exponentials involving α. We thus obtain the zero-order field passed back through the diffraction grating as

$$A(x) = 1 + \frac{1}{2} \cos 2k(1 - \cos \theta)z(x, y)$$

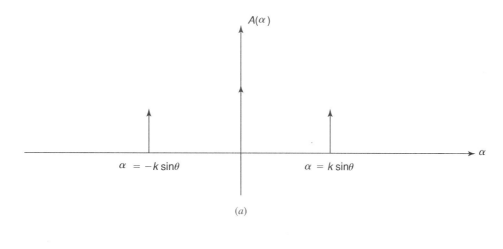

(a)

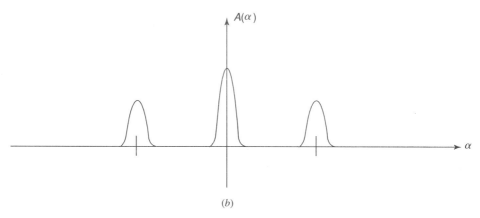

(b)

Figure 11.10 Diffracted field spectra: (a) Spectrum of field produced by line grating and
(b) spectrum of field reflected from diffusely reflecting surface.

This is the moiré for the constant-height contours. The illumination and recording
processes are illustrated in Fig. 11.11.

For gratings with very high spatial frequencies, a phase grating can yield high diffraction efficencies, without the power loss associated with amplitude gratings. Phase gratings
find widespread use in optics, in applications ranging from fiber Bragg gratings to Fourier
plane filters to distributed feedback gratings used in laser diodes.

11.5 MOIRÉ TECHNIQUES IN VIBRATION ANALYSIS

Vibrating (moving) surfaces may also be contoured using moiré techniques. Perhaps the
most intuitive procedure for doing this is to use a pulsed (strobed) optical system, wherein
the pulse length is much smaller than the period of oscillation of the object and the time
between pulses is set equal to the object's oscillation period. Contours may be generated

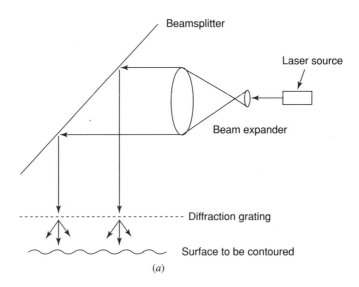

(*a*)

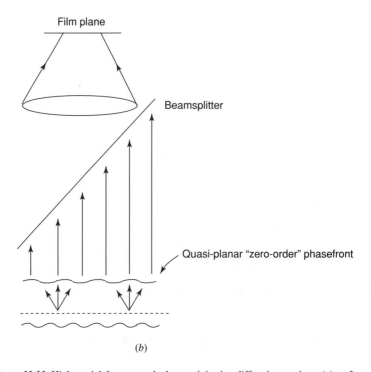

(*b*)

Figure 11.11 High spatial-frequency shadow moiré using diffraction gratings: (a) surface illumination process and (b) surface contour recording process.

by phasing the strobe with the point of maximum vibrational amplitude of the object. In this way, the object is nearly stationary during the time it is illuminated.

This process of strobing requires sophisticated electronic and optical equipment. One rather "low-tech" means of accomplishing virtually the same thing is to take a time-averaged exposure of the vibrating membrane over an integer number of vibrational cycles. (If the number of cycles is very large, it's not even strictly necessary to use exactly an integer number; the fringe pattern will not be seriously affected if this condition is not met.)

We may readily analyze the case of time-averaged moiré for sinusoidally vibrating surfaces. For example, let's revisit the case of ordinary shadow moiré contour analysis, from Section 11.2. In that section, we determined the moiré contour pattern as

$$I(x, y) = 1 + \cos\left[\frac{2\pi}{L} z(x, y)(\tan\theta_1 + \tan\theta_2)\right]$$

We'll assume an originally planar surface, which deforms in response to some sinusoidal mechanical vibration. So, all we do now is let the surface height be a function of time, that is,

$$z = z(x, y, t) = z(x, y) \sin \Omega t$$

where $z(x, y)$ is the "envelope function" for the vibrational mode of the surface. So, the equation for the moiré becomes

$$I(x, y) = 1 + \cos\left[\frac{2\pi}{L} z(x, y) \sin \Omega t (\tan\theta_1 + \tan\theta_2)\right]$$

We describe the specifics of film development more fully in Chapter 13, but for now suffice to say that for film operated in the "linear region," the transmittance of the developed film is proportional to the incident field intensity times the exposure time. When the incident field intensity is a function of time, the transmittance is equal to the integral of the intensity function over the exposure period (i.e., the total exposure is equal to the integral over all of the "incremental exposures"). Thus, for the transmittance of the film exposed to the sinusoidally varying intensity function, we have

$$\tau(x, y) = T + \int_0^T \cos\left[\frac{2\pi}{L} z(x, y) \sin \Omega t (\tan\theta_1 + \tan\theta_2)\right] dt$$

If Ωt is an integer multiple of π, then the transmittance function becomes proportional to [15–19]

$$\tau(x, y) = A + J_0\left[\frac{2\pi}{L} z(x, y)(\tan\theta_1 + \tan\theta_2)\right]$$

where A is some constant of proportionality. The Bessel function of order zero is shown in Fig. 11.12 [20]. The intensity of the moiré fringes starts out at a maximum value (for small Bessel function arguments) and passes through successive maxima and minima. The height of the successive maxima is smaller and smaller, so that the fringe visibility (i.e., contrast)

$$V = \frac{I_{max} - I_{min}}{I_{max} + I_{min}}$$

becomes less and less for the higher fringe orders.

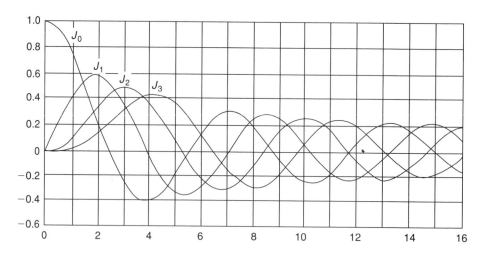

Figure 11.12 Bessel functions of the first kind. (After Harrington.)

And finally, we note that moiré, as conventional interferometry, can always be used in conjunction with holography to store individual grating patterns for subsequent re-illumination and viewing through selected reference gratings. We'll discuss this storage property of holograms further in Chapter 13.

REFERENCES

[1] Post, D., "Developments in Moiré Interferometry," *Opt. Eng.,* Vol. 21, No. 3, May/June, 1982, pp. 458–467.

[2] Oster, G., "Optical Art," *Appl. Opt.,* Vol. 4, No. 11, November 1965, pp. 1359–1369.

[3] Chiang, F-P, "Moiré Methods of Strain Analysis," *Experimental Mechanics,* Vol. 19, No. 8, August 1979, pp. 290–308.

[4] Post, D., "Moiré Interferometry for Deformation and Strain Studies," *Opt. Eng.,* Vol. 24, No. 4, July/August 1985, pp. 663–667.

[5] Tsuruta, T., Itoh, Y., and Anzai, S., "Moiré Topography for the Measurement of Film Flatness," *Appl. Opt.,* Vol. 9, No. 12, December 1970, pp. 2802–2807.

[6] Heiniger, F., and Tschudi, T., "Moiré Depth Contouring," *Appl. Opt.,* Vol. 18, No. 10, May 15, 1979, pp. 1577–1581.

[7] Chiang, C., "Moiré Topography," *Appl. Opt.,* Vol. 14, No. 1, January 1975, pp. 177–179.

[8] Browne, A. L., "Fluid Film Thickness Measurement with Moiré Fringes," *Appl. Opt.,* Vol. 11, No. 10, October 1972, pp. 2269–2277.

[9] Sciammarella, C. A., "The Moiré Method: A Review," *Experimental Mechanics,* Vol. 22, No. 11, November 1982, pp. 418–433.

[10] Meadows, D. M., Johnson, W. O., and Allen, J. B., "Generation of Surface Contours by Moiré Patterns," *Appl. Opt.,* Vol. 9, No. 4, April 1970, pp. 942–947.

[11] Halioua, M., Krishnamurthy, R. S., Liu, H., and Chiang, F. P., "Projection Moiré

with Moving Gratings for Automated 3-D Topography," *Appl. Opt.,* Vol. 2, No. 6, March 15, 1983, pp. 850–855.

[12] Allen, J. B., and Meadows, D. M., "Removal of Unwanted Patterns from Moiré Contour Maps by Grid Translation Techniques," *Appl. Opt.,* Vol. 10, No. 1, January 1971, pp. 210–212.

[13] Brooks, R. E., and Heflinger, L. O., "Moiré Gauging Using Optical Interference Patterns," *Appl. Opt.,* Vol. 8, No. 5, May 1969, pp. 935–939.

[14] Doty, J. L., "Projection Moiré for Remote Contour Analysis," *J. Opt. Soc. Am.,* Vol. 73, No. 3, March 1983, pp. 366–372.

[15] Vest, C. M., and Sweeney, D. W., "Measurement of Vibrational Amplitude by Modulation of Projected Fringes," *Appl. Opt.,* Vol. 11, No. 2, February 1972, pp. 449–451.

[16] Hung, Y. Y., et al., "Time-Averaged Shadow Moiré Method for Studying Vibrations," *Appl. Opt.,* Vol. 16, No. 6, June 1977, pp. 1717–1719.

[17] Chiang, F. P., and Lin, C. J., "Time-Average Reflection Moiré Method for Vibration Analysis of Plates," *Appl. Opt.,* Vol. 18, No. 9, May 1, 1979, pp. 1424–1427.

[18] Ritter, R., and Meyer, H-J, "Vibration Analysis of Plates by a Time-Averaged Projection Moiré Method," *Appl. Opt.,* Vol. 19, No. 10, May 15, 1980, pp. 1630–1633.

[19] Der Hovanesian, J., and Hung, Y. Y., "Moiré Contour-Sum, Contour-Difference and Vibration Analysis of Arbitrary Objects," *Appl. Opt.,* Vol. 10, No. 12, December 1971, pp. 2734–2738.

[20] Harrington, R. F., *Time-Harmonic Electromagnetic Fields,* New York: McGraw-Hill, 1960.

12

Interference and Interferometers

In this chapter, we examine the phenomenon of optical interference, particularly as it relates to the testing of optical components. Interference, as the term is used in optics, generally refers to the superposition of two optical wavefronts. As the word is applied in optical testing, one of the wavefronts is usually a perfectly planar or spherical *reference* wavefront, while the other is a *test* wavefront that has either been passed through—or reflected from—some optical component under test. The test wavefront will have acquired some distortions—wavefront aberrations of the type studied in Chapter 9—while traversing its optical path and will therefore no longer be a perfectly planar or spherical surface. The interference pattern generated when the test and reference wavefronts are added together yields both qualitative and quantitative information about the quality of the optical device.

12.1 OPTICAL COHERENCE

Mathematically speaking, the phenomenon of optical interference is fairly straightforward. Two optical fields, regarded as temporally infinite sinusoidal waveforms of the form

$$E = e^{j[\omega t - \phi(x,y,z)]}$$

where

$$\phi(x, y, z) = \text{a surface defining the wavefront}$$

are superimposed. For such single-frequency *coherent* fields, Maxwell's equations are linear in terms of the field vectors, so whenever two such fields coexist in space, the total field is given by

$$E^{\text{Tot}} = E_1 + E_2$$

We can regard the two interfering fields as infinite sinusoidal wavetrains oscillating at a constant angular frequency, ω. At microwave frequencies, it is not difficult to generate fields that are, to a very high degree of accuracy, coherent. In fact, the microwave maser predated the laser as a source of coherent radiation. In the optical regime, however, generation of coherent radiation has been a longstanding problem, which has not been completely solved, even with the advent of laser optical sources.

Historically, approximations to coherent radiation have been produced in a number of ways. Approximately temporally coherent radiation may be produced either by placing a narrow-passband optical filter in the path of a white light source or by selecting the radiation from a particular emission line of an atomic element (e.g., Cadmium vapor). Approximately spatially coherent radiation (in the plane of the wavefront) may be produced by focusing the light through a pinhole and then expanding the pinhole beam. None of these techniques produces coherent light of the quality attainable from a laser source, but the light produced is often "coherent enough" to enable the creation of certain kinds of interference patterns. The advent of the laser light source permitted a "quantum improvement" in the quality of coherent optical radiation achievable in practice, but even laser light is not perfectly coherent.

The reality of imperfectly coherent light is perhaps nowhere more important than in interferometry, where the vector addition of the optical fields absolutely requires the fields to be coherent. In order to ensure that the light in the two interfering fields is coherent, the two fields are usually obtained by splitting the light from a single source into two *equal-length* paths. The light in the two paths is operated on by different types of optical devices, but the total optical path length in both paths must be nearly equal.

A characteristic parameter of laser light sources is the *coherence length* of the emitted light. The coherence length specifies how many inches (or feet) in the longitudinal direction of the beam that the optical field can be regarded as being a perfectly coherent sinusoidal waveform. The coherence length of the laser light determines the maximum optical path length difference that can exist between the two paths of an interferometer. (Again, best results will be attained when the path length difference is as near to zero as possible.) Interference between the two fields will occur only when the difference in optical path lengths is less than the coherence length of the laser light source.

12.2 INTRODUCTION TO OPTICAL INTERFERENCE

The phenomenon of optical interference for coherent fields is governed by the *superposition theorem* [1] from electromagnetics. Maxwell's equations are linear in field strength. Therefore, if two separate electromagnetic fields coexist in the same region, the total field will be given as the sum of the two individual fields. Expressed mathematically,

$$E^{\text{Tot}} = E_1 + E_2$$

In the case of incoherent fields, field intensities add together. Thus,

$$I^{\text{Tot}} = I_1 + I_2$$

The two individual fields may be planar, cylindrical, spherical, or (as in the case of holography) the virtual image of an illuminated object. Virtually any conceivable combination of two (or more) fields can be used to create optical interference. In this section, we'll briefly illustrate some of the possible combinations of fields that can be used to form interference patterns.

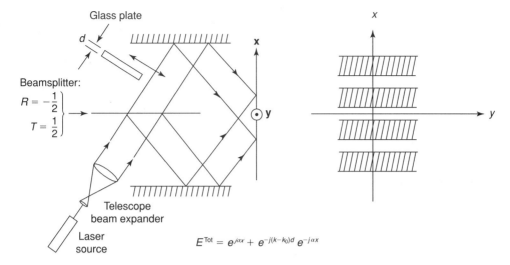

Figure 12.1 Interference of two plane wave fields.

Consider Fig. 12.1, which shows the interference of two plane wave fields. As in most of the analyses in this book, we'll ignore polarization effects (i.e, the vector nature of light) and assume that the two interfering waves have the same polarization; this will enable us to concentrate on the phase effects. It should be noted, however, that two fields cannot interfere unless they have the same polarization (or, at least have one polarization component in common, although the former case is preferable for best fringe contrast).

In Fig. 12. 1, the total field is given as

$$E^{\text{Tot}} = e^{j\alpha x} + Ae^{-j\alpha x}$$

where

A is some unit-amplitude complex number

that is,

$$A = e^{j\phi} \quad \text{for some real-valued } \phi$$

The angle ϕ represents a possible optical path length difference between the two legs of the interferometer. For the interferometer as shown, $\phi = 0$. If a plate of glass is inserted into one of the legs of the interferometer, that leg will become optically longer than the other, causing ϕ to differ from zero. We can easily show that the resultant field is given as

$$E^{\text{Tot}} = 2je^{j(\phi/2)} \sin\left(\alpha - \frac{\phi}{2}\right)$$

What we see then is that as the phase difference between the two paths of the interferometer is changed, the fringes move laterally in the plane where the two beams interfere. This interferometer is a simple example of an immensely useful interferometer known as the *Mach-Zehnder interferometer*. It can be used to detect small phase shifts by tracking fringe movement.

One of the more modern applications of this simple type of interference pattern is in the manufacture of bandpass or bandstop filters in germania-doped optical fibers [2,3]. Two interfering UV laser beams are formed as shown in Fig. 12.2, which produce a sinusoidal intensity distribution along the length of the photorefractive fiber. The sinusoidal interference pattern induces a sinusoidal index modulation along the length of the photore-fractive fiber, and this sinusoidal index distribution acts as a longitudinal diffraction grat-ing. As in the case of X-ray diffraction by atomic lattices, this longitudinal diffraction grating reflects strongly at Bragg incidence (when the spacing between successive index maxima equals a half wavelength in the fiber).

12.3 INTERFEROMETERS FOR OPTICAL TESTING: THE TWYMAN-GREEN INTERFEROMETER

In this chapter, we study several different types of interferometers for optical testing, although they all have a great deal in common in terms of their principles of operation. The basic interferometer splits a beam of light into two beams. The two beams are then passed through two different optical systems and then added together coherently on output. One leg, the *reference leg,* usually consists of nothing more than an optical delay line. The field in the reference leg ideally suffers little or no wavefront distortion and thus represents a type of ''comparison signal'' (a perfectly planar or spherical wavefront). The beam in the *test leg* is passed through the optical device under test and, in so doing, suffers a certain amount of wavefront distortion. Adding the two signals together causes the

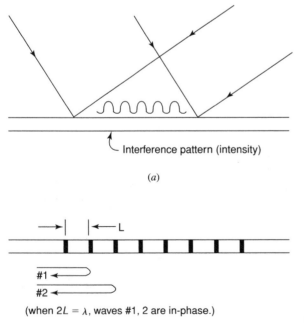

Interference pattern (intensity)

(a)

(when $2L = \lambda$, waves #1, 2 are in-phase.)

(b)

Figure 12.2 Optical fiber Bragg gratings made using two interfering fields: (a) Writing the fiber grating and (b) reflection properties of fiber grating.

wavefront phase distortion to be converted into wavefront amplitude distortions, which may then be readily observed.

This type of interferometry is closely akin to the demodulation/detection schemes employed in ordinary radio communications. The distorted "test" wavefront is analogous to a phase-modulated carrier signal, and the unmodulated reference wavefront is analogous to a "local oscillator." By adding the two signals together, the "baseband" waveform (which phase-modulates the test wavefront) may be effectively recovered and displayed.

The Twyman-Green interferometer (and its modern-day variant, the laser unequal path interferometer, or LUPI) follows the general model of this type of interferometer. Both have the form shown in Figs. 12.3 and 12.4 [4]; the only significant difference between the two is the use of a coherent laser source in the LUPI version. In Fig. 12.3, the narrow laser beam is expanded to a wider, though still collimated beam, using a telescope operated in the reverse direction. The expanded beam is passed through a beamsplitter, which directs half of the incident light energy to the reference flat and half to the object under test (in this case, a parabolic mirror). Lens L_1 focuses the expanded laser beam to the rear focal point of the lens. The rear focal point of L_1 is placed two focal-lengths in front of the reflector, at its center of curvature. The light is reflected back from both the reflector under test and the reference flat. The beam reflected back from the reference flat has flat phase, and that reflected back from the mirror under test, having passed back through lens L_1, is a distorted plane wave. Both reflected fields are directed

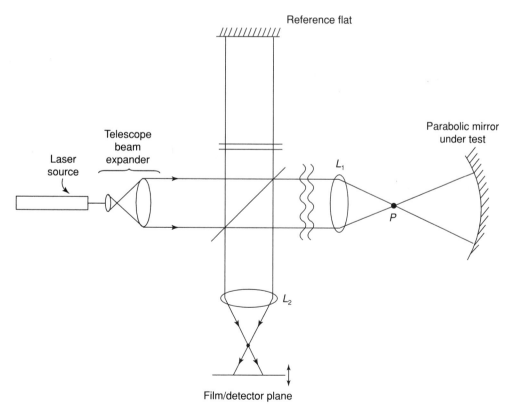

Figure 12.3 Laser unequal path interferometer for testing a parabolic mirror.

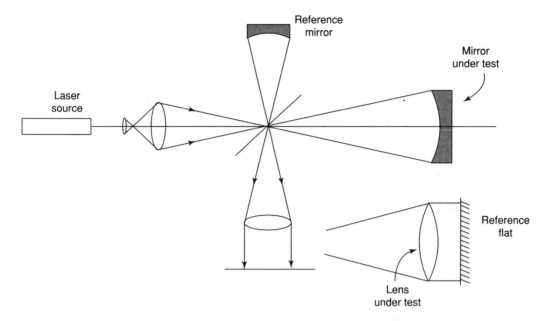

Figure 12.4 Twyman-Green/LUPI interferometer.

back toward the beamsplitter, where they are combined and directed to the lens L_2. Note that both reflected beams are again split at the beamsplitter; one beam from each is directed toward L_2, and the other is directed back toward the laser. In practical LUPI systems, an isolator may be necessary to protect the laser from this back-reflected light energy.

The light energy incident upon lens L_2 is in the form of an intensity-modulated plane wave, of the type shown in Fig. 12.1. Lens L_2 is optional and merely serves to magnify or reduce the size of the interference pattern in the image plane.

In this type of interferometer, the distortion of the reflected wavefront is equal to twice the surface distortion (due to the two-way path involved). The nature of the interference pattern between the planar reference beam and the distorted ''test'' beam may be inferred in a straightforward fashion, using an analysis of the reference and test wavefront phases. If the wavefront reflected back from the mirror (and passed back through L_1) has the quasi-planar form

$$E^{\text{test}} = e^{j\psi(x,y)}$$

and the (assumed planar) reference wavefront has the form

$$E^{\text{ref}} = 1$$

then the total field is given by

$$E^{\text{Tot}} = 1 + e^{j\psi(x,y)}$$

and the intensity (which is the quantity recorded in an interferogram recorded onto ordinary film or photodetectors) is

$$I^{\text{Tot}} = \left|1 + e^{j\psi(x,y)}\right|^2$$

Clearly then, dark/bright contour lines (fringes) will occur when

$$\psi(x, y) = (2n + 1)\pi$$

$$= 2n\pi, \quad \text{respectively}$$

The interference fringes form a contour plot showing contours of constant displacement from a perfectly spherical surface. These contour plots are no more difficult to read than a U.S. Geological Survey topographic map, though in this case, there can be some ambiguity in the direction of the displacement (i.e., is it into, or out of, the plane of the test mirror?).

Many other variations on the LUPI concept are possible [5,6], which may have advantages for certain applications. A version of the interferometer that is similar to the original Twyman-Green patent is shown in Fig. 12.4.

The Twyman-Green/LUPI interferometer also has uses beyond the testing of optical components. As shown in Fig. 12.5, this interferometer, along with the Mach-Zehnder interferometer, has use in the measurement of phase objects [7,8]. (This application is similar to the use of the interferometer for testing lenses.) Phase objects arise in a number of scientific applications, including biological slide transparencies and aerodynamic fluid flows. The phase object is a basically transparent object that primarily alters the wavefront phase of an optical field passing through it. This is, of course, exactly the same type of phenomenon that arises in the optical mirror testing applications described previously. By superimposing the fields in the test and reference paths, interference fringes consisting of contours of constant phase are formed. A contour plot showing the temperature contours around a candle flame is shown in Fig. 12.6 [7]. This interferogram was obtained using a Twyman-Green interferometer.

One additional aspect of interferogram construction needs to be mentioned. In the figures showing both the Twyman-Green and the Mach-Zehnder interferometers, we've indicated areas in the diagrams where the fields were presumed to be perfectly planar and others where the fields were presumed to be perfectly spherical. In some of those areas, we've also indicated that the total field (the sum of the test and reference fields) contained dark and bright fringes, which constituted the interference pattern. An example of such a fringe pattern is shown in Fig. 12.6. In other words, we've assumed the existence of perfectly planar (and spherical) wave fields that were amplitude-modulated. However, as we saw in Chapter 2, the only allowable type of propagating *plane wave* field is a uniform-amplitude plane wave field (containing no modulation whatever). Any field on some planar surface, which is amplitude or phase modulated, must necessarily contain a *spectrum* of plane wave fields. Therefore, the optical fields that exist in those "plane wave" portions of the interferometer (which contain the distorted "test" wavefronts) must be composed of some finite spectrum of plane wave fields, propagating at various angles with respect to the optic axis of the system. In other words, these fields contain a finite bandwidth in the plane wave spectrum domain.

In spherical coordinates, the same situation holds. A perfectly spherical wave must have uniform amplitude. An amplitude or phase-modulated "spherical" wave will, in reality, consist of a spectrum of spherical wave harmonics; it will not be a purely spherical wave. In both cases (spherical and plane wave), each spatial mode will propagate at a unique propagation velocity, and the disparity in propagation velocities between the different plane/spherical wave modes causes signal distortion. (This type of distortion also takes place in fiberoptic and microwave waveguide systems.) Image distortion is a function of both the propagation distance and the plane/spherical wave bandwidth. That is, the greater the modal bandwidth, the greater the image distortion, for a given propagation distance.

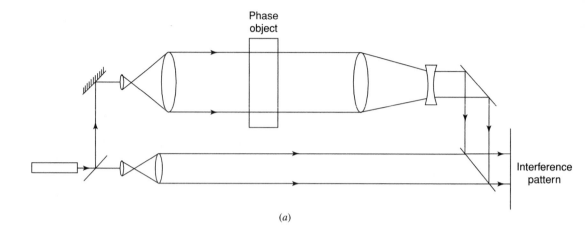

(a)

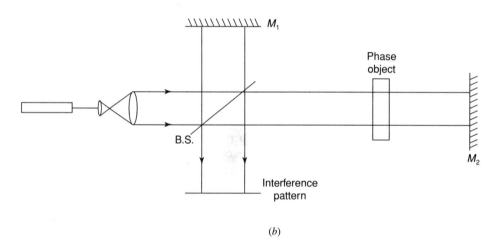

(b)

Figure 12.5 Interferometers for wind tunnel testing: (a) Mach-Zehnder and (b) Twyman-Green/LUPI.

Conversely, the greater the propagation distance, the greater the image distortion for a given plane/spherical wave bandwidth. Thus, the total image distortion is proportional to the product of the propagation distance and the modal bandwidth of the image. In the fiberoptic industry, this "figure-of-merit" is sometimes called the *distance-bandwidth product*.

As we saw in Chapter 2, the "spatial bandwidth" of an image is related to the fineness of the image features. Therefore, an interferogram with fine fringes has a broad-band plane wave spectrum. This interferogram will degrade rather rapidly with propagation distance. Therefore, it is important to use a fairly compact interferometer system in that case, in order to minimize distortion in the interferogram. The spherical wave is often preferred over the quasi-planar wave for carrying amplitude modulation information. This

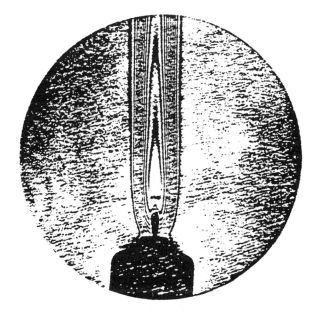

Figure 12.6 Temperature field around the flame of a candle (after Grigull and Rottenkolber).

is because an AM spherical wave—located in the far field of its image distribution—will propagate without any type of wavefront distortion, as was shown in Chapters 1 and 2.

12.4 INTERFEROMETERS FOR OPTICAL TESTING: THE FIZEAU INTERFEROMETER

In Chapter 2, we looked briefly at the planar Fabry-Perot interferometer, which consisted of two semireflecting plane sheets. That interferometer was shown to possess certain filtering characteristics in the frequency domain. The Fabry-Perot interferometer also has applications in the testing of optical flats. The modified version of this interferometer is known as the Fizeau interferometer, an example of which is shown in Fig. 12.7.

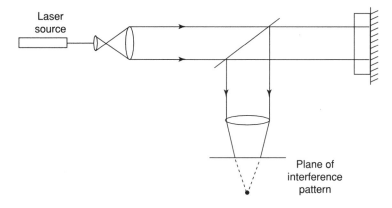

Figure 12.7 Interferometer for testing surface flatness.

In Fig. 12.7, light from a laser source is expanded and made incident on a beamsplitter. Half the beam is discarded, and half is directed toward the Fabry-Perot cavity. In this case, the Fabry-Perot cavity consists of a glass-covered mirror [9]. As in the case of the Fabry-Perot device, the interference in this case takes place between fields reflected from the front and rear faces of the transparent slab (the rear face being backed by a mirror). This interferometer may be used to determine the flatness of either the slab or the mirror (with the other being the reference surface). For example, if the transparent slab is the test object, the reference mirror may be a mercury mirror that conforms to the back face of the slab.

This interferometer may be operated in two distinct modes. In the traditional mode, the interferometer is operated in the same fashion as the Mach-Zehnder and LUPI interferometers discussed previously. That is, when the reflections from the front and back faces are in phase, bright fringes will appear, and when the reflections are out of phase, dark fringe contours will appear. In the traditional mode, accuracy is obtained by using laser light of very short wavelength. This mode of operation is successful for testing smooth reflecting surfaces.

For the testing of diffusely reflecting surfaces, a different approach has to be taken [10]. In this second approach, longer wavelength (near infrared) laser light is used. The longer wavelength light is reflected specularly from the surface, whereas shorter wavelength visible light will be reflected diffusely, forming a speckled interference pattern. The specular reflection allows for better quality interferograms. Rather than simply constructing an interference pattern by summing the test and reference fields at a single frequency (in a single measurement), the frequency of the laser light is swept. The interferogram, rather than being recorded on ordinary film, is recorded by an array of photodetectors. The photodetector images recorded are each referenced to the frequency of the laser light illumination.

For each point in the pixelated detector image, the difference in laser frequency for successive bright or dark spots is recorded; the optical path length from the back face to the front face of the Fabry-Perot cavity is proportional to this difference in frequency.

With the help of Fig. 12.8, we may perform simple first-order analyses of the two modes of operation (considering only the first bounce). An exact solution, considering multiple bounces, is a straightforward extension. In the traditional, ''short-wavelength'' mode, the two fields are as shown. Field 1 is regarded as having zero phase. The sum of the two optical fields is given as

$$E_1 + E_2 = 1 - e^{-j2kd}$$

(assuming equal-amplitude division at the front interface). In this equation, the minus sign

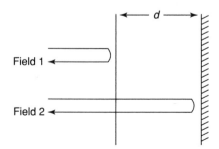

Figure 12.8 Single bounce analysis of the Fizeau interferometer.

preceding the second term on the RHS refers to the phase reversal at the perfect mirror, and k is the wavenumber in the slab medium. Clearly, bright fringes occur when

$$2kd = (2n + 1)\pi \qquad (12.1)$$

and dark fringes occur when the phase is a multiple of 2π.

In the modern, "long-wavelength" case, (12.1) still holds for each point in the image, except now the laser frequency is varied so that each point in the image goes through at least one transition from light to dark to light again. Say a dark fringe occurs at a laser frequency corresponding to wavelength λ_1 in the medium, and the next dark fringe occurs at a higher frequency corresponding to a smaller wavelength λ_2 in the medium. Thus,

$$2k_1 d = 2n\pi$$

$$2k_2 d = 2(n + 1)\pi$$

and

$$d = \frac{\pi}{k_2 - k_1}$$

This "frequency sweep" technique, though perhaps a recent development in the field of optics, has been in use for many years in testing microwave circuits via the automated network analyzer.

12.5 INTERFEROMETERS FOR OPTICAL TESTING: THE CONFOCAL MICROSCOPE

In Chapter 6, we looked at the confocal microscope as a device for obtaining images of optically transparent slides. The confocal microscope may also be operated in a reflective mode for obtaining surface roughness data [11]. The principle of operation is shown in Fig. 12.9.

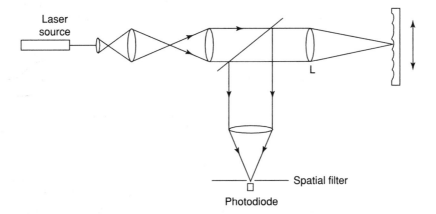

Figure 12.9 Reflective confocal microscope for interferometrically measuring surface roughness.

Light from a laser source is expanded and made incident on a beamsplitter. Half of the beam is discarded, and the other half is focused onto a diffraction-limited spot on the rough surface. (Recall from Chapters 2 and 4 that both the lateral and longitudinal extent of the spot is inversely proportional to the angle of the focused cone of rays.) If the surface happens to coincide with the location of the spot, the optical energy incident on the surface will retrace its path back to the beamsplitter (as a collimated wave) and will be focused onto the single photodiode shown.

If the surface does not coincide with the spot location, the optical field will be reflected back to the lens L with a different radius of curvature than the incident field had. Thus, the field transmitted by the lens L, incident on the beamsplitter, will no longer be collimated; it will be slightly spherical (diverging or converging, depending on whether the surface is in front of or behind the spot). This slightly spherical wave will not be focused onto the photodetector; thus, the point will not register as in focus. By raster scanning the sample, the entire sample may be contoured in this fashion.

12.6 INTERFEROMETERS FOR OPTICAL TESTING: THE JAMIN INTERFEROMETER

Heretofore, we've looked at interferometers in which the properties of the "test" wavefront were deduced by summing this unknown wavefront with a known reference wavefront. In this section, we begin to look at a new type of interferometer—known as a *wavefront shearing interferometer*—in which the properties of an unknown test wavefront are obtained by comparing the wavefront with itself. Figure 12.10a shows a type of shearing interferometer—known as a Jamin interferometer—useful in testing right-angle prisms [12,13]. In this case, the interferometer is used to determine how closely the sidewalls of the prism make a 90° angle with each other.

In this device, light from a laser source is expanded and made incident on the beamsplitter B.S. Half the beam is discarded, and the remainder is directed to a thick transmissive slab. In a first-order analysis, wherein we only consider the singly reflected fields from the front and back faces, we may regard the front face of the slab as a beamsplitter. One beam is directed downward toward the prism, and the other beam is transmitted into the thick slab, where it is reflected off the mirrored back face. This beam then exits the slab, so that the two parallel beams propagate down to the prism, trace out each other's path in opposite directions, are recombined at the thick slab, and are deflected by the beamsplitter B.S. downward toward the plane of interference.

If the two sidewalls of the prism make a perfect 90° angle, both fields will be collimated and parallel to each other. Thus, no fringe pattern will be formed in the plane of observation. If, however, the sidewalls are not at a perfect right angle, then the two fields that trace through the prism in opposite directions will be inclined with respect to each other when combined. In this case, a set of linear bright and dark fringes will be formed in the observation plane.

With the help of Fig. 12.10b, the width of the fringes may be obtained as a function of prism angle deviation from 90°. We can trace the path of the ray reflected from the front of the thick slab. This ray traverses the interferometer in a clockwise direction. The ray enters the prism (of index n) at normal incidence and reflects from the right-hand wall (assumed to be the tilted wall). In the prism, the ray traverses the path shown, finally

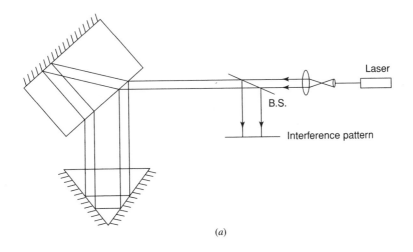

(a)

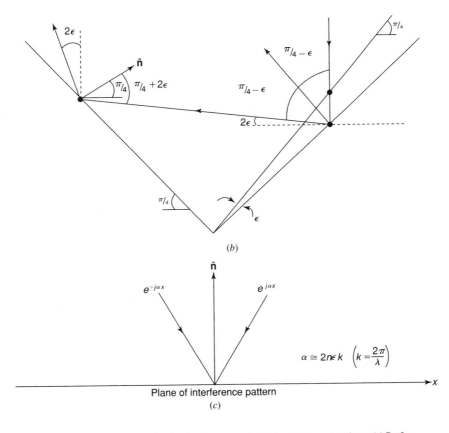

(b)

(c)

Figure 12.10 Jamin-type shearing interferometer for testing right-angle prisms: (a) Perfect
right-angle prism; (b) angle error of ϵ radians in one side; and (c) superposition of two tilted plane waves to produce sinusoidal field distribution in the
measurement plane.

traveling upward at an angle of 2ϵ, where ϵ is the deviation angle of the right-hand wall. The ray exits the prism, whereupon its deviation angle changes as

$$2\epsilon \rightarrow 2n\epsilon$$

The ray field traversing the interferometer in the counterclockwise direction is deviated by the same amount in the opposite direction. The two plane wave fields are combined in the plane of the interference pattern as shown in Fig. 12.10c. The resultant field takes the form

$$E^{\text{Tot}} = \cos(2n\epsilon x)$$

Alternate fringes are spaced a distance d, where

$$2n\epsilon d = \pi$$

The wavefront shearing interferometer may also be used to evaluate the quality of lenses [14]. Wavefront shearing interferometers for lens testing are shown in Fig. 12.11. (We'll shortly discuss how to interpret sheared interferograms for lenses.) Light from a laser source is expanded and divided into two paths. (The larger lens in the beam expander is considered the lens under test.) Each path is directed through a thick shearing plate that is rotated in opposite directions in the two optical paths, as shown. This causes the upper beam to be laterally displaced into the plane of the paper and the lower beam to be laterally displaced out of the plane of the paper.

The superposition of two laterally sheared beams from a lens can yield information on lens quality. The interpretation of the interference pattern due to laterally sheared wavefronts is not as straightforward as in the case of the Mach-Zehnder or LUPI interferometers. However, accurate determination of wavefront characteristics is still possible. Herein, we'll follow the method of Rimmer [15] for interpreting laterally sheared interferograms.

Assume that two sheared fields from a test lens have been collimated and superposed as shown in Fig. 12.12. An interference pattern will be formed in the region of overlap of the two beams. The two fields are assumed to have uniform amplitude functions and phase functions of the form

$$E_1 = e^{j\psi(x,y)}$$
$$E_2 = e^{j\psi(x-s,y)}$$

In the region of overlap, these two beams interfere to produce the intensity distribution

$$I^{\text{Tot}} = (E_1 + E_2) \cdot (E_1 + E_2)^*$$
$$= 2\{1 + \cos[\psi(x, y) - \psi(x - s, y)]\}$$

The maxima and minima of this intensity pattern form the bright and dark fringes of the sheared interferogram. In the method of Rimmer, the wavefront is calculated at points along the wavefront separated by the shearing distance, s. In the case of a one-dimensional wavefront, we would proceed by arbitrarily setting the phase at point 1 equal to zero (i.e., point 1 is the phase reference). The total intensity of the interference pattern is assumed to be known from measurements. Therefore, we may deduce the arguments of the cosine function (we'll call these arguments $x_1, x_2, \ldots$). Thus,

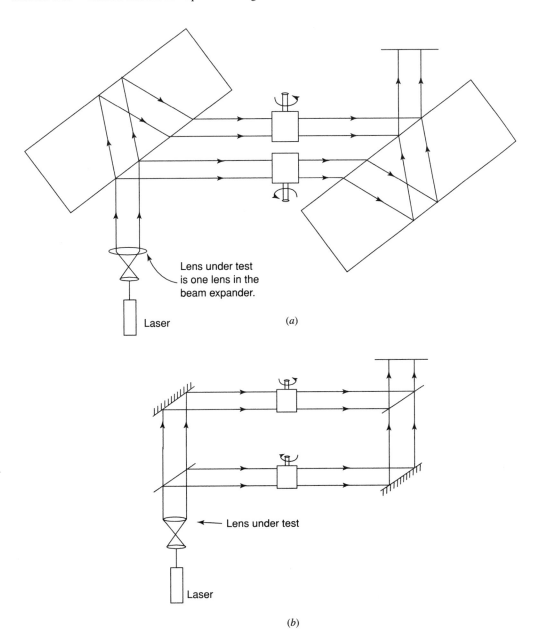

Figure 12.11 Jamin-type shearing interferometers for testing lenses: (a) Using thick slabs and (b) using beamsplitters in a typical Mach-Zehnder configuration.

$$\psi_2 = x_1{}'$$

$$\psi_3 - x_1 = x_2{}'$$

and so on. In this way the points along the wavefront may be recursively calculated.

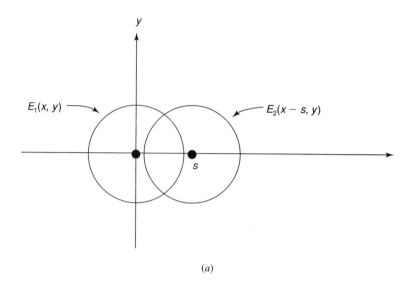

(a)

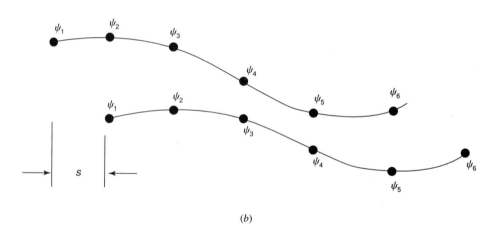

(b)

Figure 12.12 On the interpretation of shearing interferograms: (a) Two superposed
sheared wavefronts and (b) reconstruction of a one-dimensional wavefront.

Obtaining the x_i from the intensity plots is strictly possible only up to modulo π. In other
words, since $\cos x = \cos(2\pi - x)$, it is impossible to know whether

$$0 \leq x < \pi \quad \text{or} \quad \pi \leq x < 2\pi$$

In this case, enforcing the continuity of x can generally resolve the ambiguity.

In the more practical case of a two-dimensional wavefront, shearing in two orthogonal directions allows reconstruction of the wavefront. This is shown in Fig. 12.13. In this
case, we obtain four equations in the four unknowns, one of which is chosen as the
reference:

$$x_1 = \psi_2 - \psi_1$$
$$x_2 = \psi_4 - \psi_3$$
$$y_1 = \psi_3 - \psi_1$$
$$y_2 = \psi_4 - \psi_2$$

When working with two-dimensional arrays larger than 2×2, evaluation of all possible shears in the two orthogonal directions yields many more equations than unknowns. Rimmer shows how this redundancy can be used to create an $n \times n$ system of equations stable with respect to noisy measurements. Rather than adding new equations for each shearing measurement, each shearing measurement involving a particular phase term (say the mth) is summed, and the result forms the mth row of the matrix equation.

Figure 12.14 shows a shearing interferogram for a lens with a focal length of 10 cm and a diameter of 1.428 cm. Figure 12.15 shows the mathematically reconstructed wavefront from the lens.

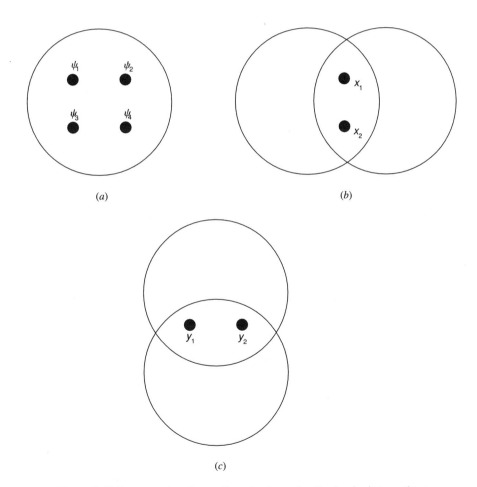

(a)

(b)

(c)

Figure 12.13 Reconstruction of a two-dimensional wavefront by shearing in two orthogonal directions: (a) Original wavefront, (b) horizontal shear, and (c) vertical shear. (After Rimmer.)

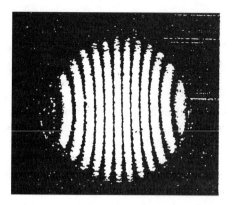 

Figure 12.14 Measured shearing interferograms of a 10 cm focal length, F/D = 7 lens. (After Rimmer [15].)

12.7 OTHER SHEARING INTERFEROMETERS

As mentioned earlier, a two-dimensional wavefront may be reconstructed from wavefront shearing data when the shearing takes place in two orthogonal directions. One easy way to do this with the interferometer in Fig. 12.11 is simply to rotate the test lens 90° and

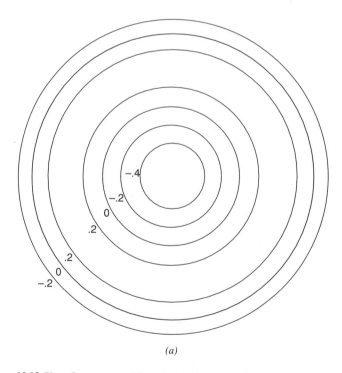

(a)

Figure 12.15 Phase front computed from the interferograms of Fig. 12.14. (After Rimmer [15].)

take two sets of measurements. This technique works with any type of interferometer in which shearing takes place in one plane only.

One classic type of shearing interferometer that shears in one plane is the cube beamsplitter shown in Fig. 12.16 [16,17]. A cube is cut along the diagonal, with the two

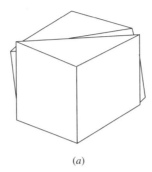

(a)

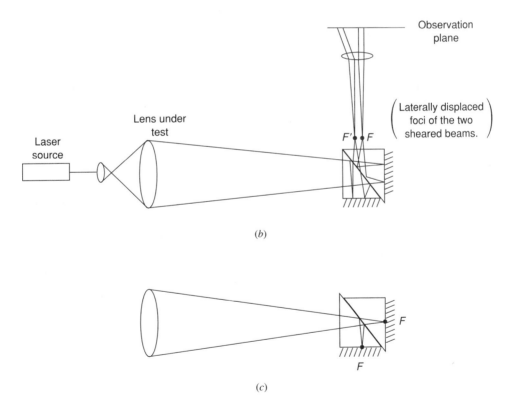

(b)

(c)

Figure 12.16 Compact shearing interferometer using a cube beamsplitter: (a) A cube beamsplitter; (b) interferometer based on the cube beamsplitter (F is focal point for reflection off the back wall and F′ is the focal point for reflection off the bottom wall.); and (c) condition for sheared beams having coincident focal points (i.e., the focus of convergent spherical wave coincides with the back and bottom walls of cube).

halves slightly rotated as shown in Fig. 12.16a, and then cemented back together again to form a stable unit. As shown in Fig. 12.16b, the light from a laser source is made divergent using a small objective lens, and the divergent beam is made incident on the lens under test. The field from the lens under test is convergent and is made incident on the cube beamsplitter.

One beam is reflected off the cube diagonal (which acts as a beamsplitter) and is also reflected off the lower cube wall, as shown. This beam is reflected off the lower wall, passed back up through the beamsplitter (half is reflected off the beamsplitter back to the source), focused at the point F, and then collimated for viewing on the observation plane. We'll assume that the lower half of the cube is aligned with the laser beam and that the upper half is rotated.

The second interfering beam passes through the beamsplitting diagonal and is incident on the rotated right-hand face of the cube. This beam is reflected off the right-hand wall back to the beamsplitting diagonal. The beam will be reflected upward, off the diagonal, and will be directed slightly out of the plane of the paper and slightly off the optic axis of the test lens in the plane of the paper. This beam will be focused to the laterally displaced focal point F', and will be sheared with respect to the other beam. This is a rather inexpensive type of shearing interferometer for lens testing.

In the cube beamsplitter shown, two spherical waves are produced which focus to two laterally displaced foci F, F'. The two spherical waves are transformed by the collimating lens into two tilted plane waves that interfere in the observation plane. Since the two plane waves are tilted with respect to each other, even perfect test lenses will give rise to a linear fringe pattern. This linear fringe pattern can be eliminated if the field from the test lens is made convergent on the rear/bottom faces of the cube beamsplitter as shown in Fig. 12.16c. In this case, both diverging spherical waves will emerge from a common focal point, sheared with respect to each other.

A slightly more complex shearing interferometer, capable of producing shear in two orthogonal directions, is shown in Fig. 12.17 [18]. This device is basically a Mach-Zehnder configuration, wherein the two beamsplitters are connected so as to rotate in unison (they are shown as a single, connected beamsplitter). The beamsplitters control beam rotation out of the plane of the paper, whereas the mirror M_1 controls beam rotation in the plane of the paper. Thus, shear in two independent planes is produced.

One type of interferometer not generally regarded as a shearing interferometer is shown in Fig. 12.18 [19]. In this interferometer, the two sheared beams are created not by a beamsplitter but by a diffraction grating. The two interfering beams represent two different diffraction orders from the grating. In this instrument, light from a laser source is made divergent with a well-corrected objective lens. The diverging field is made incident on the lens under test. The lens converts the diverging wave into a converging wave, whose focus lies just behind the diffraction grating. The convergent field is converted into three convergent spherical wave fields, and the zero-order field is blocked by an opaque dot in the focal plane. The ± 1 diffraction orders are passed through the focal plane and allowed to interfere in the observation plane as shown.

Since the diffraction grating gives rise to two spherical waves that have laterally displaced centers of curvature, even a perfect test lens will produce linear fringes in the observation plane. Lens aberrations are determined as a function of deviation of the fringes from perfect linearity.

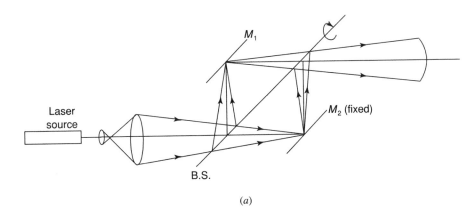

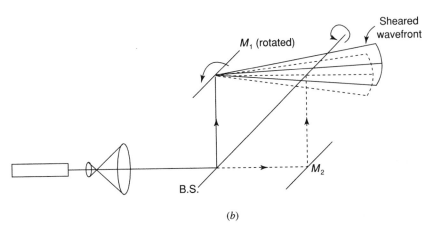

Figure 12.17 Wavefront shearing interferometer (shearing in two planes): (a) Interferometer in nominal state (beams overlapping) and (b) interferometer with mirror M_1 rotated.

12.8 EXAMPLES OF SHEARING INTERFEROGRAMS

We've already seen how to reconstruct a wavefront from its shearing interferogram. In this section, we look at some simple wavefront aberrations and the resulting shearing interferograms they produce. The two aberrations to be examined in this section are the defect of focus and spherical aberration.

To begin, consider the defect-of-focus aberration defined by

$$E(x, y) = e^{ja(x^2 + y^2)}$$

We consider the superposition of two sheared wavefronts of this form. The sum of these two sheared fields is given as

$$E^{\text{Tot}} = e^{jar_1^2} + e^{jar_2^2}$$

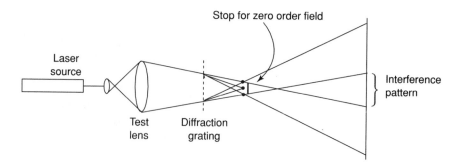

Figure 12.18 Shearing interferometer using diffraction grating.

where

$$r_1^2 = (x - \Delta)^2 + y^2 = x^2 + y^2 - 2x\Delta + \Delta^2$$
$$r_2^2 = (x + \Delta)^2 + y^2 = x^2 + y^2 + 2x\Delta + \Delta^2$$

That is, one wavefront is centered at $x = \Delta$, and the other is centered at $x = -\Delta$. So,

$$r_1^2 = r^2 - 2x\Delta + \Delta^2$$
$$r_2^2 = r^2 + 2x\Delta + \Delta^2$$

where

$$r^2 = x^2 + y^2$$

Adding the two complex exponentials gives for the total field

$$E^{\text{Tot}} = e^{ja(r^2 + \Delta^2)} \cdot 2 \cos(a\Delta x)$$

The field intensity is proportional to the square of the field strength, thus this field gives rise to linear fringes perpendicular to the line connecting the centers of the two wavefronts. (*Note:* This analysis also indicates why the Ronchi test gives rise to linear fringes even for perfect test lenses that produce quadratic wavefronts.)

In the second case, that of spherical aberration, the total field is given as

$$E^{\text{Tot}} = e^{jar_1^4} + e^{jar_2^4}$$

Performing the usual operations yields

$$E^{\text{Tot}} = e^{ja(\bar{r}^4 + 4x^2\Delta^2)} \cdot 2 \cos(4\bar{r}^2 x \Delta)$$

where

$$\bar{r}^2 = x^2 + y^2 + \Delta^2$$

and it's evident that spherical aberration produces curved interference fringes.

It's possible to go through all of the primary aberrations, calculate the fringe patterns, and plot them out to determine their corresponding shearing patterns.

12.9 INTERFEROMETERS FOR SURFACE TESTING: THE MIREAU INTERFEROMETER

The interferometer shown in Fig. 12.19 has been discussed extensively in the literature [20,21] as a means for surface profiling. An incoherent light source is focused by lens L_1 onto the aperture A_1, the extent of which will determine the area of surface to be profiled on the sample. The beam passed through the aperture is collimated by lens L_2 and passed through the narrowband (frequency) filter F, as shown. These two steps give approximate spatial and temporal coherence, as described in Section 12.1. The beam is deflected by a beamsplitter and directed toward the sample. Prior to reaching the sample, the beam is passed through the Mireau interferometer, which moves linearly.

The Mireau interferometer is similar in principle to the Fizeau interferometer depicted in Fig. 12.7. The Mireau interferometer consists of three stages: a focusing lens, a reflective dot stage, and a beamsplitter stage. The lens focuses the collimated light beam onto a spot in the vicinity of the surface being profiled. The spot may be focused at

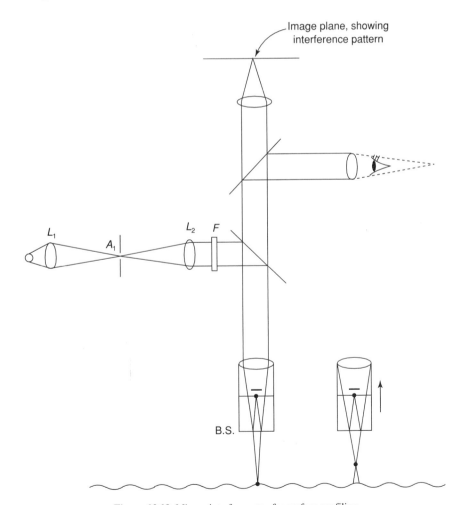

Figure 12.19 Mireau interferometer for surface profiling.

the surface or slightly in front or behind the surface, depending on the location of the Mireau on its carriage. The converging beam passes through a glass plate, which carries a small reflective dot on it. (The small dot does not significantly affect the converging beam.) The beam then hits a beamsplitter, pictured at the bottom of the Mireau interferometer. All three pieces of the interferometer move as a unit, so that the focal spot moves from a point slightly behind the surface to a point slightly in front of the surface.

At the beamsplitter, one beam is transmitted through to the sample, and the other is reflected back to the reflective spot. The beam reflected by the beamsplitter is directed back toward the second stage of the Mireau (the reflective spot), where it is focused and re-reflected back toward the third stage (the beamsplitter). The reflective spot is located exactly at the focal point of the converging beam, so that it always produces a strong reflection. The beam transmitted by the beamsplitter is focused near the vicinity of the surface. If the cone of the converging beam of rays is fairly narrow, the depth-of-field of the device will be fairly long in extent (see Chapter 4). This means that even if the spot is not exactly focused at the surface, a fairly strong optical reflection will still occur at the sample. (This device operates on a phase—not an amplitude—basis, unlike a confocal system.)

The two beams (now diverging spherical waves) are re-combined at the beamsplitter and directed back toward the first stage of the Mireau (the focusing lens). The two beams are re-collimated at this lens and then directed upward toward the observation plane. In this plane, a photodetector array receives the interference pattern (for each value of the Mireau position) and the set of interference patterns is analyzed. If the intensities of the reflected and transmitted beams are denoted I_r and I_t, respectively, the intensity interference pattern of the two has been shown in previous sections of this chapter to be given as

$$I^{\text{Tot}} = I_r + I_t + 2\sqrt{I_r I_t}\cos\phi(x, y)$$

where ϕ is the relative phase difference between the field reflected by the dot and the field reflected by the surface. If ϕ has a linear, time-varying phase attached to it (due to the movement of the Mireau on its carriage), and if both field intensities are assumed relatively constant (due to the long depth-of-field), then we may write the interference pattern as a function of time as

$$I^{\text{Tot}} = I_1 + I_2 \cos[\phi(x, y) + \alpha(t)]$$

where $\alpha(t)$ is a phase term introduced by movement on the carriage.

The technique for analyzing the interference patterns for the various carriage positions is known as the integrated bucket technique. In this approach, the detector outputs are integrated over quarter-wavelength intervals of detector movement. For example, say the cosine term above is integrated from 0 to $\pi/2$ in alpha as

$$\int_0^{\pi/2} \cos[\phi(x_0, y_0) + \alpha]\, d\alpha = \cos\phi_0 - \sin\phi_0$$

This quantity is measured by the photodetector and associated electronics, and will be denoted by A. The corresponding measurements for the intervals $\pi/2$ to π, π to $3\pi/2$ and from $3\pi/2$ to 2π will be denoted as B, C, D. Thus, we obtain

$$A = I_1 + I_2\,[\cos\phi_0 - \sin\phi_0]$$

$$B = I_1 + I_2\,[-\cos\phi_0 - \sin\phi_0]$$

$$C = I_1 + I_2\,[\cos\phi_0 + \sin\phi_0]$$

From the measured quantities A, B, C, and the previous three relations, we may obtain the phase function as

$$\phi(x_0, y_0) = \tan^{-1} \frac{C(x_0, y_0) - B(x_0, y_0)}{A(x_0, y_0) - B(x_0, y_0)}$$

and from this phase, distance can be determined. Keep in mind that the phase is due to a two-way path length difference between the reference spot and the surface.

Our initial assumption was that the depth-of-focus was long enough that intensity variations in the surface scattered field could be ignored. However, the intensity modulation due to focusing effects can also be used to improve the sensitivity of this technique when large surface height discontinuities are being measured [22]. In addition, measurements taken at more than wavelength can be used to increase measurement sensitivity.

We can easily show how multiple wavelength interferometry can be used to increase sensitivity to large surface discontinuities (greater than a quarter wavelength). For example, say the path length difference between the reflection path to the dot and the transmission path to the surface is d. Then, phase measurements may be taken at two different frequencies to yield

$$k_1 d = \phi_1 \quad \text{and} \quad k_2 d = \phi_2$$

where

$$\phi_1 = n_1(2\pi) + \phi_{0,1}$$

and

$$\phi_2 + n_1(2\pi) + \phi_{0,2}$$

and

$$\phi_{0,1}, \phi_{0,2} < 2\pi$$

In other words, the optical path length at either wavelength may extend well beyond one wavelength (and may actually consist of several wavelengths), but the optical path at both wavelengths must consist of the same integer number of wavelengths, whatever that integer number happens to be.

Subtracting ϕ_1 and ϕ_2 yields

$$\Delta\phi = \phi_2 - \phi_1 = 2\pi d \left(\frac{1}{\lambda_1} - \frac{1}{\lambda_2} \right)$$

With this last equation, we may now solve for d, knowing that no extra factors of 2π are involved in the equation.

It should be noted in the interferometer discussed above that the aperture A_1 is not a pinhole, but rather a finite-extent aperture. Thus, the light incident on lens L_2 will be due to a number of point sources located in the aperture A_1, and each of these point sources will give rise to a plane wave of unique direction, on transmission through L_2. Thus, the "collimated" beam really consists of a (narrow) spectrum of plane waves. This spectrum is focused onto the surface of the test sample, illuminating a finite area, proportional in extent to the area of A_1.

REFERENCES

[1] Harrington, R. F., *Time-Harmonic Electromagnetic Fields,* New York: McGraw-Hill, 1960.

[2] Friebele, E. J., and Kersey, A. J., "Fiberoptic Sensors Measure Up for Smart Structures." *Laser Focus World,* May 1994, pp. 165–171.

[3] Erdogan, T., and Mizrahi, V., "Fiber Phase Gratings Reflect Advanced in Lightwave Technology," *Laser Focus World,* February 1994.

[4] Bond, C., and Pipan, C. A., "How to Align an Off-Axis Parabolic Mirror," *SPIE,* Vol. 1113, Reflective Optics II (1989), March 27–29, 1989, Orlando, FL (available from Space Optics Research Labs).

[5] Kocher, David G., "Twyman-Green Interferometer to Test Large Aperture Optical Systems," *Applied Optics,* Vol. 11, No. 8, August 1972, pp. 1872–1874.

[6] Houston, J. B., Buccini, C. J., and O'Neill, P. K., "A Laser Unequal Path Interferometer for the Optical Shop," *Applied Optics,* Vol. 6, No. 7, July, 1967, pp. 1237–1242.

[7] Grigull, U., and Rottenkolber, H, "Two-Beam Interferometer Using a Laser," *J. Opt. Soc. Am.,* Vol. 57, No. 2, February 1967, pp. 149–155.

[8] Howes, W. L., and Buchele, D. R., "Optical Interferometry of Inhomogeneous Gases," *J. Opt. Soc. Am.,* Vol. 56, No. 11, November 1966, pp. 1517–1528.

[9] Bunnagel, R., Oehring, H.-A., and Steiner, K., "Fizeau Interferometer for Measuring the Flatness of Optical Surfaces," *Applied Optics,* Vol. 7, No. 2, February 1968, pp. 331–335.

[10] de Groot, P., Smythe, R. and Deck, L., "Laser Diodes Map Surface Flatness of Complex Parts", *Laser Focus World,* February 1994, pp. 95–98.

[11] Anderson, S. G., "Confocal Laser Microsopes See a Wider Field of Application," *Laser Focus World,* February 1994, pp. 83–86.

[12] Murty, M. V. R. K., "Some Modifications of the Jamin Interferometer Useful in Optical Testing," *Applied Optics,* Vol. 3, No. 4, April 1964, pp. 535–538.

[13] Sen, D., and Puntambekar, P. N., "Shearing Interferometers for Testing Corner Cubes and Right Angle Prisms," *Applied Optics,* Vol. 5, No. 6, pp. 1009–1014.

[14] Murty, M. V. R. K., "The Use of a Single Plane Parallel Plate as a Lateral Shearing Interferometer with a Visible Gas Laser Source," *Applied Optics,* Vol. 3, No. 4, April 1964, pp. 531–534.

[15] Rimmer, M. P., "Method for Evaluating Lateral Shearing Interferograms," *Applied Optics,* Vol. 13, No. 3, March 1974, pp. 623–629.

[16] Saunders, J. B., "A Simple, Inexpensive Wavefront Shearing Interferometer," *Applied Optics,* Vol. 6, No. 9, September 1967, pp. 1581–1583.

[17] Saunders, J. B., "A Simple Interferometric Method for Workshop Testing of Optics," *Applied Optics,* Vol. 9, No. 7, July 1970, pp. 1623–1629.

[18] van Rooyen, E., and van Houten, H. G., "Design of a Wavefront Shearing Interferometer Useful for Testing Large Aperture Optical Systems," *Applied Optics,* Vol. 8, No. 1, January 1969, pp. 91–93.

[19] Ronchi, V., "Forty Years of History of a Grating Interferometer," *Applied Optics,* Vol. 3, No. 4, April 1964, pp. 437–451.

[20] Wyant, J. C., and Prettyjohns, K. N., "Three-Dimensional Surface Metrology Using a Computer Controlled Non-Contact Instrument," *Proc. SPIE* 661, 1986.

[21] Wyant, J. C., Kollopoulos, Bhushan, B., and Baslia, D., "Development of a Three-Dimensional Noncontact Digital Optical Profiler," Transactions of the ASME, *Journal of Tribology,* Vol. 108, No. 1, January 1986, pp. 1–8.

[22] Wyant, J. C., "How to Extend Interferometry for Rough Surface Tests," *Laser Focus World,* September 1993, pp. 131–135.

13

Introduction to Holography

In previous chapters, we've looked separately at a number of relatively familiar concepts in optics, including diffraction gratings, optical interference phenomena, and optical moiré phenomena. In this chapter, we'll integrate all of these concepts in order to understand optical image formation via the technique known as holography.

The discoverer of modern holography, a researcher named Dennis Gabor, referred to this technique as imaging via wavefront reconstruction. It is unfortunate that this term didn't catch on more widely, because it describes the phenomenon much accurately than the more popular—and somewhat more mystical—term, holography. In fact, a hologram is nothing more than a specialized type of diffraction grating that is produced as the recorded interference pattern between two or more different optical fields. A greatly magnified view of an actual hologram is shown in Fig. 13.1 [24], from which the contour pattern of lines can be clearly seen. When viewed macroscopically with the unaided eye a hologram simply appears as an ordinary diffraction grating or laser disk, from which various colors of the rainbow emerge when viewed from various angles.

A hologram is therefore nothing more than the recorded image of an interference pattern of the type studied in Chapter 12. However, in the case of the hologram, the fringe spacing is very fine, being on the order of the optical wavelength. Thus, the fringes are not visible to the unaided eye. So a hologram is a type of diffraction grating whose lines are curved rather than strictly linear. When properly illuminated, the hologram *reconstructs* the two optical *wavefronts* that were used to create the interference pattern in the first place. (Actually, some other fields are also created, as we'll see, but these are generally of no importance and can be easily discarded.)

This reconstruction property of holograms is quite useful, especially in interferometry. Traditional interferometry, for example, requires both interfering wavefronts to exist contemporaneously. But what happens if we wish to form an interference pattern between one field that we produced in the lab last week and one that we intend to produce sometime next month? By storing the wavefront from last week in coded hologram form, and then re-illuminating it in our experiment next month, we may then create interference between the two optical wavefronts, even though the object that created one of the wavefronts is long gone. Of even more practical importance is the case when both wavefronts correspond to the same object, under two different sets of strain conditions.

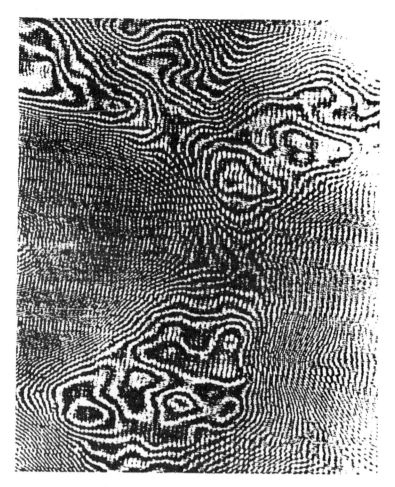

Figure 13.1 An enlarged section of a plane hologram made with $\lambda = 6328$ Å from a He-Ne gas laser. (*Conductron Corporation.*) (After Jenkins and White [24].)

The hologram does not give rise to any type of magical image that is different from anything we've studied before. The image (or wavefront) produced when the hologram is illuminated is simply a virtual image of the type we've been studying all along so far in this book. Thus, it's not so surprising that it should have some of the three-dimensional properties of an ordinary virtual image produced by any other kind of imaging system. The only "weird" thing about the holographically generated image is that it is produced without an actual object being present. But the image itself is simply an ordinary virtual image.

13.1 THE LEITH-UPATNEIKS TRANSMISSION HOLOGRAM

The original Gabor hologram, though interesting from a historical standpoint, has been obsolete for decades and will not be described here. (The interested reader can consult the literature [1–3].) Much more useful in modern holography is the *Leith-Upatneiks* hologram, which employs spatial heterodyning techniques to spatially separate into sepa- rate diffraction orders the many images that are produced in the wavefront reconstruction process.

As mentioned earlier, a hologram is nothing more than a recorded interference pattern between (typically) two optical fields. In this context, a hologram is very much like the interference patterns generated by the LUPI or Mach-Zehnder interferometers described in Chapter 12. Generally, one of the interfering fields is a ''reference field,'' a planar or spherical wave field similar to the reference field used in the LUPI. The other field is an ''object field'' similar to the ''test field'' used in the LUPI. Typical configurations used for recording the interference pattern between the two fields are shown in Fig. 13.2. In Fig. 13.2a, a simple Mach-Zehnder interferometer is depicted, and in Fig 13.2b, a ''spheri- cal wave Mach-Zehnder'' interferometer is shown. Both create a recorded interference pattern between the test and reference wavefronts.

We may readily find an expression for the recorded interference pattern between the reference and virtual image fields. Referring to Fig. 13.3, assume a planar reference wave of the form

$$E_{\text{ref}} = e^{j\alpha x}$$

and a virtual image (scattered) wave of the form

$$E^{\text{scat}}(x, y)$$

in the hologram plane. The waves are assumed to be co-polarized. The total electric field intensity in the plane of the hologram is given by

$$I^{\text{Tot}}(x, y) = (E_{\text{ref}} + E_{\text{obj}})(E^*_{\text{ref}} + E^*_{\text{obj}})$$

or

$$I^{\text{Tot}}(x, y) = \left|E^2_{\text{ref}}\right| + \left|E^2_{\text{obj}}\right| + E_{\text{ref}}E^*_{\text{obj}} + E^*_{\text{ref}}E_{\text{obj}}$$

Substituting in the expressions for the reference and object virtual image fields, we have for the total intensity in the x, y plane

$$I^{\text{Tot}}(x, y) = 1 + \left|E^2_{\text{obj}}\right| + E^*_{\text{obj}}e^{j\alpha x} + E_{\text{obj}}e^{-j\alpha x}$$

If a negative photographic plate is placed in the x, y plane, its transmittance after develop- ment will be of the following form [4] (for a more exact representation for the transmittance function, see the next section):

$$\tau(x, y) = \tau_0 - (Bt) I^{\text{Tot}}(x, y)$$

where τ_0 is the transmittance of the transparency without any exposure, B is a constant, and t is the exposure time. We'll talk about the significance of this equation in more detail in the next section. For now, the important thing to note is that this equation states that the transmittance of the developed transparency depends on the intensity of the total field, as well as on the exposure time—both in a linear fashion. In reality, both types of linearity only hold over a certain range of illumination intensities and exposure times (i.e., if illumi- nated long enough or with enough intensity, the film eventually saturates, with further exposure producing no further changes in the transmittance of the developed film).

Thus, in the linear region of the $t - E$ curve, the transmittance of the developed hologram takes the form

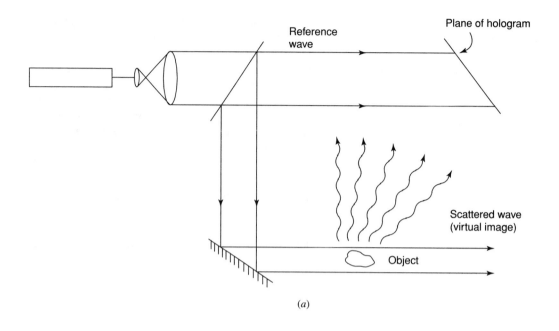

(a)

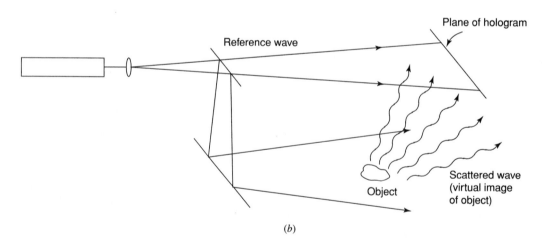

(b)

Figure 13.2 Two common configurations for reconstructing spatially separated wavefronts: (a) Mach-Zehnder interferometer configuration and (b) spherical-
wave Mach-Zehnder configuration.

$$\tau(x, y) = \tau_0 - Bt\{1 + I_{\text{obj}}(x, y) + E_{\text{obj}}(x, y)e^{-j\alpha x} + E^*_{\text{obj}}(x, y)e^{-j\alpha x}\}$$

Creating the hologram transmittance function is only half the process of wavefront
reconstruction, or holography. The second step involves re-illuminating the hologram to
reconstruct the virtual image of the original illuminated object. If the recorded interference
pattern between the reference field and the virtual image field of the object (i.e., the
hologram) is now illuminated by the original reference wavefront, the field transmitted

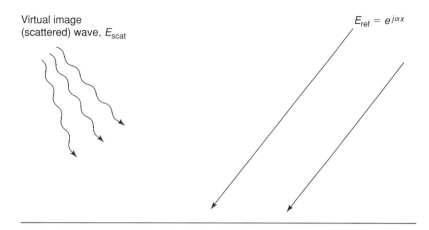

Figure 13.3 Detail of interference phenomenon in hologram formation.

by the hologram is obtained simply by multiplying the incident field by the hologram transmittance:

$$E_{\text{trans}} = \tau(x, y) \cdot E_{\text{inc}}(x, y)$$

$$= (\tau_0 - Bt)e^{j\alpha x} - Bt\, I_{\text{obj}}(x, y)e^{j\alpha x} - Bt\, E_{\text{obj}}(x, y) - Bt\, E_{\text{obj}}^*(x, y)e^{j2\alpha x}$$

These various fields are shown in Fig. 13.4. The term of interest in the expression above is the third term on the RHS. This is the reconstruction of the original virtual image field of the scattering object.

We'll make some assumptions about the terms in the equation above, which are often more valid for reconstructed wavefronts than for ordinary virtual images. The first assumption involves the spatial bandwidth of the scattered fields. When the object in Fig. 13.2 is illuminated by a coherent plane wave, only one side of the object is illuminated; the other side is shadowed. This illumination function tends to make reflections from the object rather directional, so that the object is seen in only certain viewing directions. In addition, if the object is reflective, the scattered field will tend to be even more directional, radiating in the specular directions determined by Snell's law of reflection. Therefore, the virtual image fields scattered by reflective objects under coherent plane wave illumination tend to be fairly directional, visible in rather limited fields of view and invisible elsewhere. In many instances, this field of view is quite limited. Thus, the fields shown in Fig. 13.4 tend to be restricted to limited fields of view in angular space. If the propagation constant α (related to the offset angle of the reference wave) is large enough, there will be no overlap of the virtual images shown in Fig. 13.4. Thus, the various virtual images will be separated by a viewing angle. When this is the case, the reconstructed virtual image will be of good quality.

In addition to the virtual image field discussed above, the hologram also produces the conjugate of this field, although it is offset in a different viewing direction. As we saw repeatedly in Chapter 5, the conjugate of a diverging virtual image field is a converging real image field. This real image may be recorded by placing film in the image focal plane.

Since the reconstructed virtual image field is indeed a true virtual image, it will exhibit all of the qualities of such images, including the same three-dimensional nature

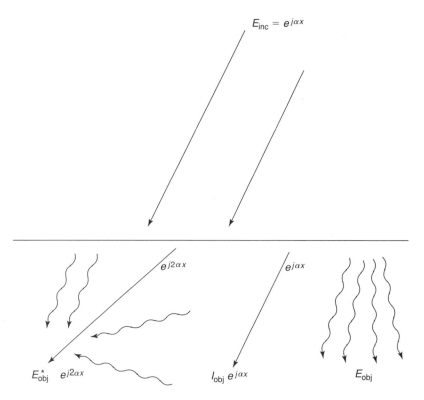

Figure 13.4 The fields produced on reconstruction.

as the original virtual image directly scattered by the actual object. That is, if the reconstructed virtual image is viewed from different directions, it will display the various "different" views that would have been observed when actually viewing the original object. So, a hologram is capable of storing and reading out numerous planar images, whereas an ordinary photograph may only store a single planar image.

The implications for information storage are immediately evident. For example, instead of storing multiple views of a single object, in principle it seems possible to store multiple unrelated images (for example, successive pages in a book) on a single hologram. Hence, viewing the hologram from different directions will yield the images of the various pages. This type of "heterodyned" storage could yield an immense image storage capacity. Alternatively, the viewing direction could be held constant, and the image to be read out could be determined by the direction of the reconstructing plane wave field. As it turns out in practice, thick holograms are most practical for storing large amounts of data, due to the narrow range of angles over which the reconstructed image is visible.

13.2 HOLOGRAM RECORDING MEDIA

Typically, film properties are discussed in connection with descriptions of holography. This is because the film properties directly affect the way in which the incident light intensity during exposure is converted into the hologram during the film processing. In this section, we look at film properties from a qualitative perspective in order to understand some of the ways in which the film affects the recording of the hologram.

The most significant phenomenon affecting hologram formation is the relationship between the developed film transmittance and the incident light intensity. For a negative transparency under coherent (monochromatic) illumination, this relationship is given by the $t - E$ curve shown in Fig. 13.5 [2,5,6]. In the figure, the exposure, E, is defined as the optical field intensity times the exposure time. Since intensity has units of power, exposure has units of energy. The hologram is exposed in the linear region of the $t - E$ curve (starting just below the knee of the curve, as shown), wherein amplitude transmittance is linearly related to field intensity during development. Note that since exposure (intensity) is always positive, exposing the film will only cause the transmittance to decrease. Thus, the film is always operated at the knee on the *left-hand side* of the linear region in the figure.

The presence of this linear region on the $t - E$ curve justifies our assumption in the previous section that amplitude transmittance is directly proportional to total field intensity (apart from a constant bias term). Following [5], we may write the expression for this linear region of the curve as

$$\tau(x, y) = a - bt\, i(x, y) \tag{13.1}$$

where b is the slope of the $t - E$ curve and a is the bias offset. The variable t is the exposure time.

It's tempting to think that this relation might hold at each and every point on the photographic film, independent of what's going on at the neighboring points. Unfortunately, this is not the case. Consider for example, the chemical emulsion shown in Fig. 13.6. The emulsion contains chemicals that undergo a reaction when exposed to light. Say

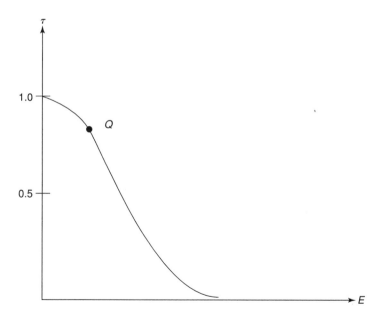

Figure 13.5 Typical curve of amplitude transmittance versus exposure for a negative transparency.

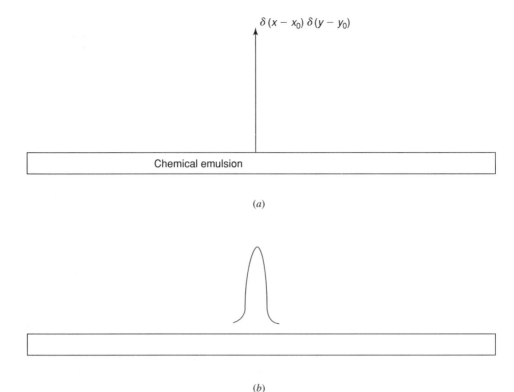

$$\delta(x - x_0)\,\delta(y - y_0)$$

Chemical emulsion

(a)

(b)

Figure 13.6 Simple model of chemical processes in the exposure of photographic film:
(a) Plot of light intensity and (b) plot of transmittance of developed emulsion.

the emulsion is exposed to an incident light beam that is (somehow) perfectly focused to a point on the emulsion, producing the impulse illumination function shown in Fig. 13.6a. The chemicals at x_0, y_0 will begin to react, producing the chemical changes that lead to changes in emulsion transmittance. This reaction, however, initiated at the point x_0, y_0, will begin to spread out and involve chemicals in neighboring regions of the emulsion as well. In other words, it is impossible to completely localize this chemical reaction to a single point. The reaction will always expand over some finite radius in the x, y plane, since the chemicals are relatively mobile, and a chemical reaction at one point will invariably draw chemicals from neighboring points into the reaction.

Thus, the chemical reaction at one point is dependent on both the illumination at that point and on the chemical reactions taking place at neighboring points in the emulsion. Therefore, the simple point-by-point description given in (13.1) is not general enough to describe the film development process accurately.

The reader will notice the similarity between the phenomenon of film development described above and the phenomenon of the point spread function of optical imaging systems, discussed in Chapter 5. In that discussion, we noted that a point source of light in the object plane gives rise to a blurred spot in the image plane, wherein the blurred spot was referred to as the point spread function. Therefore, the blurred impulse shown in Fig. 13.6b is a type of point spread function, except this time it is a point spread function for photographic film, not a lens.

If the incident light intensity in the x, y plane is given as a convolution of the form

$$I^{\text{inc}}(x, y) = \iint_{-\infty}^{\infty} I^{\text{inc}}(x, y)\, \delta(x - x_0)\, \delta(y - y_0)\, dx_0 dy_0 \qquad (13.2)$$

then the developed transmittance will be expressed in the form

$$\tau(x, y) = a - bt \iint_{-\infty}^{\infty} I^{\text{inc}}(x_0, y_0) PSF(x - x_0, y - y_0)\, dx_0 dy_0 \qquad (13.3)$$

that is, in convolution form.

We've seen how lens imaging action may be viewed either in convolutional terms (via the point spread function) or in spatial frequency terms (via the modulation transfer function). The same is true of photographic film; that is, film action may also be viewed in terms of an MTF. Note that since film responds only to optical field intensity, whether or not the incident illumination is coherent, film does not possess coherent transfer function properties, only MTF properties.

Consider Fig. 13.7, which shows a film emulsion illuminated by a square-wave illumination function. If the period of the square wave is much longer than the extent of the film point spread function, then the developed transparency will have substantially the same appearance as the intensity distribution shown. If, however, the period of the square wave begins to become comparable with the width of the film point spread function, the chemical reactions taking place in the illuminated regions of the film will begin to encroach upon the shadowed regions, causing a loss in contrast as shown. If the period of the square wave becomes finer still, a point will eventually be reached where the entire film gets developed and all contrast is lost.

The reduction in contrast is described in terms of a modulation transfer function that is analogous to the MTF of optical lenses. A typical curve for film MTF is shown in Fig. 13.8 [3]. Expressed in terms of MTF, the film transmittance is

$$\tau(\alpha) = K\delta(\alpha) - bt\, I(\alpha) \text{MTF}(\alpha) \qquad (13.4)$$

which is merely the Fourier transform of (13.3).

Note that the film MTF, as defined, is directly proportional to the film point spread function—not its square, as in the case of lens MTF.

Vander Lugt and Mitchell [5] describe one technique for evaluating film MTF. There are many others [6], but this one is useful to study because it involves some of the concepts of Fresnel region fields and transforms that we've looked at in previous chapters. With reference to Fig. 13.9, the Vander Lugt/Mitchell approach involves creating an off-axis Fresnel zone plate by forming an interference pattern between an on-axis spherical wave (of curvature $R/2$) of the form

$$E_{\text{spherical}} = e^{-jk(x^2 + y^2)/R}$$

and an obliquely incident plane wave of the form

$$E_{\text{planar}} = e^{j\alpha x}$$

The field intensity due to the sum of these two waves is

$$I^{\text{Tot}} = |E_p^2| + |E_s^2| + E_p E_s^* + E_p^* E_s$$

or

$$I^{\text{Tot}} = 2 + e^{j\alpha x} e^{-jk(x^2 + y^2)/R} + e^{-j\alpha x} e^{jk(x^2 + y^2)/R}$$

This last equation is plugged into (13.3)

Chapter 13 ■ Introduction to Holography

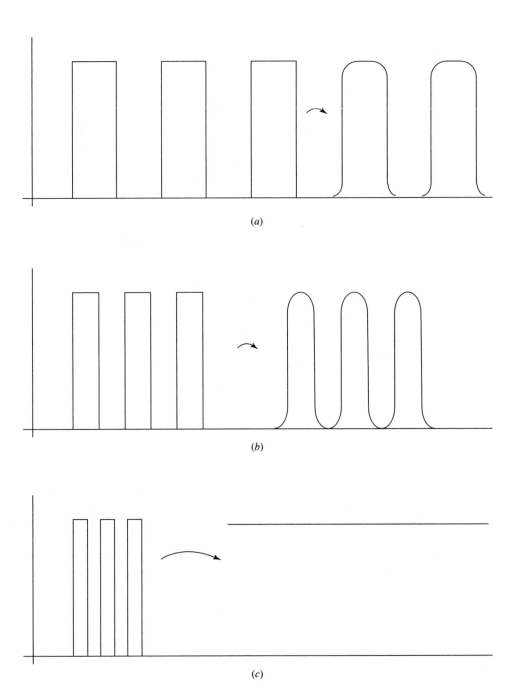

(a)

(b)

(c)

Figure 13.7 Loss of film contrast with increasing spatial frequency.

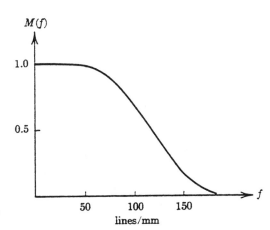

Figure 13.8 Typical modulation transfer function of film. (After Goodman [3].)

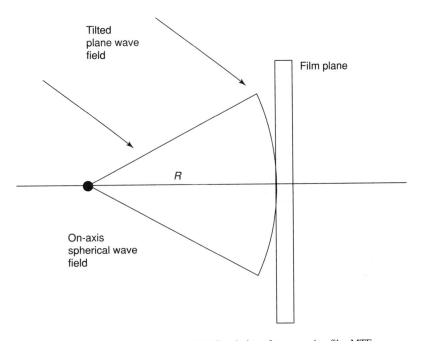

Figure 13.9 The Vander Lugt-Mitchell technique for measuring film MTF.

$$\tau(x, y) = a - bt \iint_{-\infty}^{\infty} I^{\text{inc}}(x_0, y_0) \, PSF(x - x_0, y - y_0) \, dx_0 dy_0 \qquad (13.5)$$

to produce

$$\tau(x, y) = a - bt \iint_{-\infty}^{\infty} \{2 + e^{j\alpha x_0} e^{-jk(x_0^2 + y_0^2)/R} + e^{-j\alpha x_0} e^{jk(x_0^2 + y_0^2)/R}\}$$

$$\times PSF(x - x_0, y - y_0) \, dx_0 dy_0 \qquad (13.6)$$

This developed transparency can now be placed in the front focal plane of a lens and illuminated by a normally incident plane wave. The plane wave is assumed to have uniform amplitude; hence, the field transmitted by the transparency is identically equal to the transmission function given above. The field in the back focal plane of the lens will, of course, be the Fourier transform of the field in the front focal plane, which is the transmittance function from (13.6). We can use two well-known Fourier transform properties to obtain this Fourier transorm at once, without using any math at all.

The first Fourier transform property we'll use is the convolution theorem, which states that the FT of a function that is expressed in convolution form is merely the product of the FTs of the two individual functions being convolved. In this case, the FT of the film PSF is the film MTF. The FT of the intensity function is given as the sum of the FTs of the 3 individual functions (in this case, we'll only be concerned with the third term; the other two will propagate in directions away from this term. By the shifting theorem, the FT of a function multiplied by a linear exponential is merely a translated FT of the function. Thus, the FT of the third term will appear in the $x > 0$ halfplane, laterally displaced in the rear focal plane by a distance proportional to α, the tilt of the plane wave field. The FT of the quadratic phase term is again a quadratic phase term, of different radius of curvature. (We note that this quadratic phase term has unit amplitude.)

Putting all of this together, we see that the FT of the third term on the RHS of (13.6) has a *magnitude* directly proportional to the film MTF, along with a spherical wave phase that won't concern us. If a photodetector is scanned throughout the rear focal plane of the lens, the signal level generated by the detector will then be directly related to the value of the film MTF. In this way, film MTF may be directly read out. Thus, this simple technique is useful for determining the MTF of a photographic film.

13.3 INTERFEROMETRIC VIBRATION ANALYSIS VIA WAVEFRONT RECONSTRUCTION

In the first section of this chapter, we indicated how a hologram may be produced by forming an interference pattern between two optical wavefronts on a photographic film. Implicit in the discussion is the necessity that the film and the two beams remain perfectly stationary during the exposure period, so that a high-contrast interference pattern can be produced (i.e., the transmittance in the transparent areas is nearly unity, and the transmittance in the opaque areas is nearly zero). At first glance, it would appear that any movement occurring during the exposure time would completely ruin the developed hologram.

In general, it's true that movement of various portions of the test setup does indeed tend to reduce the quality of the developed hologram. However, in one particular circumstance (namely, in the case of sinusoidal linear vibration), the quality of the resultant hologram is not ruined. Indeed, the hologram assumes a new form that reconstructs an interference pattern of contour plots of vibrational intensity.

This rather remarkable property of holograms was discovered early on in the development of modern holography, having first been published in 1965 by Powell and Stetson [7]. This process is somewhat similar to moiré contouring, in that constant-amplitude contours of a linearly vibrating object are mapped out, allowing the shape of the vibrational modes to be immediately visualized.

A sketch of one type of test setup for making the holograms is shown in Fig. 13.10. This setup is based on the Twyman-Green interferometer, whereas we've used the Mach-Zehnder interferometer in previous sections of this chapter. In the original Powell-Stetson paper, the vibrating object was the bottom of a film can driven by a solenoid located inside the can. We may gain a quick understanding of the hologram and reconstruction fields based on the discussions of the previous sections.

We'll assume that the hologram produces no spatial frequency components that lie outside the flat region of the film MTF curve in Fig. 13.8. Thus, we may drop the convolutional form (13.3) and write the transmittance versus exposure as

$$\tau(x, y) = \tau_0 - bt\, I(x, y)$$

where b is a constant and t is the exposure time.

Now, in an incremental period of time, an incremental exposure is produced as

$$d\tau(x, y) = \tau_0 - b\, I(x, y)\, dt$$

This equation forms the basis of our analysis of time-varying recording of holograms.

In this analysis, the reference field will be represented in the generic form

$$E_{\text{ref}}(x, y, z) = A(x, y, z)e^{j\psi_{\text{ref}}(x,y,z)}$$

and the time-varying field scattered by the vibrating object is approximated as a phase-modulated field of the form

$$E_{\text{scat}}^t(x, y, z) = E_{\text{scat}}(x, y, z)\, e^{j\psi_{\text{scat}}(x,y,z,t)}$$

where the superscript t indicates "time-varying." In other words, the amplitude of the scattered field is effectively constant over time, and the only characteristic of the field

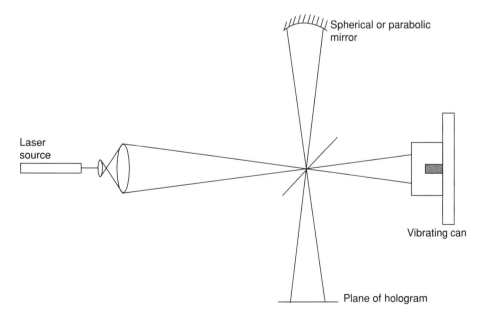

Figure 13.10 One possible type of optical setup (based on the Twyman-Green interferometer) for recording holograms for vibration analysis.

that is changing with time is the phase, which changes as the shape of the vibrating object changes (causing the path length to the object to change).

As shown in Fig. 13.11, the time-varying phase function is given as

$$\psi_{\text{scat}}(x, y, z, t) = k_0(\cos \theta_1 + \cos \theta_2) \, m(x, y) \cos[\Omega t + \psi_0(x, y)] \quad (13.7)$$

when the angles of incidence and reflection are not zero. (*Note:* We assume diffuse reflection from the scattering object, so that the angles of incidence and reflection may not necessarily be equal.) In the equation above, Ω = vibration frequency.

The instantaneous value of light intensity on the hologram film plate during recording is given as

$$I(x, y, t) = \left| E_{\text{ref}}(x, y) + E_{\text{scat}}^t(x, y, t) \right|^2$$

$$= \left| E_{\text{ref}} \right|^2 + \left| E_{\text{scat}}^t \right|^2 + E_{\text{ref}} E_{\text{scat}}^{t*} + E_{\text{ref}}^* E_{\text{scat}}^t \quad (13.8)$$

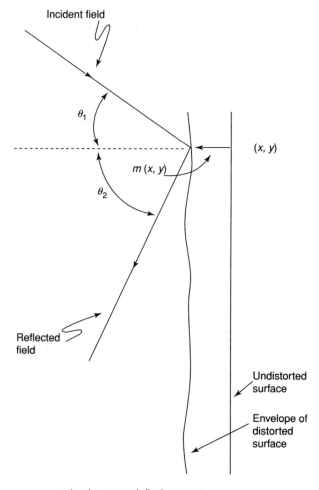

$m(x, y)$ = normal displacement

Figure 13.11 On the geometrical relationships for producing a hologram of a vibrating skin.

The transmittance of the developed hologram is given as the integral of the time-varying light intensity taken over the total exposure period. Thus, the total transmittance is given as

$$\tau(x, y) = \int_0^{t_{exp}} d\tau(x, y)$$
$$= \int_0^{t_{exp}} I(x, y) \, dt \qquad (13.9)$$

neglecting the constant bias term in the transmittance and the slope coefficient. Substituting (13.8) into (13.9) gives the transmittance of the developed hologram (neglecting the "baseband" terms not containing the reference field phase) as

$$I(x, y) = E_{ref}(x, y)E_{scat}^*(x, y) \int_0^{t_{exp}} e^{-j\psi_{scat}(x,y,t)} \, dt$$

$$+ E_{ref}^*(x, y)E_{scat}(x, y) \int_0^{t_{exp}} e^{j\psi_{scat}(x,y,t)} \, dt$$

We've explicitly written the time-varying phase above in (13.7); thus the two integrals take on the generic form

$$I = \int_0^{t_{exp}} e^{\pm jk_0(\cos\theta_1 + \cos\theta_2) \, m(x,y) \cos[\Omega t + \psi_0(x,y)]} \, dt$$

If the exposure time is an integral number of cycles of the vibrational frequency, Ω, then the generic integrals above take on the form of the Bessel function of order zero:

$$I = J_0[k_0(\cos\theta_1 + \cos\theta_2)m(x, y)] = J_0(\eta)$$

so that the hologram transmittance is readily obtained as

$$\tau(x, y) = K + [E_{ref}(x, y)E_{scat}^*(x, y) + E_{ref}^*(x, y)E_{scat}(x, y)]J_0(\eta)$$

where K is a constant term.

Thus, we see that the ordinary hologram transmittance is now modulated by the oscillatory Bessel function. The effect off this Bessel function modulation is to map contour lines onto the virtual image of the vibrating object, indicating contours of constant vibrational amplitude.

The function $J_0(u)$ is shown in Fig. 13.12 [23]. Thus, the transmittance of a point that does not move at all (i.e., for which $m(x, y) = 0$) will be much greater than the transmittance at a point where the deflection $m(x, y)$ corresponds to any of the successive extrema of the Bessel function. Therefore, the greater the magnitude of the motion, the less the contrast of the successive contour lines. (The light contour lines become darker and darker with increasing vibrational amplitude.) Hence, the further out the contour line, the less its visibility will be, until a limit is reached where the contours are no longer visible at all.

Typically, contrast is defined locally as

$$C = \frac{\text{Maximum intensity } - \text{ Minimum intensity}}{\text{Maximum intensity } + \text{ Minimum intensity}}$$

Therefore, the fringe contrast so defined is a maximum near $\eta = 0$ and decreases as η increases. This reduction of fringe visibility with increasing fringe order is shown in Fig.

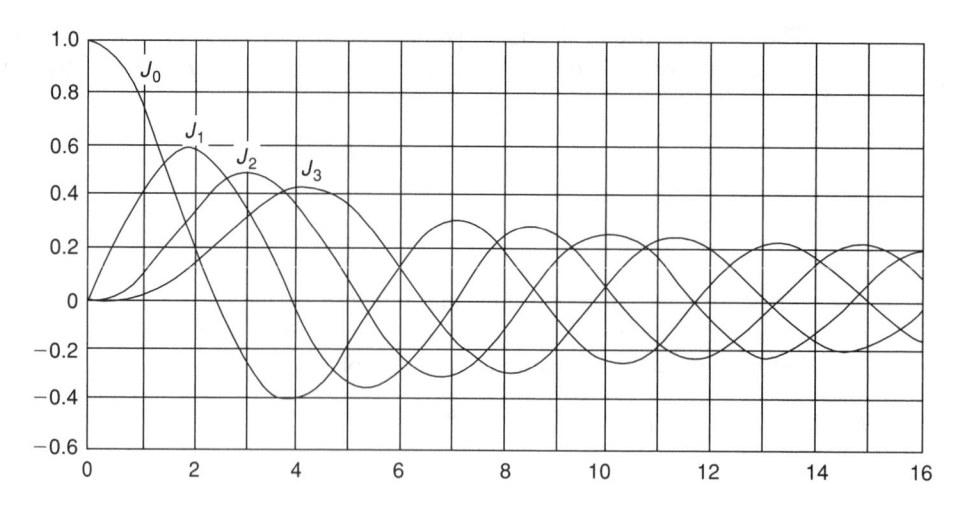

Figure 13.12 Bessel functions of the first kind. (After Harrington.)

13.13, which is taken from the original paper of Powell and Stetson [7]. (Note that fringe visibility varies as the *square* of the Bessel function.)

A clever variation on this theme has been proposed by Levitt and Stetson [8] wherein the phase $\psi_0(x, y)$ of vibration at a given point on the surface may be extracted by phase-modulating the field incident on the vibrating object (i.e., the field which, on reflection from the vibrating object, will become the "scattered field" for the hologram). The setup is shown in Fig. 13.14 (taken from the Levitt and Stetson article).

The optical phase due to the independently controlled vibrating mirror (which is assumed to be free of mechanical deformation and hence vibrates as a rigid body) is

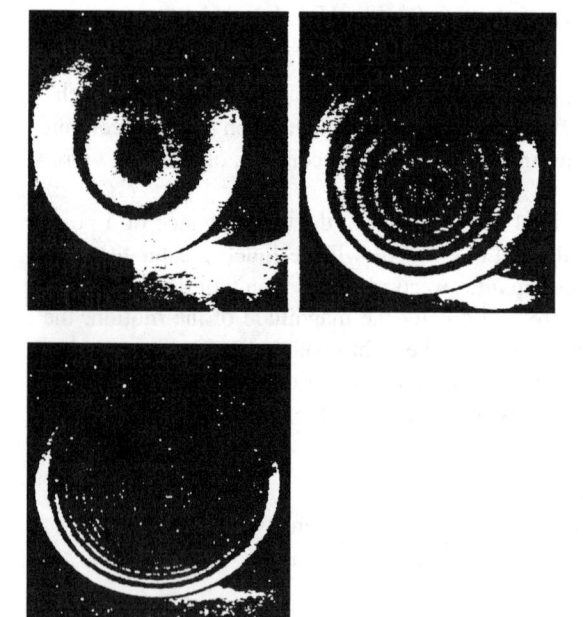

Figure 13.13 Reconstructions of three holograms of a 35 mm film-can bottom with a progressive increase in amplitude of excitation at the lowest resonance frequency of the can bottom. (After Powell and Stetson.)

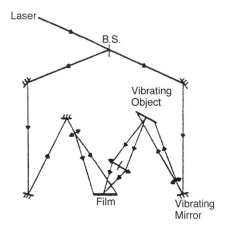

Figure 13.14 Hologram camera with a vibrating mirror used to phase modulate the object illumination. (After Levitt and Stetson [8].)

$$\phi_{mod} = \phi_{max} \cos(\Omega t + \psi_{mod})$$

and the optical field phase due to the vibration of the nonrigid body-under-test is

$$\phi_{vib} = k(\cos \theta_1 + \cos \theta_2) m(x, y) \cos[\Omega t + \psi_0(x, y)]$$

Adding the two phasor quantities—$\exp(j\phi_{mod})$ and $\exp(j\phi_{vib})$—together (e.g., using a phasor diagram) gives a new phase quantity

$$e^{jA\cos(\Omega t + B)}$$

where A is given by the law of cosines:

$$A = \phi_{max}^2 + \phi_{vib}^2 - 2\phi_{max}\phi_{vib} \cos(\psi_0 - \psi_{mod})$$

where

$$\phi_{vib} = k_0(\cos \theta_1 + \cos \theta_2)m(x, y)$$

So, using the original Powell/Stetson process, we obtain the hologram transmittance as a function of the total phase excursion A, going through various peaks and nulls as A increases. The light/dark fringe contours are now due not only to $m(x, y)$ but also to the phase of the vibrating mirror. Thus, by keeping $m(x, y)$ constant and varying the phase of the signal to the vibrating mirror, contours of constant phase may be mapped out on the vibrating object.

13.4 HOLOGRAPHIC INTERFEROMETRY

At the beginning of this chapter, we mentioned that one application of holography was in the area of interferometry, wherein interference patterns may be obtained between two objects that do not exist simultaneously, by storing the virtual image of one object in the form of a hologram and comparing it with the virtual image of another object at another time. This may be accomplished in a number of ways, using double-exposure holography, by cascading two holograms together, or by illuminating the hologram with multiple source and reference fields. In this section, we'll study some of these techniques.

By the square-law $t - E$ response function of photographic film, it is immaterial whether an interference pattern between two distinct virtual images is recorded simultaneously (during a single exposure) or sequentially (during two distinct exposures). This is true as long as the $t - E$ curve is never driven into saturation and the relationship between transmittance and intensity remains linear. A comparison between the two types of recording is shown schematically in Fig. 13.15.

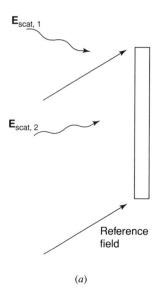

(a)

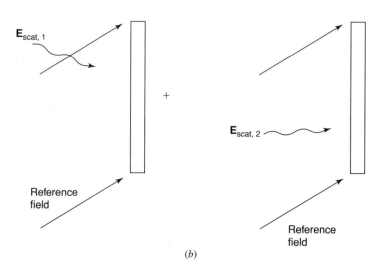

(b)

Figure 13.15 (a) Simultaneous recording versus (b) sequential recording of interference
patterns.

The case of sequential recording is analyzed simply by summing two intensity functions of the type derived in Section 13.1. Thus, the total transmittance of a twice-exposed hologram is a special case of the time-varying exposure discussed in the previous section and given in the usual way as (where we assume linear recording and neglect the constant bias term in the expression for the $t - E$ curve)

$$\tau(x, y) = (E_1 + E_{ref})(E_1^* + E_{ref}^*) + (E_2 + E_{ref})(E_2^* + E_{ref}^*).$$

When

$$E_{ref} = e^{j\alpha x}$$

then

$$\tau(x, y) = \text{on-axis field} + e^{j\alpha x}(E_1^* + E_2^*) + e^{-j\alpha x}(E_1 + E_2)$$

For simultaneous holographic recording of the interference pattern between the two fields, the total transmittance of the singly exposed hologram is

$$\tau(x, y) = (E_1 + E_2 + E_{ref})(E_1^* + E_2^* + E_{ref}^*).$$

When

$$E_{ref} = e^{j\alpha x}$$

then

$$\tau(x, y) = \text{on-axis field} + e^{j\alpha x}(E_1^* + E_2^*) + e^{-j\alpha x}(E_1 + E_2)$$

So, as far as the spatially heterodyned fields propagating in the directions exp $(\pm j\alpha x)$ are concerned, there is no difference between the simultaneous and sequential recording (as long as the film is never driven into saturation). As a result, an interference pattern between two optical fields may be recorded in two steps, at two different points in time, between two fields that need not exist simultaneously.

Many other types of optical interference phenomena may be recorded holographically. (Historically, one of the major application areas of holography has been in interferometry.) For example, when two reference fields are used, a Mach-Zehnder interferometer may be synthesized for holographically recording the interference pattern between a plane wave and a phase-distorted wave [9]. The conventional Mach-Zehnder is shown in Fig. 13.16 and the holographic simulation of the interferometer is shown in Figs. 13.17 and 13.18. In the first case, the hologram transmittance is obtained by expanding the expression

$$\tau(x, y) = \left| E_s + e^{j\alpha x} + {}^{-j\alpha x} \right|^2$$

with the resultant diffracted fields shown in Fig. 13.18a. In the second case,

$$\tau(x, y) = \left| E_s + e^{j\alpha x} + e^{j2\alpha x} \right|^2$$

and the resulting diffracted fields are shown in Fig. 13.18b. In the second case, a Mach-Zehnder has been effectively produced holographically, using two reference beams in the recording process. In the first case, the object wave and its conjugate have been produced, creating an interference pattern having twice the sensitivity of an ordinary Mach-Zehnder.

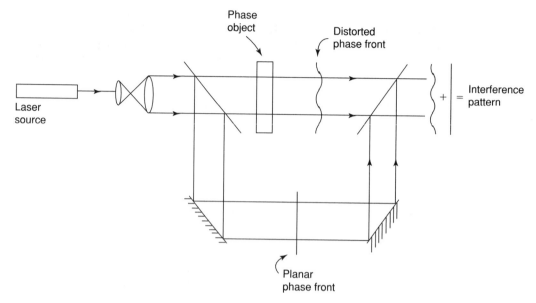

Figure 13.16 Conventional Mach-Zehnder interferometer for observation of phase objects.

13.5 DIFFERENCE AND SUBTRACTIVE HOLOGRAPHY

An additional application of holography in the area of interferometry involves the generation of holograms that indicate differences between two different virtual images. Generally in this context, it is assumed that certain portions of the image will change, while other portions remain unchanged. The idea is to isolate those portions of the image that have changed from those that have not. One way of accomplishing this is via difference holography [10]. The principle is quite straightforward. Difference holography is a form of double-exposure holography wherein a half-wave plate is inserted in the reference arm between exposures. Thus, the transmittance of the twice-exposed hologram is

$$\tau(x, y) = (E_1 + E_{\mathrm{ref}})(E_1^* + E_{\mathrm{ref}}^*) + (E_2 - E_{\mathrm{ref}})(E_2^* - E_{\mathrm{ref}}^*).$$

When

$$E_{\mathrm{ref}} = e^{j\alpha x}$$

then

$$\tau(x, y) = \text{on-axis field} + e^{j\alpha x}(E_1^* - E_2^*) + e^{-j\alpha x}(E_1 - E_2)$$

Hence, the *difference* between the two virtual images is produced on reconstruction.

Another technique of obtaining differences between images has been termed *holographic subtraction* and involves the use of two beams on reconstruction. This is another type of three-beam holography of the type discussed in the previous section in connection with interferometer synthesis. The technique described in [10] is useful for performing

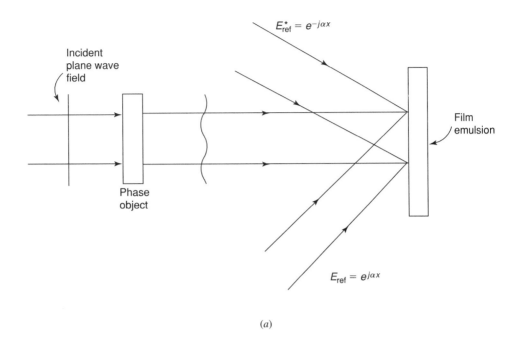

(a)

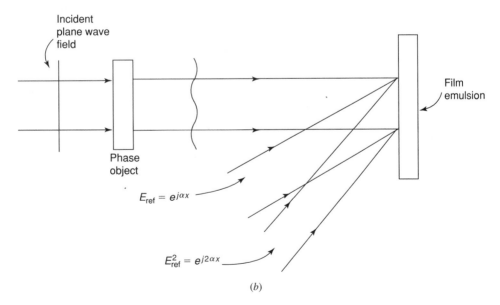

(b)

Figure 13.17 Holographic recording of Mach-Zehnder interferometer: (a) Symmetric reference fields and (b) offset reference fields.

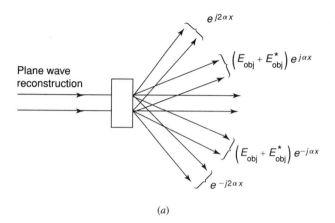

(a)

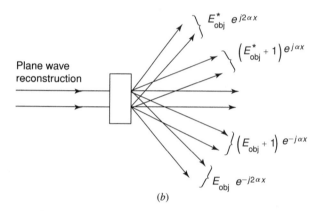

(b)

Figure 13.18 Holographic reconstruction of Mach-Zehnder interferometer: (a) Symmetric reference beams and (b) offset reference beams.

interferometry between a holographically stored virtual image and a real-time virtual image. The principle is relatively straightforward. The hologram recording process is performed in the usual way, as shown in Fig. 13.19. The transmittance of the developed hologram is given as usual as (here, we use the exact expression for the $t - E$ curve, since the constant bias term comes into play in the reconstruction process):

$$\tau(x, y) = \tau_0 - Bt\, I(x, y)$$

or

$$\tau(x, y) = \tau_0 - Bt[1 + E_1 E_1^* + E_1 e^{-j\alpha x} + E_1^* e^{j\alpha x}]$$

If the developed hologram is illuminated using the plane wave field $\exp(j\alpha x)$, as shown in Fig. 13.19b, then the field E_1 is produced on-axis. If, in addition, this hologram is illuminated by the field E_2—the virtual image field scattered from a second object—then, the on-axis field consists not only of the first-order diffracted field E_1 arising from the off-axis planar reference wave, but it also consists of the undiffracted field E_2 from the second object passing straight through the hologram. If the magnitude of the reference wave is much greater than the magnitude of the scattered object wave E_1, then the total on-axis field is given approximately by

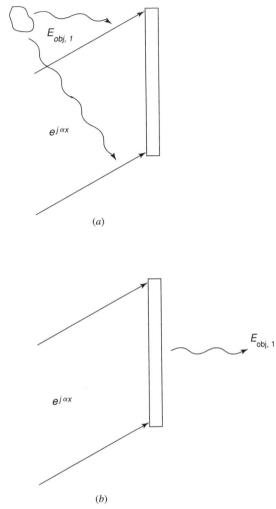

$E_{obj, 1}$

$e^{j\alpha x}$

(a)

$e^{j\alpha x}$

$E_{obj, 1}$

(b)

Figure 13.19 Recording/reconstruction process for real-time holographic interferometry: (a) Recording and (b) reconstruction.

$$E^{\text{Tot}} = (-Bt)E_1 + (\tau_0 - Bt)E_2$$

Clearly, by judicious choice of the exposure time, the coefficients of the two fields can be equated, so that a (subtractive) interference pattern between them may be obtained. This subtractive pattern allows for real-time comparison between stored virtual image and a real-time virtual image.

13.6 SANDWICH HOLOGRAMS

Interference patterns may be obtained not only by adding holographically reconstructed fields but also by multiplying them. This is done by overlaying two holograms. This process has been discussed in a series of articles [11–14].

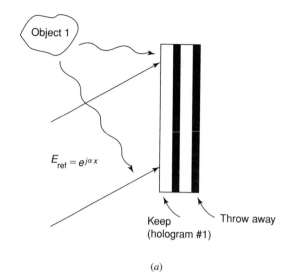

(a)

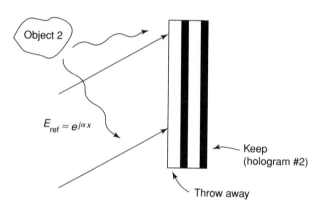

(b)

Figure 13.20 On the concept of sandwich holography: recording the two halves of the sandwich hologram: (a) First exposure and (b) second exposure (with two new plates).

The process is straightforward and is illustrated in Figs. 13.20 and 13.21. Sandwich holography is a type of double-exposure holography in which two overlayed film plates are first exposed by the reference wave and first object field. One of these plates is then saved and the other discarded. Now two more plates are placed in the setup and are illuminated by the reference wave and the second object field. Again, one plate is saved from the second exposure and one discarded, in such a way that opposite plates are saved from the two exposures. When the two complementary plates from the two exposures are placed together and illuminated by a reconstruction beam (equal to the reference beam during recording), the on-axis field is the sum (i.e., interference pattern) of the two object fields. This is shown mathematically as follows. The total transmittance of the two developed plates is given as

$$\tau^{\text{Tot}} = \tau_1 \tau_2 = \{1 + |E_1^2| + E_1 e^{-j\alpha x} + E_1^* e^{j\alpha x}\}$$
$$\cdot \{1 + |E_2^2| + E_2 e^{-j\alpha x} + E_2^* e^{j\alpha x}\}$$

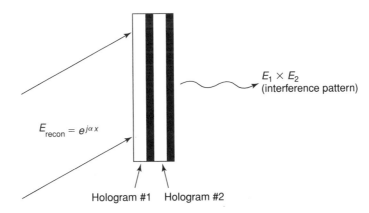

Figure 13.21 On the concept of sandwich holography: Reconstructing the interference pattern.

Illumination of this hologram by the reference field: $\exp(j\alpha x)$ produces the on-axis interference pattern between the two virtual images.

13.7 HOLOGRAPHIC OPTICAL ELEMENTS: THE HOLOGRAPHIC TWYMAN-GREEN INTERFEROMETER

We've seen how a Mach-Zehnder interferometer may be realized using holography. In this section, we'll examine a holographic Twyman-Green interferometer [15]. The central element in the interferometer is a holographically generated off-axis Fresnel zone plate. This zone plate is conceptually similar to the zone plates discussed in Chapter 2. In fact, that plate is an example of a *binary hologram* having a transmittance function equal to 0 or 1. The off-axis Fresnel zone plate (OFZP) is generated by recording the interference pattern between an off-axis plane wave field and an on-axis spherical wave field, as shown in Fig. 13.22 (or, vice versa, as in the original article). Upon re-illumination by the reference field, this plate produces the usual three waves: an approximately planar undiffracted wave, the object wave (in this case, a converging spherical wave) and a conjugate object wave (in this case, a diverging spherical wave). In the present implementation of the Twyman-Green, we're interested in the undiffracted field and the converging spherical wave, as shown in Fig. 13.23. The OFZP is illuminated by a tilted plane wave, producing a tilted plane wave and an on-axis converging spherical wave on transmission. The essentially planar undiffracted wave is reflected from a planar mirror, while the converging spherical wave field is passed through a focal point and reflected from an imperfect spherical test mirror, as shown in Fig. 13.23a. When both reflected fields re-illuminate the hologram from the back, each produces an on-axis, diverging spherical wave field, as shown in Fig. 13.23b. The on-axis spherical wave field consists of a perfect spherical wave summed with the aberrated spherical wave field reflected from the mirror, and thus a Twyman-Green interferometer is produced.

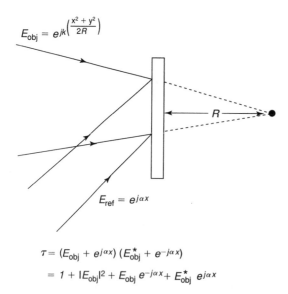

$$E_{obj} = e^{jk\left(\frac{x^2 + y^2}{2R}\right)}$$

$$E_{ref} = e^{j\alpha x}$$

$$\tau = (E_{obj} + e^{j\alpha x})(E_{obj}^* + e^{-j\alpha x})$$
$$= 1 + |E_{obj}|^2 + E_{obj}\, e^{-j\alpha x} + E_{obj}^*\, e^{j\alpha x}$$

Figure 13.22 Off-axis Fresnel zone plate: Recording.

13.8 CORRECTION OF LENS ABERRATIONS VIA HOLOGRAPHY

Since an ordinary two-beam hologram generates a conjugate image of the object field, this conjugate image may be used for the purpose of canceling out lens aberrations. If, for example, the object field is the field passed through an optical system having aberrations, this aberrated field may be recorded holographically using an offset planar reference field as shown in Fig. 13.24 (using a telescope beam expander as the optical system under test). The transmittance is

$$\tau(x, y) = 1 + |E_{aber}|^2 + e^{-j\alpha x} E_{aber} + e^{j\alpha x} E_{aber}^*$$

If this hologram is then re-illuminated by the (aberrated) field passed by the system, the field in the direction of $\exp(j\alpha x)$ is perfectly corrected by virtue of the multiplication of the aberration function by its conjugate.

13.9 PHASE-ONLY HOLOGRAMS

So far in this chapter, we've only dealt with amplitude holograms that have a transmittance function modulated by the intensity of the optical fields incident on the hologram plate. It turns out that if an amplitude hologram is recorded on certain types of bleachable emulsions, the amplitude variations may be transformed into phase (index) variations. In the type of negative exposure films we've been dealing with, in which increased intensity translates into increased color change (reduced transmittance), the color changes may be altered via bleaching techniques into refractive index changes; the darker the original film, the higher the index of the bleached film.

These types of phase-only holograms have certain applications in image processing

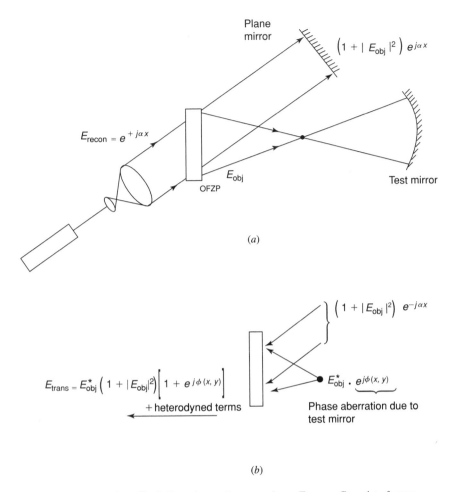

Figure 13.23 Using off-axis Fresnel zone plate to produce a Twyman-Green interferometer: (a) Illumination of OFZP by laser and (b) second illumination of OFZP by reflected fields.

applications wherein the loss in transmitted light intensity due to the darkened portions of the hologram is significant. When transmitted light intensity is important, phase holograms can be useful. If an amplitude hologram is analogous to an AM radio system, in which an amplitude-modulated signal wave is heterodyned onto a sinusoidal carrier, then a phase hologram is analogous to a phase, or frequency-modulated system, in which a sinusoidal carrier is phase modulated by the signal waveform.

The standard amplitude hologram generates three optical fields: the strong undiffracted field and two first-order diffracted fields consisting of the object field and its conjugate. In the case of a phase hologram, the transmittance phase takes on the form

$$\tau(x, y) = e^{j[1 + E_{obj}E_{obj}^{\pm} + 2E(x,y)\cos(\alpha x + \phi(x,y))]}$$

where it is assumed that

$$E_{obj}(x, y) = E(x, y)\, e^{j\phi(x,y)}$$

and

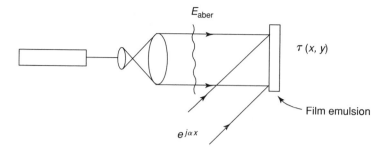

$$\tau\,(x,\,y) = 1 + |E_{aber}|^2 + E_{aber}\,e^{-j\alpha x} + E^*_{aber}\,e^{j\alpha x}$$

(a)

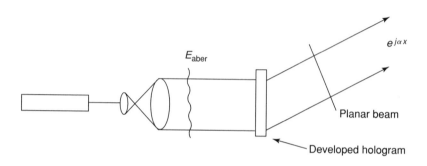

If $E_{inc} = E_{aber}$, $E_{trans} = e^{j\alpha x} +$ fields propagating in other directions.

(b)

Figure 13.24 Holographic correction of aberrations in a telescope beam expander: (a) Recording the hologram and (b) using the hologram for lens correction.

$$E_{ref}(x,\,y) \,=\, e^{j\alpha x}$$

If we now re-illuminate this hologram by multiplying this transmittance function by the original reference wave, we no longer obtain the three nicely separated fields from amplitude holography. What we have to do now is Fourier-analyze the resulting transmitted field (in α space) in order to obtain the many diffracted fields. What we find is that integrals of the form studied in Chapters 1 and 2 arise, which may be solved (as before) in terms of Bessel functions. The resulting expressions are not very informative, except in the case of small index variations, wherein linearity can be used to separate the contributions due to the three terms [16]. We won't undertake a mathematical analysis of phase holograms in this book, other than to say that the theoretical diffraction efficiency of a phase hologram is roughly five times that of a thin amplitude hologram [17]. The mathematical analysis for small (linear) phase deviations proceeds in essentially the same fashion as for amplitude holograms.

13.10 EVANESCENT WAVE HOLOGRAMS

In addition to using propagating plane waves for hologram recording and reconstruction, it is also possible to use surface wave fields of the type that propagate in thin film wave-guides and evanescent mode fields (see Chapter 2). In this section, we give a brief overview of holograms produced and viewed in this fashion.

In the production of the Leith-Upatnieks hologram, we've seen the necessity of combining the virtual image field of an illuminated object with a "reference" field of the form $\exp(j\alpha x)$—that is, a field that propagates in the plane of the film emulsion. The formation of the Leith-Upatnieks hologram, however, does not at all depend on the functional nature of this field in the direction *normal* to the plane of the film. In the normal direction, the field may be a uniform propagating plane wave, or it may be an exponentially decaying evanescent mode field; it doesn't matter.

Thus, there is absolutely no theoretical reason why a hologram couldn't be just as well produced with an evanescent or thin film waveguide optical field as with an ordinary propagating plane wave field inclined at some angle with respect to the plane of the hologram. In actual practice, there are numerous ways in which to record and view a hologram using evanescent mode fields [18,19,20].

One of the earliest such techniques is shown in Fig. 13.25 [18]. The film plate consists of a glass base covered with a thin emulsion film. The index of the base is less than the index of the emulsion and the entire plate is immersed in a high-index liquid. An obliquely incident propagating plane wave field is made incident on the emulsion side of the plate as shown. The angle of incidence is chosen in such a way that the field propagates in the high-index liquid, but is just beyond cutoff in the emulsion.

In the case of the setup shown in Fig. 13.25 (taken from [18]), the index of the surrounding liquid was given as 1.73 and the angle of inclination from normal in the liquid was given as 68.5°. This was the angle at which the reflected field magnitude suddenly became nearly unity, indicating a cutoff condition in the emulsion. Phase matching at the interface (i.e., enforcing Snell's law) gives the experimentally determined value of the index of the emulsion as

$$n_1 \sin \alpha_1 = n_2 = 1.61$$

At this angle of incidence, the incident optical field (which is a propagating plane wave in the high-index fluid) is an exponentially decaying optical field in the lower-index film emulsion. As in the ordinary Leith-Upatnieks hologram, the transmittance of the developed hologram depends on the intensity of the total interference field in the emulsion (which, for the reference field, is a decaying function versus distance into the emulsion). Thus,

$$
\begin{aligned}
\tau(x, y) &= (e^{j\alpha x}e^{-\sigma z} + E_{\text{obj}}) \cdot (e^{-j\alpha x}e^{-\sigma z} + E_{\text{obj}}^*) \\
&= e^{-2\sigma z} + E_{\text{obj}}E_{\text{obj}}^* + E_{\text{obj}}e^{-j\alpha x}e^{-\sigma z} + E_{\text{obj}}^*e^{j\alpha x}e^{-\sigma z}
\end{aligned}
$$

and we see that, apart from the second term in the expression, the entire hologram is

Recording

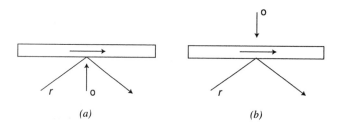

Reconstruction

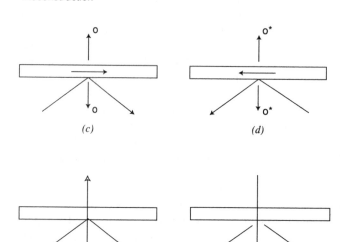

Figure 13.25 Recording and reconstruction schemes for holograms formed by interference between an ordinary object wave o and an evanescent reference wave r. The hologram recorded in (a) as well as (b) gives the same reconstructed waves—true wave field o and complex-conjugate wave field o^*—using the arrangements shown in (c), (d), (e), and (f). (After Bryngdahl.)

confined to the region near the outer surface of the emulsion. In this way, a thin hologram is always produced, regardless of the thickness of the emulsion.

On reconstruction, the developed hologram plate is re-illuminated with the original reference wave field. Thus, the field transmitted by the hologram illuminated in this way is

$$E^{\text{trans}}(x, y) = e^{j\alpha x}(e^{-2\sigma z} + E_{\text{obj}}E^*_{\text{obj}}) + E_{\text{obj}}e^{-\sigma z} + E^*_{\text{obj}}j^{2\alpha x}e^{-\sigma z}$$

We note that the term involving E_{obj} is the only one that is not multiplied by a term of the form $\exp(\pm j\alpha x)$. Hence, only that term can propagate out the back side of the film plate. The other terms can still propagate out the front side of the film plate, which is in direct contact with the high-index liquid. When the hologram is illuminated by the conjugate of the original reference field, the conjugate object field is produced, as in the case of the ordinary Leith-Upatnieks hologram. Other combinations of recording and readout waves are shown in Fig. 13.26 from [18].

Rather than using a propagating plane wave field in a high-index liquid to produce an evanescent mode field in the film emulsion, the reference and readout waves may be produced using the thin film slab waveguide shown in Figs. 13.27 and 13.28 from [19].

Recording

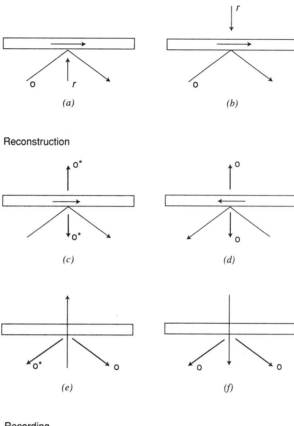

Reconstruction

Recording

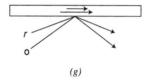

Reconstruction

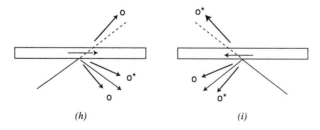

Figure 13.26 Recording and reconstruction schemes for a hologram formed by interference between an evanescent object wave o and an evanescent reference wave r. Parts (b) and (c) show the directions of the reconstructed true wave field o and complex-conjugate wave field o^* for opposite directions of the illuminating evanescent wave. (After Bryngdahl.)

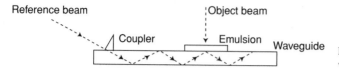

Figure 13.27 The principle of recording a waveguide hologram. (After Putilin et al.)

The advantage of this waveguide reference field over the plane wave field in a high-index liquid is that in the former case, *all* extraneous fields are cut off from propagation in both directions away from the plane of the hologram, whereas in the latter, they are cut off only in the transmission direction, as mentioned in the previous paragraph.

13.11 VOLUME HOLOGRAMS

Volume amplitude holograms cannot be analyzed mathematically in closed form, and volume phase holograms can only be analyzed for the case of planar object and reference waves [21]. Even then, the analysis is extremely involved (and incidentally, closely related to the analysis of bulk acousto-optic cells [22]). The easiest way of visualizing the effect of thick holograms is to regard them as a type of three-dimensional periodic lattice wherein strong diffracted fields are produced at the Bragg angles. The effect of the planar object and reference fields is to produce a volume diffracting lattice. When illuminated by the reconstruction beam, the lattice produces a strong diffracted field in the direction of the

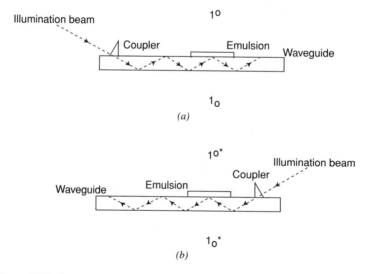

Figure 13.28 The reconstruction of thin waveguide holograms. (*a*) Two ordinary object waves *O* are generated on both sides of the waveguide with the guided reconstruction beam the same as the reference beam for recording. (*b*) Two conjugate object waves *O** are generated with the guided wave traveling in the opposite direction of the reference beam for recording. (After Putilin et al.)

original object beam. The complex analysis [21] indicates a theoretical diffraction efficiency of 100% for volume phase holograms.

13.12 COLOR HOLOGRAPHY

A recent article [25] describes a new process for creating full-color holograms having true color reproduction and wide viewing angles. The system uses three lasers (a 632.8 nanometer laser for red, a 488 nm laser for blue and a 532 nm laser for green light) to create a ''white light'' reflection hologram in three sequential exposures.

REFERENCES

[1] Born, M., and Wolf, E., *Principles of Optics,* New York: Pergamon, 1980.

[2] Stroke, G. W., *An Introduction to Coherent Optics and Holography,* New York: Academic Press, 1969.

[3] Goodman, J., *Introduction to Fourier Optics,* New York, McGraw-Hill, 1968.

[4] Brandt, G. B., ''Hologram-Moire Interferometry for Transparent Objects,'' *Applied Optics,* Vol. 6, No. 9, September 1967, pp. 1535–1540.

[5] Vanderlugt, A., and Mitchell, R. H., ''Technique for Measuring the Modulation Transfer Functions of Recording Media,'' *J. Opt. Soc. Am.,* Vol. 57, No. 3, March 1967, pp. 372–379.

[6] Grumet, A., ''MTF Evaluation with Optical Matched Filters,'' *Applied Optics,* Vol. 16, No. 1, January 1977, pp. 154–159.

[7] Powell, R. A., and Stetson, K. A., ''Interferometric Vibration Analysis by Wavefront Reconstruction,'' *J. Opt. Soc. Am.,* Vol. 55, No. 12, December 1965, pp. 1593–1598.

[8] Levitt, J. A., and Stetson, K.A., ''Mechanical Vibrations: Mapping Their Phase with Hologram Interferometry,'' *Applied Optics,* Vol. 15, No. 1, January 1976, pp. 195–199.

[9] De, Manoranjan, and Sevigny, Leandre, ''Three-Beam Holographic Interferometry,'' *Applied Optics,* Vol. 6, No. 10, October 1967, pp. 1665–1671.

[10] Collins, Leland, ''Difference Holography,'' *Applied Optics,* Vol. 7, No. 1, January 1968, pp. 203–205.

[11] Abramson, N., ''Sandwich Hologram Interferometry: A New Dimension in Holographic Comparison,'' *Applied Optics,* Vol. 13, No. 9, September 1974, pp. 2019–2025.

[12] Abramson, N., ''Sandwich Hologram Interferometry 2: Some Practical Calculations,'' *Applied Optics,* Vol 14, No. 4, April 1975, pp. 981–984.

[13] Abramson, N., ''Sandwich Hologram Interferometry 3: Contouring,'' *Applied Optics,* Vol. 15, No. 1, January 1976, pp. 200–205.

[14] Abramson, N., Bjelkhagen, H., and Skande, P., ''Sandwich Holography for Storing Information Interferometrically with a High Degree of Security,'' *Applied Optics,* Vol. 18, No. 12, June 15, 1979, pp. 2017–2021.

[15] Chen, C. W., Wyant, J. C., and Breckenridge, J. B., "Holographic Twyman-Green Interferometer," *Applied Optics,* Vol. 24, No. 6, March 15, 1985, p. 736.

[16] Cathey, W. T., "Spatial Phase Modulation of Wavefronts in Spatial Filtering and Holography," *J. Opt. Soc. Am.,* Vol. 56, No. 9, September 1966, pp. 1167–1171.

[17] Urbach, J. C., and Meier, R. W., "Properties and Limitations of Hologram Recording Materials," *Applied Optics,* Vol. 8, No. 11, November 1969, pp. 2269–2281.

[18] Bryngdahl, O., "Holography with Evanescent Waves," *J. Opt. Soc. Am.,* Vol. 59, No. 12, December 1969, pp. 1645–1650.

[19] Putilin, A. N., Morozov, V. N., Huang, O., and Caulfield, H. J., "Waveguide Holograms with White Light Illumination," *Opt. Engr.,* Vol. 30, No. 10, October 1991, pp. 1615–1619.

[20] Metz, M., "Edge-Lit Holography Strives for Market Acceptance," *Laser Focus World,* May 1994, pp. 159–163.

[21] Burckhardt, C. B., "Diffraction of a Plane Wave at a Sinusoidally Stratified Dielectric Grating," *J. Opt. Soc. Am.,* Vol. 56, No. 11, November 1966, pp. 1502–1509.

[22] Scott, C. R., *Field Theory of Acousto-Optic Devices,* Norwood, MA: Artech House, 1992.

[23] Harrington, R. F., *Time-Harmonic Electromagnetic Fields,* New York: McGraw-Hill, 1960.

[24] Jenkins, F. A., and White, *Fundamentals of Optics,* New York, McGraw-Hill, 1976.

[25] Murray, J., "Laboratory Focuses on New True-Color Holography," *Laser Focus World,* December 1994, pp. 23–25.

PART V
INTRODUCTION TO OPTICAL INFORMATION PROCESSING

14

A Sampling of Optical Information Processing Systems Based on Fourier Plane Filtering

In Chapter 5, we introduced the Fourier transforming property of thin lenses in the Fresnel region. We showed that if an object transparency is placed in the front focal plane of a lens, the Fourier transform of that transparency function is formed in the rear focal plane (with the correct phase relationships between the various Fourier components). That relatively simple property of thin lenses contains within it immense implications for designing optical processing systems. For example, since the Fourier transform of a two-dimensional object scene can be produced optically (in parallel, and for all spatial frequency components of the object scene *simultaneously, at the speed of light*), this immediately suggests the possibility of producing an optical Fourier transform machine capable of incredible processing speeds. If the Fourier transform amplitudes only were required, these could be read out directly by a detector array (followed by A/D converters) in the Fourier transform plane. The accuracy of the Fourier transform would be determined by the number of bits of precision in the A/D's. If Fourier transform phases were required as well, these could be determined interferometrically, using techniques discussed later in this chapter.

Perhaps more important than the operation of Fourier transforming, however, are the twin operations of optical convolution and correlation. These operations are widely used in character and pattern recognition, object detection and identification, feature extraction, machine vision, optical associative memories and optical neural networks. These operations can be realized optically by Fourier transforming one transparency function, multiplying that transform by the transform of a second transparency function, and then inverse Fourier transforming the resultant product of the two transforms. We'll look at two separate ways of performing optical correlations using the traditional "4f" correlator (4 focal lengths long) and the joint transform correlator.

The field of optical filtering dates back more than a hundred years to early experiments performed by Ernst Abbe in 1893 and even today remains a topic of significant research interest in such diverse areas as optical information processing (correlation and convolution devices), optical associative memories, and optical neural networks. Even optical techniques for evaluating wavelet transforms are being studied using optical filtering technology. The sustained interest in optical filtering stems from the development of modern spatial light modulators which allow optical filters to be synthesized under digital control in real time.

14.1 HISTORY OF FOURIER PLANE FILTERING

The first experiments into the effects of Fourier plane filtering were conducted by Ernst Abbe over a century ago—long before the advent of coherent light sources. Yet they clearly illustrated the optical signal processing powers of Fourier plane filtering. Goodman [1] gives a good account of these experiments, and the discussion in this section closely follows his description.

Abbe was evidently the first to recognize the equivalence of decomposing an object transmittance function into a Fourier transform spectrum and into a spectrum of propagating and evanescent plane waves. From that observation, it was a small step to the realization that if the object was regarded as comprised of a spectrum of plane waves, one could model the imaging process not by tracing rays (as in the classical picture of optical imaging), but by following the progress of the various constituent plane wave components of the transparency through the optical system. Tracing plane wave components is a relatively straightforward matter since tilted plane wave components must all (to a first order) converge at various laterally displaced points in the focal plane (which, by the lens law, always lies between the object and image planes).

Abbe was apparently not aware of the Fourier transform relationship between objects and their corresponding "images" when the two are located in the front and back focal planes, respectively, of a lens. Nevertheless, he still pioneered the development of the · plane wave spectrum viewpoint in optics. A sketch of his experimental setup is shown in Fig. 14.1 [1]. In this setup, the various spectral components that are focused in the rear focal plane will not have the true phase relationships of the Fourier components contained in the original image. (There is a residual quadratic phase distribution added to the true phase distribution, as was shown in Chapter 5.) But then Abbe's experiments of a hundred years ago used both ordinary incoherent light and amplitude-only filters anyway, so the quadratic phase wouldn't have affected his experiments.

In the figure, Abbe's object transparency was chosen to be a mesh grating. As discussed in Chapter 10, the transmittance function is periodic in the x- and y-directions; hence, it is expressible in terms of a two-dimensional Fourier series. In other words,

$$\tau(x, y) = \sum_{m,n} F(\alpha_m, \beta_n)\, e^{j(\alpha_m x + \beta_n y)}$$

where

$$\alpha_m = 2m\pi/\Delta x$$
$$\beta_n = 2n\pi/\Delta y$$

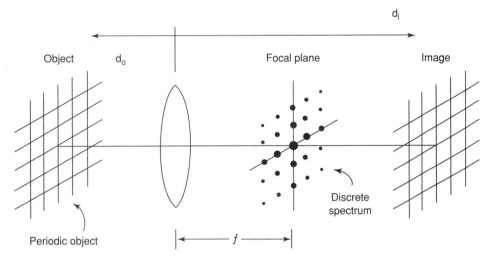

Figure 14.1 The Abbe-Porter experiment. (After Goodman.)

and

Δx, Δy are the periodicities of the mesh in the x-, y-directions, respectively.

Clearly, from the Fourier transform representation of the transmittance function, the Fourier spectrum of the mesh is a discrete function (a 2-D sequence of delta functions) in the x-, y-directions. Pictures of the mesh and its Fourier transform amplitudes are shown in Fig. 14.2 [1]. Unlike the mathematical Fourier transform, the optical Fourier transform does not consist of a sequence of delta functions, but rather a grid of spots of finite extent. These spots are caused by the finite aperture (point spread function) of the lens.

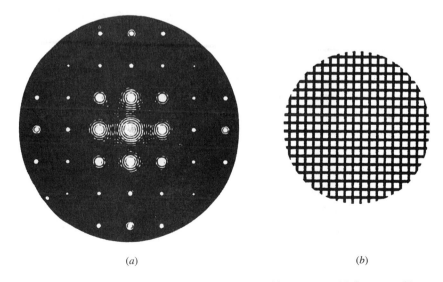

(a)   (b)

Figure 14.2 Photograph of the unmodified mesh and its spectrum: (a) Spectrum; (b) image. (After Goodman.)

Abbe placed one-dimensional, binary, amplitude-only (slit) filters in the Fourier transform plane (FTP) and then observed the resultant image-plane patterns that were formed from the *filtered* object spectra. In effect, he was multiplying the 2-D FT of the mesh by a 1-D pulse transmittance function, to remove all periodicity in the direction orthogonal to the slit and create a 1-D spectrum and image. Such 1-D filtered FTP spectra are shown in Figs. 14.3 and 14.4 along with their corresponding images. As we saw in Chapter 10, the images and spectra in Figs. 14.3 and 14.4 are indeed FT pairs.

Mathematically, Abbe reduced the 2-D periodic mesh transmittance function to the 1-D periodic function,

$$\tau(x, y) = \sum_m F(\alpha_m, \beta_0)e^{j(\alpha_m x + \beta_0 y)}$$

by removing all Fourier orders corresponding to $\beta \neq 0$.

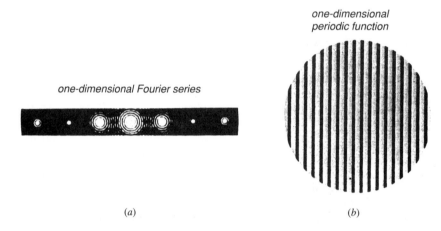

(a)   (b)

Figure 14.3 Mesh filtered with a horizontal slit: (a) Spectrum; (b) image.

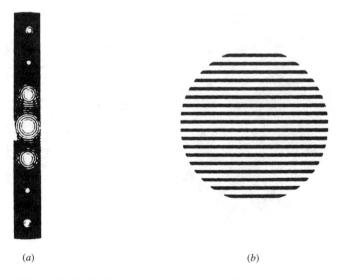

(a)   (b)

Figure 14.4 Mesh filtered with a vertical slit: (a) Spectrum; (b) image.

Abbe was very clever in using a doubly periodic transmittance function for his experiments, since the discrete spectral components would be in sharp focus in the FTP. Nonperiodic object transmittance functions give rise to *continuous* Fourier spectra that generally have little definition or visible structure. Had Abbe not used the doubly periodic 2-D mesh function, his ideas would not have been so clearly demonstrated.

Amplitude-only filters similar to Abbe's still find use today, for example, in one type of image processing function known as *edge enhancement*. We'll look at some more sophisticated techniques in edge enhancement later, but one simple way to remove noise in an image and emphasize the outlines (or edges) of various objects is to place an opaque circular disk in the FTP of Fig. 14.1. Since the edges correspond to rapid changes in brightness, and hence involve high-frequency components in the input scene plane wave spectrum, filtering out the low spatial-frequency components eliminates the low-frequency noise in the input scene and leaves the edges intact, creating an edge-enhanced output scene in which object outlines are more clearly delineated.

In 1935, Zernike proposed an early technique of image enhancement (now known as the Zernike *phase contrast* method) which used spatial filtering techniques. Using a system of the type shown in Fig. 14.1, he demonstrated a means for improving the contrast visibility of certain *phase-only* object transparencies (biological slides, wind tunnel air flows, etc.). The problem with viewing thin phase objects is that they possess little visible contrast structure, since that kind of structure arises primarily from amplitude variations (except in the case of thick or layered phase objects wherein Bragg interactions arise). So, Zernike needed a way to view thin phase objects that had little visible structure. For such objects, the transmittance of the object in the object plane is given as

$$\tau(x, y) = e^{j\phi(x, y)} \cong 1 + j\phi(x, y)$$

as a result of the assumption that the slide is thin and $\phi \ll 1$. In the FTP, the following distribution is obtained (neglecting quadratic phase factors)

$$F(\alpha, \beta) = PSF(\alpha, \beta) + j\Phi(\alpha, \beta)$$

If a quarter wave dot (a type of *phase-only filter*, or *POF*) is placed at the optic axis of the system, so as to multiply the PSF function by a factor of j and make it co-phasal with the FT of the exponential, then the field distribution in the image plane is

$$E_{\text{image}}(x, y) = 1 + \phi(x, y)$$

The quarter wave dot has little effect on the FT of the phase function, affecting only the near-zero frequency components lying near the optic axis. These are generally of small amplitude and are not significant contributors to the spectrum (and even if they were, the eye is insensitive to such long-wavelength variations in an image anyway). The resultant intensity distribution (which would be detected by film or photodetectors in the image plane) is

$$I_{\text{image}}(x, y) = E_{\text{image}}E_{\text{image}}^* \cong 1 + 2\phi(x, y)$$

when $\phi \ll 1$. In this case, the phase function interferes directly in-phase with the constant amplitude function—rather than in quadrature—producing good contrast images.

As is shown in Chapter 12, phase can also be recovered by adding a plane wave to the phase-modulated wave to produce an interference pattern consisting of alternating bright and dark contours, from which phase variations generally can be readily inferred. In this chapter, we'll examine ways of recovering phase via the technique of Fourier plane filtering, that is, by placing slides of various types in the Fourier transform plane of a lens, so as to enhance the visibility of the phase variations by altering the spectrum of the object. This type of operation falls into the category of optical image processing and is directly analogous to the well-known techniques of electronic signal processing which involve filtering of signal spectra for the purpose of extracting information from the signal.

14.2 OPTICAL SIGNAL PROCESSING (CORRELATION) USING THE 4F CORRELATOR

Both optical and electronic signal processing are based on the convolution theorem for Fourier transforms. In both cases, the spectrum of an input function is multiplied by some type of mask function, thereby altering the spectrum. In the case of signal/image processing, the mask is usually some type of "window" function (we'll look at windows later, in the section on wavelets), whereas in correlation, the mask function is (the FT of) another signal or image function, similar in form to the input function. In signal processing, the goal is to produce some type of enhanced or improved version of the image, as in the case of the Zernike phase contrast method and the edge enhancement technique based on high-pass spatial filtering. In correlation, the goal is to produce a sharp, bright spot that indicates the presence of a particular shape in the input object scene.

As long as the system configuration of Fig. 14.1 remained the only one available for processing image spectra, the field of optical signal processing made only modest progress. It wasn't until the realization [2] of the exact Fourier transform relation that exists between field distributions in the front/back focal planes of a lens that optical signal processing systems suddenly became possible. This realization became the basis for the 4f correlator, described below.

We now consider the optical system shown in Fig. 14.5. This system is based on the Fourier transforming property of lenses described in Chapters 5 and 6. This is given as

$$E_{\text{FTP}}(x_2, y_2) = \iint_{A_1} E_{\text{inp}}(x_1, y_1) e^{-j(k/f)(x_2 x_1 + y_2 y_1)} \, dx_1 \, dy_1$$

$$= \tilde{E}_{\text{inp}}(\alpha, \beta); \qquad \alpha = \frac{kx_2}{f}, \qquad \beta = \frac{ky_2}{f}$$

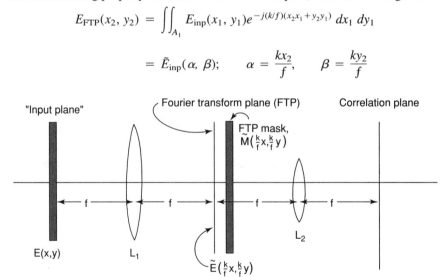

Figure 14.5 Traditional 4f optical correlator.

A transmittance mask is placed in the Fourier transform plane (FTP) of the system so that the FTP field (to the physical optics approximation) becomes

$$\tilde{E}_{\text{FTP}}^{\text{Tot}}\left(\frac{k}{f}x_2, \frac{k}{f}y_2\right) = \tilde{E}_{\text{FTP}}\left(\frac{k}{f}x_2, \frac{k}{f}y_2\right)\tilde{M}\left(\frac{k}{f}x_2, \frac{k}{f}y_2\right)$$

where the tilde is placed over the mesh transmittance M to indicate that the mask is a Fourier transform quantity. (The classic Vander Lugt filter is sometimes even referred to as a *Fourier transform* hologram, because it is a hologram created using the Leith-Upatneiks "offset reference" configuration; it is the hologram of a *Fourier transform* rather than the hologram of a *scene.*) So, by the Fourier transforming property of the second lens, the field in the correlation plane of the system in Fig. 14.5 is

$$E_{\text{corr}}(x_3, y_3) = \iint_{A_2} \tilde{E}_{\text{FTP}}\left(\frac{k}{f}x_2, \frac{k}{f}y_2\right)\tilde{M}\left(\frac{k}{f}x_2, \frac{k}{f}y_2\right) e^{-j[(k/f)x_3x_2 + (k/f)y_3y_2]} \, dx_2 \, dy_2$$

Throughout the following discussion, we'll assume that all light is co-polarized (whether that means linear, circular, elliptical, or whatever), so that scalar field quantities may be used in the equations from now on. Substituting for the Fourier transform expressions in the equations above, that is,

$$\tilde{E}_{\text{FTP}}(x_2, y_2) = \iint_{A_1} E_{\text{inp}}(x_1, y_1) e^{-j[(k/f)x_2x_1 + (k/f)y_2y_1]} \, dx_1 \, dy_1$$

and

$$\tilde{M}(x_2, y_2) = \iint_{A_1} m(x_1', y_1') e^{-j[(k/f)x_2x_1' + (k/f)y_2y_1')} \, dx_1' \, dy_1'$$

and making use of the equation,

$$\int_{-\infty}^{\infty} e^{\pm j\omega t} \, dt = 2\pi\delta(\omega)$$

(where δ is the unit impulse function/Dirac delta function), we obtain the correlation plane field as

$$E_{\text{corr}}(x_3, y_3) = \iint_{A_1} E_{\text{inp}}(x_1, y_1) m(-x_1 - x_3, -y_1 - y_3) \, dx_1 \, dy_1$$

(which is not a correlation, but a *convolution,* similar to the convolutional equations we encountered in Chapter 1). Defining a new function,

$$m'(x_1', y_1') = m(-x_1', -y_1')$$

that is, obtained by reversing the signs of the axes in the "input plane" for the Fourier transform mask, we obtain the correlation expression,

$$E_{\text{corr}}(x_3, y_3) = \iint_{A_1} E_{\text{inp}}(x_1, y_1) m'(x_1 + x_3, y_1 + y_3) \, dx_1 \, dy_1$$

This expression is central to understanding the operation of the classical 4f optical correlator. The correlator is generally operated in one of two different modes. In the "filtering" mode, the mask in the FTP alters the spectrum of the input image for the purpose of enhancing or suppressing certain spatial frequency components, in much the same way that an electrical bandpass filter operates on radio signals to enhance signal quality. In the "correlation" mode, the correlator uses a mask whose transmittance is "the conjugate of the Fourier transform of" an object transmittance anticipated to lie in the input scene. When the object Fourier transform multiplies the conjugate Fourier transform in the FTP, they produce a planar phase front, which in turn produces a high-intensity spot in the correlation plane (indicating the presence of the object in the input scene). We'll examine both modes of operation in the following two sections.

14.3 IMAGE ENHANCEMENT USING FTP FILTERING TECHNIQUES

When 4f correlator systems are used for image enhancement, the purpose of the system is generally to enhance certain features in an image that may have been degraded due to poor photographic conditions. In this mode of operation, the FTP mask literally acts as a filter, passing certain Fourier components contained within the image, while blocking others (or altering the relative phases of the various components). As in electrical signal processing, one major reason for using Fourier plane filtering is to remove unwanted "out-of-band" noise. In the case of an image, this noise may take the form of "high-frequency" *speckle noise* which might be obscuring certain image features. FTP filters are also used to improve image sharpness by filtering out low-frequency image components and passing high-frequency components that contribute to sharp edges. Unfortunately, this type of filtering often accentuates high-frequency noise of the type described above; thus, *edge enhancement* filters must be used with some care.

High-pass filtering is not the only possible means for sharpening an image. As is seen in Chapters 5 and 6, image sharpness varies dramatically simply when the film is axially defocused from the focal plane of the imaging system. However, all plane wave components of the image are present in the defocused planes as well as in the plane of focus. The only difference is that the plane wave components are not in proper phase away from the focal plane. They only possess their proper phase relationships in the focal plane. This fact immediately suggests one means of improving the sharpness of a blurred image—namely, rather than blocking certain plane wave components while enhancing others, perhaps the existing plane wave components in the degraded image simply need to be re-phased with respect to each other in order to enhance the sharpness of the image.

In this section, we are concerned with amplitude-only techniques for enhancing images, rather than the phase-only technique described in the previous paragraph. These techniques have direct analogues with both analog electrical bandpass filtering techniques and digital filtering techniques.

In the previous section of this chapter, we proved the *convolution theorem* for Fourier transforms—that is that multiplication in the FT domain coresponds to convolution/correlation (depending on the orientation of the mask input transparency) in the spatial domain. Since many of the properties of FTP masks are more easily visualized and understood in the spatial domain than in the spectral domain, we should keep both viewpoints in mind when studying FTP filters. In view of the convolutional effect that these FTP filters have on input images, they are sometimes referred to as *convolution masks,* although this name can cause confusion when applied to correlation systems.

Convolution masks can be used to sharpen the focus of images that have been blurred or otherwise degraded in either known or unknown ways. For example, consider Fig. 14.6, which shows a sharp line grating transmittance function and a line grating transmittance function wherein the grating was moved while a photographic exposure was taken. (See Chapter 13 for an in-depth discussion of film development effects.) The "perfect step" edge contrast function has been blurred into a "finite ramp" function, producing a linear transmittance taper between the perfectly transmissive and opaque lines of the grating. This creates smeared, nonlocalized edges. The exact position of the true edge is no longer precisely known; all that is known is that the edge lies somewhere between the light and dark areas of the grating.

One traditional means of "sharpening" the linear ramp function is to differentiate the blurred image luminance function. Differentiating the blurred edge function from Fig. 14.6 produces the "edge-enhanced" image shown in Fig. 14.7. In the edge-enhanced image, the transmissive (light) portions of the grating have become opaque (dark), and only the two edges of each strip are now light. Often, the "image" shown in Fig. 14.7*b* appears sharper to the eye than the one in Fig. 14.7*a*.

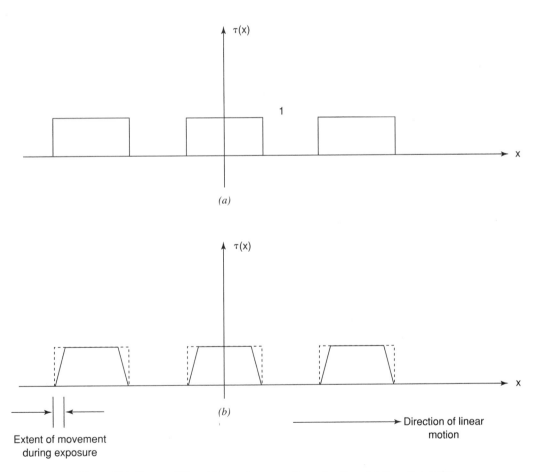

Figure 14.6 Sharp and blurred line grating transmittance functions: (a) Sharp line grating transmittance function and (b) blurred line grating transmittance function due to linear motion during photographic exposure.

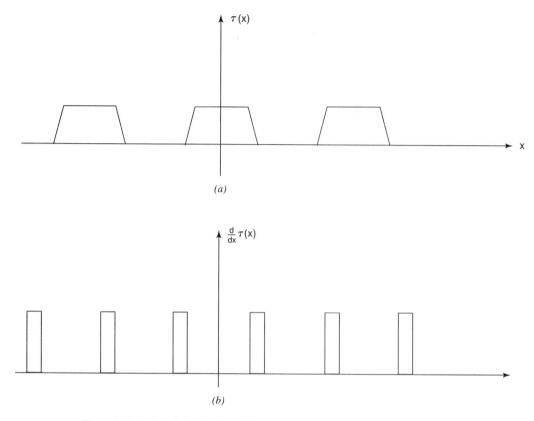

Figure 14.7 Differentiating the blurred line grating function to produce an edge-enhanced
image: (a) Line grating function blurred due to linear motion during exposure,
and (b) edge-enhanced line grating function.

We may look at the effect of the differentiation operator in both the spatial and
spectral domains. By the 1-D FT relation,

$$f(x) = \int_{-\infty}^{\infty} F(\alpha)\, e^{j\alpha x}\, d\alpha$$

we immediately see by differentiating both sides of this equation that

$$\frac{d}{dx} f(x) = \int_{-\infty}^{\infty} [j\alpha F(\alpha)]\, e^{j\alpha x}\, d\alpha$$

Hence, differentiation in the spatial domain corresponds to multiplication by $(j\alpha)$ in the
spectral domain. Thus, an FTP filter that produces this image differentiation operation has
a transmittance function linearly proportional to α. This type of filter is known as a gradient
filter since it causes a gradient operation to be performed on the input image.

The gradient filter has a transmittance function

$$M(\alpha) = j\alpha$$

which corresponds to a convolution function of

$$m(x) = \int_{-\infty}^{\infty} j\alpha \, e^{j\alpha x} \, d\alpha = \frac{d}{dx} \int_{-\infty}^{\infty} e^{j\alpha x} \, d\alpha$$

$$= \frac{1}{2\pi} \frac{d}{dx} \, \delta(x) = \text{doublet function}$$

The doublet function is shown in Fig. 14.8.

Convolving the 1-D doublet function with the 1-D image transmittance function $f(x)$ yields

$$g(x) = \int_{-\infty}^{\infty} f(x') \, m(x - x') \, dx'$$

$$= \int_{-\infty}^{\infty} f(x') \, \frac{d}{dx'} \, \delta(x - x') \, dx'$$

Integrating this expression by parts yields

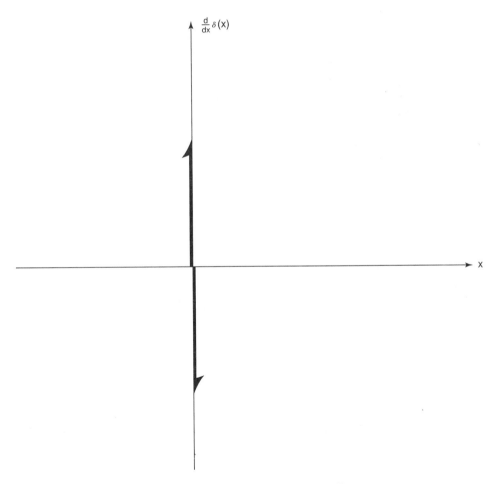

Figure 14.8 Doublet function for continuous gradient filter.

$$g(x) = \frac{d}{dx} f(x)$$

So we see that convolving an input image with a doublet function produces a continuous analogue derivative function in the spatial domain.

In virtually all modern electro-optical systems, images are carried not in a continuous fashion on analogue photographic film, but rather on pixelated displays having the "waffle" form of the SLMs described in Chapter 8. For these devices, the mask transmittance functions may be in the pixelated form shown in Fig. 14.9. For such a pixelated FTP mask, the transmittance function will be of the form

$$M(\alpha, \beta) = \sum_{m,n} M(\alpha_m, \beta_n)\, p_{\Delta\alpha\Delta\beta}(\alpha_m, \beta_n)$$

where the function $p(\cdot)$ on the RHS of the equation above represents a unit-height pulse function of length $\Delta\alpha$, $\Delta\beta$, respectively, in the α, β directions, centered at α_m, β_n. We may calculate the convolution function corresponding to this pixelated mask function in the usual way as

$$m(x, y) = (\Delta\alpha\Delta\beta\, \text{sinc}\Delta\alpha\, x/2.0\, \text{sinc}\Delta\beta\, y/2.0) \sum_{m,n} M(\alpha_m, \beta_n)e^{j(\alpha_m x + \beta_n y)}$$

This is the input plane mask transmittance function that is convolved with the original input image function.

More pertinent to the problem at hand, however, is the case where the input image is pixelated and the FTP mask is continuous. In this instance, the input image function is represented as

$$E_{\text{inp}}(x, y) = \sum_{m,n} E_{\text{inp}}(x_m, y_n) p_{\Delta x,\, \Delta y}(x_m, y_n)$$

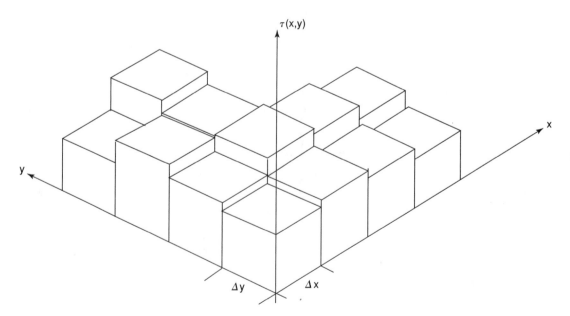

Figure 14.9 A typical pixelated display.

Now, the continuous derivative used previously becomes a finite difference. Going back to the example of the 1-D line grating, we see that the convolution function for the finite difference operator is the following finite approximation to the doublet function:

$$m(x) = \frac{1}{\Delta} [\delta(x + \Delta) - \delta(x)]$$

To show that this operator corresponds to a finite difference operation, we may convolve it with a 1-D image function $f(x)$ to obtain

$$g(x) = \int_{-\infty}^{\infty} f(x - x') \, m(x') \, dx'$$

or

$$g(x_n) = \frac{1}{\Delta} [f(x_n + \Delta) - f(x_n)]$$

This is the discrete, digital, pixelated finite-difference approximation to the continuous derivative function. In the spectral domain, the FTP filter function (corresponding to the finite difference operator above) is given as

$$M(\alpha) = \frac{1}{2\pi} \int_{-\infty}^{\infty} m(\mathrm{x}) e^{-j\alpha x} \, dx$$

so that

$$|M(\alpha)| = \frac{1}{\pi\Delta} \sin \alpha\Delta/2.0$$

This is a sinusoidal FTP mask function that periodically emphasizes and deemphasizes the spatial frequency components of the image plane wave spectrum.

The spatial domain function $m(x)$ is referred to in the digital signal processing (DSP) literature as a *digital filter* [4]. In the optics literature, it is referred to as the impulse response of the filter.

For the purpose of edge enhancement, it may be more effective to correlate the input image with the following digital filter:

$$m(x) = \frac{1}{2\Delta} \{[\delta(x + 3\Delta) + \delta(x + 2\Delta)] - [\delta(x + \Delta) + \delta(x)]\}$$

This is known as *scaling*. Such scaled finite-difference operators are known as *Haar wavelets* and are useful in edge detection. We'll consider these operators again in the section on wavelet transforms later in this chapter.

The gradient operator considered above responds to derivatives in the underlying function, producing *outline images,* as explained earlier. Other types of differential operators are possible, however. One popular differential operator, which responds to changes in image intensity in either Cartesian direction, is the Laplacian operator and its variants.

The various discrete Laplacian operators are derived from the continuous Laplacian in rectangular coordinates. This operator responds to image intensity *curvature* and generally responds twice at an edge (at each side). The continuous Laplacian operator in rectangular coordinates is given as

$$\nabla^2 f = \frac{\partial^2 f(x,\,y)}{\partial x^2} + \frac{\partial^2 f(x,\,y)}{\partial y^2}$$

where each second derivative is approximated as

$$\frac{\partial^2 f}{\partial x^2} = \frac{1}{\Delta^2}\,[f_1 - 2f_0 + f_{-1}]$$

for the function $f(x)$ shown in Fig. 14.10. So, the 2-D discrete Laplacian may be written as

$$\nabla^2 f \cong \frac{1}{\Delta^2}\,[f(x + \Delta,\, y) + f(x - \Delta,\, y) + f(x,\, y + \Delta) + f(x,\, y - \Delta) - 4f(x,\, y)]$$

From our previous work in 1-D in this section, it's evident that the convolution function that produces this discrete Laplacian is

$$m(x, y) = \frac{1}{\Delta^2}[\delta(x + \Delta, y) + \delta(x - \Delta, y) + \delta(x, y + \Delta) + \delta(x, y - \Delta) - 4\delta(x, y)]$$

and the FTP mask is

$$M(\alpha,\,\beta) = \frac{1}{4\pi^2\Delta^2}\,[e^{j\alpha\Delta} + e^{-j\alpha\Delta} + e^{j\beta\Delta} + e^{-j\beta\Delta} - 4]$$

This 2-D digital filter is often written in the matrix form:

$$m(x,\,y) = \begin{bmatrix} 0 & 1 & 0 \\ 1 & -4 & 1 \\ 0 & 1 & 0 \end{bmatrix}$$

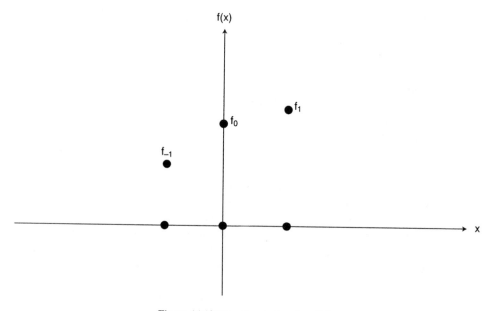

Figure 14.10 The discrete function $f(x)$.

Many variations of Laplacian-type operators exist and can be used for edge enhancement. Examples include [5]:

$$m(x, y) = \begin{bmatrix} -2 & 1 & -2 \\ 1 & 6 & 1 \\ -2 & 1 & -2 \end{bmatrix}$$

$$m(x, y) = \begin{bmatrix} -2 & 1 & -2 \\ 1 & 4 & 1 \\ -2 & 1 & -2 \end{bmatrix}$$

$$m(x, y) = \begin{bmatrix} -1 & -1 & -1 \\ 1 & 3 & -1 \\ 1 & 1 & -1 \end{bmatrix}$$

Other types of "Laplacian" masks include [6]

$$m(x, y) = \begin{bmatrix} -1 & -1 & -1 \\ -1 & 8 & -1 \\ -1 & -1 & -1 \end{bmatrix}$$

its biased version,

$$m(x, y) = \begin{bmatrix} -1 & -1 & -1 \\ -1 & b+8 & -1 \\ -1 & -1 & -1 \end{bmatrix} \quad \text{where } b > 0$$

and an adaptive biased version

$$m(x, y) = \begin{bmatrix} -1 & -1 & -1 \\ -1 & b(m, n) + 8 & -1 \\ -1 & -1 & -1 \end{bmatrix} \quad \text{where } b > 0$$

In the last equation, the adaptive bias might be related to some local property of image intensity, such as the average intensity of the eight surrounding pixels.

Another choice of digital filter is the Sobel operator [7] given by

$$m(x, y) = (1 + j)\delta(x - \Delta, y - \Delta) + 2\delta(x - \Delta, y)$$
$$+ (1 - j)\delta(x - \Delta, y + \Delta) + 2j\delta(x, y - \Delta)$$
$$- 2j\delta(x, y + \Delta) - (1 - j)\delta(x + \Delta, y - \Delta)$$
$$- 2\delta(x + \Delta, y) - (1 + j)\delta(x + \Delta, y + \Delta)$$

which can be shown to implement the operation,

$$\left[\left(\left\langle \frac{\partial f}{\partial x} \right\rangle_y \right)^2 + \left(\left\langle \frac{\partial f}{\partial y} \right\rangle_x \right)^2 \right]^{1/2}$$

where $\langle \ \rangle_x$ and $\langle \ \rangle_y$ denote averages over x, y, respectively. This edge enhancement operator utilizes first derivatives like the gradient operator, but is sensitive to intensity changes in both orthogonal directions.

We previously showed that the FTP mask for the Laplacian filter is given as the sum of the exponential functions

$$M(\alpha, \beta) = \frac{1}{4\pi^2\Delta^2} [e^{j\alpha\Delta} + e^{-j\alpha\Delta} + e^{j\beta\Delta} + e^{-j\beta\Delta} - 4]$$

This type of FTP mask function can be realized using the Leith-Upatneiks offset holographic techniques discussed in Chapter 13 [1,3]. This mask function may be produced using the experimental setup shown in Fig. 14.11. The field in the plane of the FTP mask is

$$E(\alpha, \beta) = M(\alpha, \beta) + e^{j\alpha_0 x}$$

As shown in the figure, the center hole is four times as large as the others. In reality, the center hole might be four times as transmissive. As is shown in Chapter 13, the transmittance of the developed FTP mask is proportional to the *intensity* of the total field at the mask plane during exposure. Thus, the FTP mask transmittance will be proportional to

$$I = |EE^*| = 1 + |M(\alpha, b)|^2 + M(\alpha, \beta)e^{-j\alpha_0 x} + M^*(\alpha, \beta)e^{j\alpha_0 x}$$

If the developed mask is then illuminated by a normally incident plane wave, the field transmitted by the mask will be proportional to $M(\alpha, \beta)$ and will be tilted in the $\alpha = \alpha_0$ direction. The Laplacian FTP mask could also be generated by successive exposures with tilted plane wave fields having tilt angles corresponding to the hole locations in the input digital filter mask [8].

An optical system incorporating this "Vander Lugt" FTP mask for image enhance-

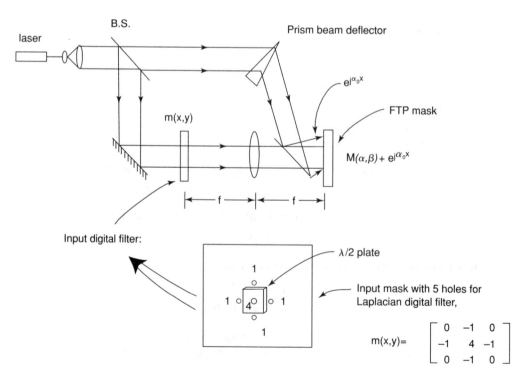

Figure 14.11 Modified Mach-Zehnder interferometer for producing a Vander Lugt FTP mask (same setup as used for Leith-Upatneiks hologram formation).

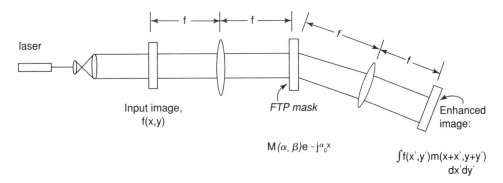

Figure 14.12 Experimental setup for image enhancement using Vander Lugt FTP masks.

ment is shown in Fig. 14.12. Once again, this mask was produced using a Leith-Upatnieks "offset reference" holographic system of the type described in Chapter 13. This system is commonly known as a "4f" correlator. Note that in addition to the desired correlation signal heterodyned to the spatial frequency α_0, there is a conjugate signal heterodyned at the spatial frequency $-\alpha_0$ and an on-axis signal. In other words, the Vander Lugt FTP mask produces three different diffraction orders, resulting in considerable loss of light intensity in the correlation plane. In the following section, we investigate a mask designed to mitigate this loss of light intensity.

14.4 OPTICAL CORRELATION USING THE PHASE-ONLY FILTER

The previous section covered optical techniques in image enhancement. These techniques involved straightforward modification of the image plane wave spectrum for the purpose of bringing out certain features of the image. In general, image enhancement modifications are not severe; hence, the processed images generally resemble the input images quite closely. In this and the following section, we examine techniques in optical correlation, an operation that is in principle the same as optical image enhancement. For example, the same 4f correlator architecture as above can be used for correlation. The philosophy inherent in the operation of correlation is much different, however, from image enhancement.

Correlation is a type of generalized transform operation. In the same way that the sinusoidal Fourier kernel "filters" or "selects" the various sinusoidal components of an image (when that image is multiplied by the sinusoid and integrated against it), kernels of every conceivable functional nature can be used to "select" the appearances of those functional forms in an input image. This is done by the very same operation used in Fourier transform theory. Namely, we multiply our "operand function" by some kernel, and then we integrate the product of the two functions over the domain of the operand function and the kernel function. The process is exactly the operation of correlation.

The similar operations of Fourier transforming and correlation are thus seen to be similar in principle to the dot product from vector algebra. The dot product (or inner product) is used to decompose a vector into its components along a set of orthogonal coordinate axes. The Fourier transform decomposes a function into its sinusoidal components. Correlation is also a type of integral inner product, which allows a continuous function to be decomposed into a set of functions of specified form.

The goal of the optical correlator is to indicate correlation by producing a planar phase front in the FTP, which in turn forms a sharp spot in the output (correlation) plane. The intensity of the spot indicates the magnitude of the inner product between the input image function and a mask function. The Vander Lugt 4f correlator (using a holographic-type complex amplitude and phase FTP mask) was developed in the early 1960s and represented a milestone in optical signal processing technology. Since its discovery, however, two primary types of correlation technologies have emerged as having superior correlation plane performance. These are the binary phase-only FTP mask (called the binary phase-only filter, or BPOF) and the binary joint transform correlator (BJTC). The first of these new technologies is discussed in this section and the second in the next.

In 1982, Joseph Horner [9] devised a figure-of-merit (modified by Caulfield [10] and now known as the *Horner efficiency*) for measuring the ratio of light intensity in the correlation plane spot to the light intensity in the input image plane. Mathematically, the Horner efficiency is given by the equation

$$\eta_H = \frac{\iint_{\text{corr.plane}} \left[\iint_{x',y'} |f(x', y')\, m(x' + x, y' + y)|^2\, dx'dy' \right] dxdy}{\iint_{P_1} |f(x, y)|^2\, dxdy}$$

When the input function matches the mask function, the result of the optical correlation operation is a bright spot on-axis in the correlation plane. The definition above is Horner's original version, and Caulfield's modification considers the integration to take place only over the correlation spot.

The actual FTP mask transmittance is assumed given as $M(\alpha, \beta)$. Thus, if this function is realized using a Leith-Upatnieks *amplitude hologram,* it directly attenuates the signal transmitted by the input plane transparency. Light attenuation through this amplitude transparency—as well the production of unneeded diffraction orders that it may cause—is the principal cause of Horner efficiencies less than 100%. Now, it turns out [11] that in optical image processing, the *phase* part of the plane wave spectrum coefficients of an image (i.e., the phase part of the image Fourier transform) is far more important in reconstruction than is the magnitude part. In fact, an image can often be well reconstructed (at least in edge-enhanced form) using only the phase of the Fourier transform. It also turns out that the phase part of the FTP mask function $M(\alpha, \beta)$ is the most important part in producing optical 4f correlators.

The importance of phase in an optical correlator is easily seen. In an optical correlator, the mask function $M(\alpha, \beta)$ is traditionally given as the complex conjugate of the FT of the input function. When this is the case, the 4f correlator yields the correlation of the input transparency function with the impulse response of the FTP mask. The result of this correlation is a spot in the correlation plane. A small, well-defined spot is produced in the correlation plane when a perfectly plane wave is produced just behind the mask in the FTP. But a perfect plane wave is produced only when the FTP mask phase is exactly the conjugate of the FT of the input plane transparency. Now, the spatial extent of the correlation plane spot is also determined by the amplitude taper of the field in the FTP. A uniform amplitude, uniform phase field in the FTP produces an Airy function spot in the correlation plane. Any amplitude distribution other than uniform will produce a wider spot, though possibly with smaller sidelobes. However, the size of the correlation plane spot is much more sensitive to phase errors than amplitude errors in the FTP.

Using an FTP mask equal to the conjugate of the FT of the input image results in an amplitude taper in the FTP that is equal to the *square* of the amplitude of the FT. A phase-only FTP filter, on the other hand, produces a less severe amplitude taper that is just proportional to the amplitude itself (not its square). Therefore, the phase-only FTP mask results in less amplitude taper in the FTP, and hence a sharper spot in the correlation plane.

Horner himself has done considerable work on the phase-only filter (POF). In [12], he compared the performance of phase-only, amplitude-only, and complex Vander Lugt matched filters. His results confirmed the work of Oppenheim and Lim [11], showing that in optical correlation, as well as in optical image processing, the Fourier transform phase is of primary importance. Figure 14.13 shows computer simulation results for the three types of FTP filters, given the input figures shown. In the simulations, if the input function is $f(x, y)$ and its Fourier transform is $F(\alpha, \beta)$, then the mask transmittance of the Vander Lugt filter is

$$M(\alpha, \beta) = F(\alpha, \beta) = |F(\alpha, \beta)| e^{-j\phi_F(\alpha, \beta)}$$

(ignoring the other diffraction orders produced by the Leith-Upatnieks amplitude holo-gram). The mask transmittance of the amplitude-only filter is

$$M(\alpha, \beta) = |F(\alpha, \beta)|$$

and the mask transmittance of the phase-only filter is

$$M(\alpha, \beta) = e^{-j\phi_F(\alpha, \beta)}$$

The results shown in Fig. 14.13 were produced using these various types of FTP masks. Clearly, the POF is superior for correlation plane performance, since it produces a highly visible, highly localized correlation plane spot, indicating the exact location in the input plane of the reference shape.

Evaluation of the Horner efficiency using the formula above results effectively in a four-dimensional integral in the numerator, which is very time-consuming for numerical integration. As a result, Gianino and Horner [13] suggested rewriting the Horner efficiency (using Parseval's theorem for Fourier transforms and the convolution theorem) in terms of FTP quantities as

$$\eta_H = \frac{\iint_{\text{FTP}} |F(\alpha, \beta) M(\alpha, \beta)|^2 \, d\alpha d\beta}{\iint_{\text{FTP}} |F(\alpha, \beta)|^2 \, d\alpha d\beta}$$

In this way, the four-dimensional integral in the numerator has been reduced in dimensionality by a factor of 2.

Continuous POFs may be constructed using various types of film materials, but for real-time applications, the masks must be displayed on high-speed electronic devices capable of rapid image formation and erasure. Spatial light modulators of the type mentioned in Chapter 8 are well suited for this purpose. A binary POF may be synthesized from a binary phase SLM having two transmission states, $\tau = \pm 1$. Although such a drastic simplification over the ordinary phase-only filter (now a binary phase-only filter, or BPOF)

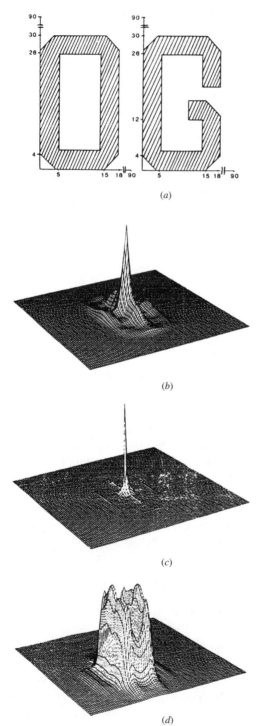

(a)

(b)

(c)

(d)

Figure 14.13 FTP filter comparisons: (a) Letters O and G used in correlation experiment. Numbers refer to points on the 90 × 90 input plane of the FFT. The letter O contains 356 points of unit height; the letter G 355 points. (b) Autocorrelation $|g * g^*|^2$ using classical matched filter (full phase and amplitude). (c) Autocorrelation with phase-only filter $|g * g_\varphi^*|^2$. (d) Autocorrelation with amplitude-only filter $|g * g_A^*|^2$. (After Horner and Gianino.)

might seem to be completely impractical, such a filter has very good correlation performance.

Horner and Leger [14] have shown the utility of such binary phase-only FTP filters. Using the tank input image shown in Fig. 14.14a, a BPOF was calculated; the binarization can be chosen on the basis:

$$\phi_{\text{BPOF}} = 1; \qquad -\pi/2 < \phi_F < \pi/2$$

$$\phi_{\text{BPOF}} = -1; \qquad \pi/2 < \phi_F < 3\pi/2$$

which corresponds to binarization on the basis of the sign of the cosine of the Fourier transform phase. In this case, there is no difference between a BPOF made from the conjugate of the image FT and one made from the actual FT itself, since either one would produce this same binary phase filter.

The correlation performance of the BPOF is compared in Fig. 14.15 against that obtained using a continuous POF for the tank image. The "impulse response" of the BPOF (i.e., the reconstruction of the tank input image from its binarized phase-only Fourier transform, produced by placing the BPOF in the front focal plane of a lens and then observing the field formed in the back focal plane) is shown in Fig. 14.14b. The conjugate ambiguity inherent in the binary-phase transform is manifested in the appearance of the two flipped images in the reconstruction.

Since the phase of the FT of the input image has been binarized in the BPOF to the two values 0 and π, this implies that the image that would be reconstructed from this purely real FT would necessarily be symmetric, since a symmetric image would produce a purely real Fourier transform, and vice versa. Indeed, this is the case. The image that is reconstructed from the binarized Fourier transform is symmetric in x, y, and this symmetry is obtained via the presence of an inverted image (in both the x- and y-directions), having intensity equal to that of the original upright image. This was shown in Fig. 14.14b, with the tank and its inverse appearing in the image reconstructed from the BPOF.

What this means as far as correlation is concerned is that now the input image will be effectively correlated with *two* mask functions, one being $m(x, y)$, the intended upright mask function, and the other being its inverted counterpart, $m(-x, -y)$. Depending on the symmetry properties of $m(x, y)$, there may or may not be strong correlation with $m(-x, -y)$. If the input mask function $m(x, y)$ is highly asymmetric (as in the case of the tank image), then its inverse will probably correlate poorly with the input image. What happens, however, if $m(x, y)$ is highly symmetric, and its inverse is likely to correlate strongly with the input function, producing two correlation spots?

Horner, Javidi, and Zhang [16] devised one means of eliminating the unwanted correlation with the flipped image by first coding the phase of the filter and then binarizing the resulting *coded phase* transform function. Basically, the idea here is to take a transform function having nearly real values, and to multiply that nearly real function by a highly irregular phase function prior to binarization. This process is expressed using simple Fourier series techniques. Let the input function be $f(x, y)$, the complex Fourier transform be $F(\alpha, \beta)$, and the binary phase Fourier transform be

$$\text{sign}[\cos\phi_F(\alpha, \beta)]$$

where the transmittance of the BPOF as a function of FT phase is shown in Fig. 14.16. The BPOF transmittance is a periodic square wave function of FT phase and thus may be expressed in terms of a Fourier series of the form

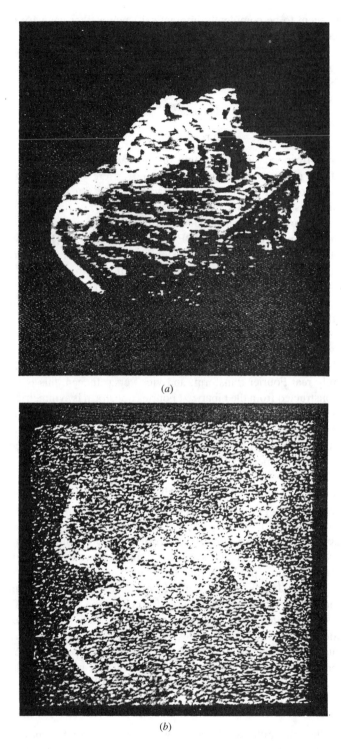

(a)

(b)

Figure 14.14 Tank images: (a) Infrared image of test object (tank) and (b) impulse response of the binary POF of test object. (After Horner and Leger.)

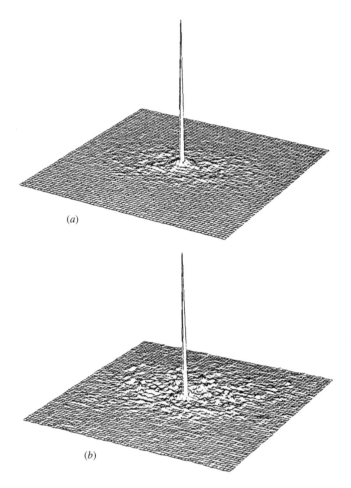

(a)

(b)

Figure 14.15 Correlation plane performance of BPOF: (a) Correlation of the test object
with the POF and (b) correlation of the test object with the binary POF.
(After Horner and Leger.)

$$\tau_{\text{BPOF}}(\alpha, \beta) = \sum_{-\infty}^{\infty} \frac{2}{n\pi} j^{n+1} e^{jn\phi_F(\alpha,\beta)}$$

where the sum includes only odd integer values of n, creating a Fourier cosine series. The
Fourier coefficients were evaluated using the usual Fourier techniques. We may rewrite
the previous equation as

$$\tau_{\text{BPOF}}(\alpha, \beta) = e^{j\phi_F(\alpha,\beta)} + e^{-j\phi^F(\alpha,\beta)} + \sum$$

We see that the binarization operation has created an infinite Fourier series, which
corresponds to higher diffraction orders similar to those created with the Vander Lugt
filter. In this case, however, the amplitudes of the higher diffraction orders are weighted
by the Fourier coefficients, so that the power in these higher orders decreases as $1/n^2$.
Hence, only the two lowest orders shown in the equation above are significant.

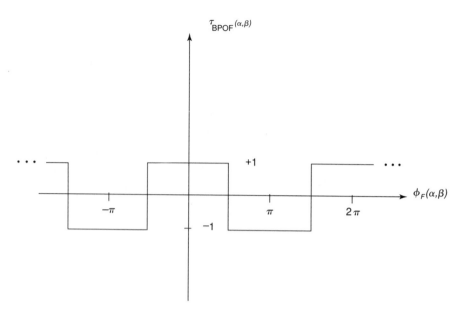

Figure 14.16 BPOF transmittance as a function of Fourier transform phase.

If the phase is multiplied by a complex phase function, $\exp([j\theta(\alpha,\beta)])$ before binarization, the BPOF transmittance becomes

$$\tau_{\text{BPOF}}(\alpha,\beta) = e^{j[\phi_F(\alpha,\beta)-\theta(\alpha,\beta)]} + e^{-j[\phi^F(\alpha,\beta)-\theta(\alpha,\beta)]} + \sum$$

So, if a phase-only mask of transmittance

$$\tau(\alpha,\beta) = e^{-j\theta(\alpha,\beta)}$$

(having a *highly* asymmetric impulse response) is then inserted in the FTP, along with this BPOF, then only one of the two inverted images in the input mask of the BPOF will correlate strongly with the input image. Figure 14.17 [16] shows correlation performance with and without the extra mask for a highly symmetric input image, a block form of the number "6."

In any real-time system, not only would the mask be displayed on some type of binary SLM, but so would the input image. One option for binarizing the input image is to use a binary amplitude SLM. Horner and Bartelt [15] have shown comparisons between correlation plane performance using continuous/binarized input images and Vander Lugt/POF/BPOF FTP filters. The continuous/binary amplitude versions of the input image are shown in Fig. 14.18 and the correlation plane performance comparisons are shown in Fig. 14.19. Despite the incredible simplicity in the binary amplitude input function and the binary phase mask, the correlation plane performance is still quite good.

One very important type of filter for practical correlation devices is the *synthetic discriminant function* (SDF) filter. An SDF filter consists of a set of several different views of an object (from different angles) superimposed on a single filter. This set of exposed images is frequently known as a *training set.* The superposition can be accomplished using multiple exposures of the film. Generally speaking, if detection of an object is needed over some angular range, the views must be very close together in angle, on

Input image used in computer simulations.

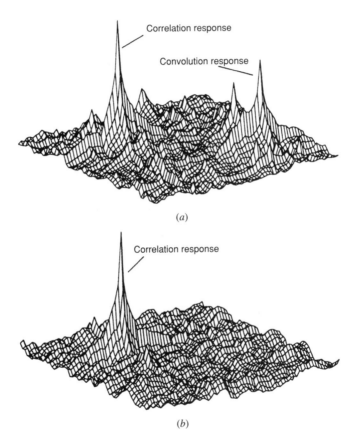

Figure 14.17 Correlation plane performance of coded-phase BPOF: (a) Normal BPOF correlation and (b) correlation using random phase-mask technique. (After Horner, Javidi, Zhang.)

the order of every 5 degrees in some cases. Figure 14.20 [17] shows a typical correlation plane response of a (phase-only) SDF filter to a target image which is a member of the training set. There is a single large spike, plus numerous smaller spikes due to correlation with images similar in form to the input image. The use of edge-enhanced input and training set images reduces the amount of cross-correlation between similar, but unequal, training set images.

(a)

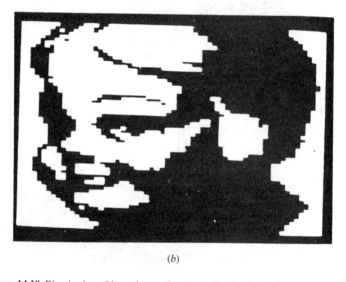

(b)

Figure 14.18 Binarization of input image. Input test signal: (a) Continuous tone and (b) binarized around median pixel value. (After Horner and Bartelt.)

The BPOF correlator has lately taken a back seat to the binary joint transform correlator (BJTC), to be described in the next section. However, with a new architecture described by Barnes et al. [18], this situation may change. The proposed correlation architecture employs an interferometer, shown in Fig. 14.21 to create the BPOF in real time. The system consists of a Mach-Zehnder interferometer similar to the type originally used by Vander Lugt, except that the plane reference wave is not necessarily tilted with respect to the (assumed untilted) "signal wave." The sum field is then made incident on a CCD detector array located in the Fourier transform plane of the lens, as shown in the figure. So, the field incident on the detector array in the FTP is

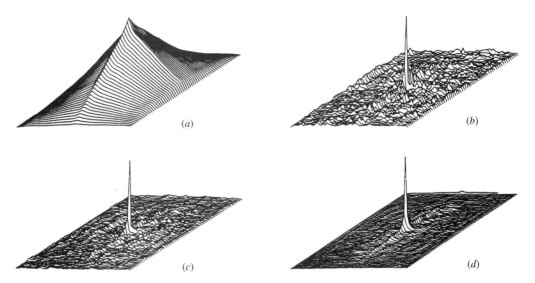

Figure 14.19 Correlation plane performance comparisons: (a) Autocorrelation of the continuous test signal with the regular matched filter. The filter was made from the continuous test signal. (b) Correlation of the binary amplitude test signal with the binary POF made from a continuous test signal. (c) Correlation of the binary amplitude test signal with the binary POF. The filter was made from a binary amplitude test signal. (d) Correlation of the binary amplitude test signal with the continuous POF. The filter was made from a binary amplitude test signal. This combination gives the highest SNR and η_H. (After Horner and Bartelt.)

$$E_1(\alpha, \beta) = 1 + F(\alpha, \beta)$$

where $F(\alpha, \beta)$ is the FT of the field in the input plane of the correlator, as shown. The total field intensity at the CCD plane is then

$$I_1(\alpha, \beta) = E_1(\alpha, \beta)E_1^*(\alpha, \beta)$$
$$= 1 + |F(\alpha, \beta)|^2 + 2 |F(\alpha, \beta)| \cos\phi_F(\alpha, \beta)$$

The frame grabber behind the detector array acquires the pixelated image intensity pattern. Now, the piezoelectric transducer (PZT) shown in the figure moves the mirror in the reference path by an optical phase angle δ, and a second detector array image is acquired by the frame grabber. (Later in this chapter we'll determine the necessary conditions on δ for obtaining a binary phase mask.) The new electric field amplitude in the plane of the detector array is

$$E_2(\alpha, \beta) = e^{j\delta} + F(\alpha, \beta)$$

and the corresponding intensity distribution is

$$I_2(\alpha, \beta) = E_1(\alpha, \beta)E_1^*(\alpha, \beta)$$
$$= 1 + |F(\alpha, \beta)|^2 + 2 |F(\alpha, \beta)| \cos[\phi_F(\alpha, \beta) - \delta]$$

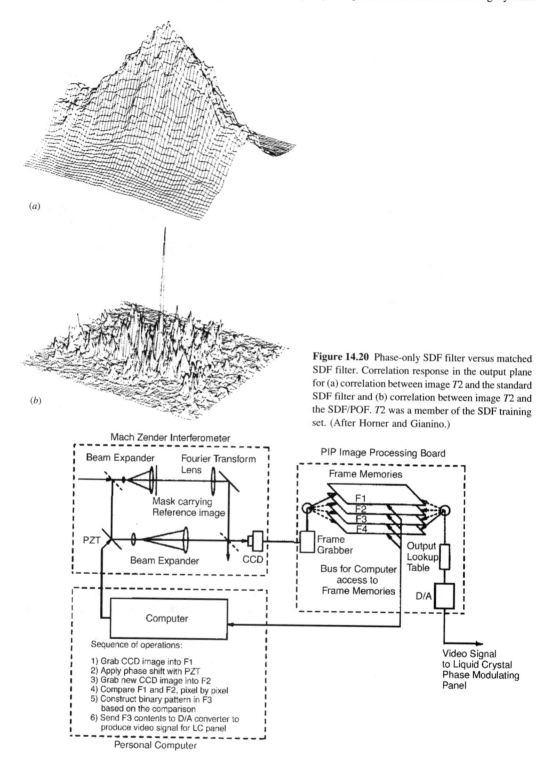

(a)

(b)

Figure 14.20 Phase-only SDF filter versus matched SDF filter. Correlation response in the output plane for (a) correlation between image *T*2 and the standard SDF filter and (b) correlation between image *T*2 and the SDF/POF. *T*2 was a member of the SDF training set. (After Horner and Gianino.)

Figure 14.21 Optoelectronic BPOF correlator. Schematic illustrating operation of the interferometric system for determining BPOFs. (After Barnes et al.)

The difference between the two intensity patterns is now taken on a pixel-by-pixel basis by a computer to yield the difference distribution

$$\Delta I = I_2(\alpha, \beta) - I_1(\alpha, \beta) \propto \cos[\phi_F(\alpha, \beta) - \delta] - \cos\phi_F(\alpha, \beta)$$

$$\propto \cos(\phi_F - \delta/2) \sin\delta/2$$

For phase binarization, it's only necessary to determine ϕ_F up to 180° (i.e., it's only necessary to divide the x, y plane into two halfplanes and decide which halfplane the phase should fall into). So, say $\delta = 180°$. Then,

$$\Delta I = \cos(\phi_F - \pi/2) = \sin\phi_F$$

Thus, when

$$\Delta I > 0, \qquad 0 \leq \phi_F < \pi$$

$$\Delta I < 0, \qquad \pi \leq \phi_F < 2\pi$$

and we see that the sign of the intensity difference gives the phase binarization for the BPOF.

Fourier transform plane correlators can also be fabricated using reflective, rather than transmissive, spatial light modulators in the Fourier transform plane. An example of one such correlator is shown in Fig. 14.22 [19].

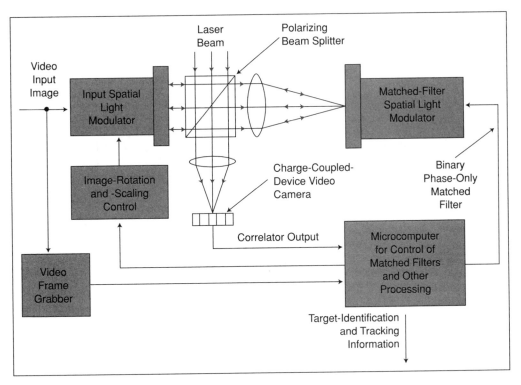

Figure 14.22 Reflective Optoelectronic correlator. This Proposed Digital-Electronic Optical Correlator would classify an unknown target depicted in the input image by performing a cross-correlation with reference images of many known targets simultaneously. (After Scholl.)

14.5 CORRELATION WITHOUT FILTERS: THE JOINT TRANSFORM CORRELATOR

In addition to the traditional 4f optical correlator, a second type of correlator architecture exists which has been the subject of considerable research in recent years. This is the joint transform correlator (JTC) introduced by Weaver and Goodman [20] and shown in Fig. 14.23. In the joint transform configuration, both 2-D transparencies to be convolved are placed side-by-side in the input plane (1f in front of the lens). The interference pattern of their two transforms is formed in the rear focal plane of the lens (the joint transform plane) located 1f behind the lens. If this "joint transform" is recorded onto square law recording media, it becomes a type of FTP mask. When this mask is placed in the front focal plane of an identical Fourier transforming lens and illuminated by coherent light, the correlation between the two input transparencies will be formed in the rear focal plane.

The operation of the JTC may be readily understood as follows. Let the total input plane transmittance be given as

$$f(x, y) = f_1(x - \Delta, y) + f_2(x + \Delta, y)$$

where the two transparencies are shifted symmetrically with respect to the $y - z$ plane

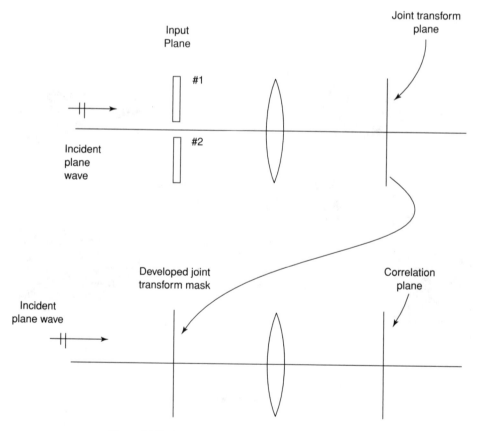

Figure 14.23 Architecture of the joint transform correlator.

and do not overlap. This combination of input functions is used for the operation of *correlation*. For *convolution*, the second function would be inverted in x and y, that is,

$$f(x, y) = f_1(x - \Delta, y) + f_2(-x + \Delta, -y)$$

as in [20]. So, the Fourier plane field is

$$F(\alpha, \beta) = \iint_{-\infty}^{\infty} f_1(x - \Delta, y)e^{-j(\alpha x + \beta y)}\,dxdy + \iint_{-\infty}^{\infty} f_2(x + \Delta, y)e^{-j(\alpha x + \beta y)}\,dxdy$$

or

$$F(\alpha, \beta) = e^{-j\alpha\Delta}F_1(\alpha, \beta) + e^{j\alpha\Delta}F_2(\alpha, \beta)$$

A negative transparency is then made of this field distribution. This transparency becomes the FTP mask for correlation. The transmittance of the developed film is proportional to the intensity of the Fourier plane field incident on the film. Thus, the transmittance of the developed film is proportional to

$$\tau(\alpha, \beta) = |F_1(\alpha, \beta)|^2 + |F_2(\alpha, \beta)|^2 + F_1(\alpha, \beta)F_2^*(\alpha, \beta)e^{-j2\alpha\Delta}$$

$$+ F^{*1}(\alpha, \beta)F_2(\alpha, \beta)e^{j2\alpha\Delta}$$

This field intensity function in the FTP is sometimes termed the *joint power spectrum* (JPS). The developed transparency is then placed into the input plane of the lens (with the original inputs #1, #2 now removed) and illuminated with a normally incident coherent plane wave. This produces four FT fields in the rear focal plane (i.e., the FTP) of the lens. These are four correlation signals given as

$$E_1 = \iint_{-\infty}^{\infty} F_1(\alpha, \beta)F_1^*(\alpha, \beta)\,e^{-j(\alpha x + \beta y)}\,d\alpha d\beta$$

$$E_2 = \iint_{-\infty}^{\infty} F_2(\alpha, \beta)F_2^*(\alpha, \beta)\,e^{-j(\alpha x + \beta y)}\,d\alpha d\beta$$

$$E_3 = \iint_{-\infty}^{\infty} F_1(\alpha, \beta)F_2^*(\alpha, \beta)\,e^{-j[\alpha(x + 2\Delta) + \beta y]}\,d\alpha d\beta$$

$$E_4 = \iint_{-\infty}^{\infty} F_2(\alpha, \beta)F_1^*(\alpha, \beta)\,e^{-j[\alpha(x - 2\Delta) + \beta y]}\,d\alpha d\beta$$

The first two terms are autocorrelations of the two input signals and are centered on the optical axis of the system (i.e., at $x = y = 0$). The third term is the cross-correlation

$$E_3(x, y) = \iint_{-\infty}^{\infty} f_1(x', y')f_2^*(x' + x + 2\Delta, y' + y)dx'dy'$$

centered at $x = -2\Delta$, $y = 0$, and the fourth term is the cross-correlation

$$E_4(x, y) = \iint_{-\infty}^{\infty} f_1^*(x', y')f_2(x' - x - 2\Delta, y' - y)dx'dy'$$

centered at $x = 2\Delta$, $y = 0$.

The JTC has been the subject of considerable research interest in recent years. The original configuration proposed by Weaver and Goodman has been modified by Javidi and Odeh [26], as shown in Fig. 14.24. Like the phase-only filters described in the previous section, this binary JTC (BJTC) correlator used a binarized (-1, $+1$) phase mask in the

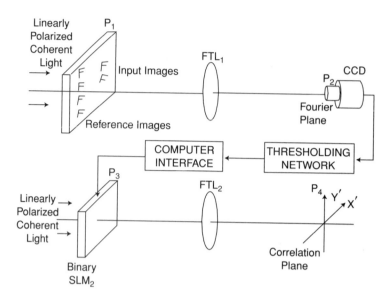

Figure 14.24 Bipolar joint transform image correlator using an electrically addressed binary SLM at the Fourier plane. FTL: Fourier transform lens. (After Javidi, Odeh.)

Fourier transform plane (in this case, the *joint transform plane*). Thus, it is to be expected that the transmitted field amplitudes in the correlation plane would be comparable to those in the phase-only filter case. In fact, Javidi reports very high power levels in the correlation spike [27], even in relation to ordinary POFs.

Javidi and Odeh [26] give an excellent description of the BJTC, which we repeat below.

> Plane P_1 is the input plane that contains the multiple reference signals and the multiple input signals displayed on SLM$_1$. The incoherent images enter the input SLM and are converted to coherent images. Either optically or electrically addressed SLMs can be used at the input plane. The images are then Fourier transformed by lens FTL$_1$, and the interference between the Fourier transforms is produced at plane P_2. The intensity of the Fourier transforms' interference is obtained by a CCD array at plane P_2 and is binarized using a thresholding network. An electrically addressed SLM operating in the binary mode is located at plane P_3 to read out the binarized intensity of the Fourier transforms' interference provided by the thresholding network. The correlation functions can be produced at plane P_4 by taking the inverse Fourier transform of the thresholded interference intensity distribution at plane P_3.
>
> The reference and input signals at plane P_1 are denoted by $\sum_{i=0}^{N} r_i(x + x_i, y)$ and $\sum_{i=0}^{M} s_i(x - x_i, y)$, respectively. The light distribution at the back focal plane of transform lens FTL$_1$ is the interference between the Fourier transforms of the two output image functions, that is,

$$G(\alpha, \beta) = \sum_{i=1}^{M} S_i\left(\frac{2\pi}{\lambda f}\alpha, \frac{2\pi}{\lambda f}\beta\right)\exp\left(-i\frac{2\pi}{\lambda f}x_i\alpha\right) \qquad (14.1)$$

$$+ \sum_{i=1}^{N} R_i\left(\frac{2\pi}{\lambda f}\alpha, \frac{2\pi}{\lambda f}\beta\right)\exp\left(i\frac{2\pi}{\lambda f}x_i\alpha\right)$$

> where (α, β) are the spatial frequency coordinates, $S_i(\cdot)$ and $R_i(\cdot)$ correspond to the Fourier

transforms of the input signals $s_i(x, y)$ and $r_i(x, y)$, respectively, f is the focal length of the transform lens, and λ is the wavelength of the illuminating coherent light. The Fourier transforms' interference intensity distribution can be written as

$$
\begin{aligned}
I(\alpha, \beta) &= |G(\alpha, \beta)|^2 \\
&= \sum_{i=1}^{N} \sum_{j=1}^{M} R_j \left(\frac{2\pi}{\lambda f}\alpha, \frac{2\pi}{\lambda f}\beta\right) R_i^* \left(\frac{2\pi}{\lambda f}\alpha, \frac{2\pi}{\lambda f}\beta\right) \exp\left[-i\frac{2\pi}{\lambda f}(x_i - x_j)\alpha\right] \\
&+ \sum_{i=1}^{N} \sum_{j=1}^{M} S_j \left(\frac{2\pi}{\lambda f}\alpha, \frac{2\pi}{\lambda f}\beta\right) R_i^* \left(\frac{2\pi}{\lambda f}\alpha, \frac{2\pi}{\lambda f}\beta\right) \exp\left[-i\frac{2\pi}{\lambda f}(x_i + x_j)\alpha\right] \quad (14.2) \\
&+ \sum_{i=1}^{N} \sum_{j=1}^{M} S_i^* \left(\frac{2\pi}{\lambda f}\alpha, \frac{2\pi}{\lambda f}\beta\right) R_j \left(\frac{2\pi}{\lambda f}\alpha, \frac{2\pi}{\lambda f}\beta\right) \exp\left[i\frac{2\pi}{\lambda f}(x_i + x_j)\alpha\right] \\
&+ \sum_{i=1}^{N} \sum_{j=1}^{M} S_i^* \left(\frac{2\pi}{\lambda f}\alpha, \frac{2\pi}{\lambda f}\beta\right) S_j \left(\frac{2\pi}{\lambda f}\alpha, \frac{2\pi}{\lambda f}\beta\right) \exp\left[i\frac{2\pi}{\lambda f}(x_i - x_j)\alpha\right]
\end{aligned}
$$

In the classical case, the inverse Fourier transform of Eq. (14.2) can produce the correlation signals at the output plane. For the proposed correlator, the Fourier transforms' interference intensity provided by the CCD array is thresholded before the inverse Fourier transform operation is applied.

The CCD array at the Fourier plane is connected to an electrically addressed SLM operating in the binary mode through a thresholding network so that the binarized interference intensity distribution can be read out by coherent light. The interference intensity is binarized according to the equation

$$
H(\alpha, \beta) = \begin{cases} +1 & I(\alpha, \beta) \geq v_t, \\ -1 & \text{otherwise} \end{cases} \quad (14.3)
$$

Here, $H(\alpha, \beta)$ is the binarized interference intensity, $I(\alpha, \beta)$ is the interference intensity given by Eq. (14.2), and v_t is the threshold value. The threshold for binarization of the Fourier transforms' interference intensity can be set by making a histogram of the pixel values of the interference intensity when the input signals match the reference signals and then picking the median. The correlation signals can be produced by taking the inverse Fourier transform of the binarized interference intensity given by Eq. (14.3).

As mentioned previously, Javidi [27] compared the performance of BJTC and POF correlators. Using the setup shown in Fig. 14.25, Javidi compared the autocorrelation of the letter F with itself and the cross-correlation of the letter F with the letter L. Autocorrelation comparisons are shown in Fig. 14.26, and cross-correlation performance is shown in Fig. 14.27. The excellent performance in relation to the POF and BPOF makes one wonder if a Vander Lugt hologram filter based on a similar binarization scheme (i.e., to create a BPOF based not on a binarized *phase* but on a binarized *thresholded amplitude*) wouldn't yield a similar performance. Javidi [30] has also compared the two types of filters for noisy input scenes.

Javidi, in his early papers, used a constant-intensity threshold value for the entire joint power spectrum. It is typical today to use adaptive thresholds, that is, thresholds that vary as a function of position in the joint power spectrum plane. Rogers et al. [28] describe a binarization scheme in which a pixel is binarized based on the intensity at the pixel in relation to the average intensity over a 3×3 pixel window surrounding the pixel in-

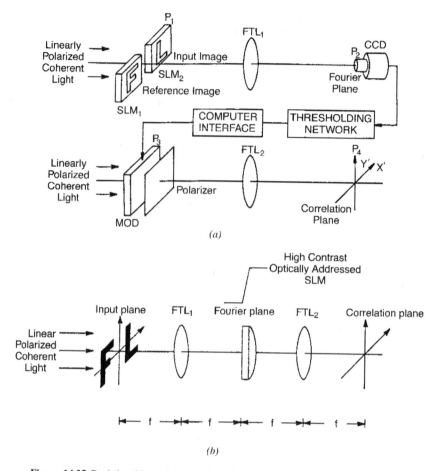

Figure 14.25 Real-time binary joint transform image correlator using (a) an electrically
addressed binary SLM and (b) a very high-contrast optically addressed SLM
at the Fourier plane. (After Javidi.)

question. This is an example of adaptive thresholding. Hahn and Flannery [29] describe
a corresponding 5×5 operator of the form

$$
\left\{
\begin{array}{ccccc}
1 & 1 & 1 & 1 & 1 \\
1 & 1 & 1 & 1 & 1 \\
1 & 1 & -24 & 1 & 1 \\
1 & 1 & 1 & 1 & 1 \\
1 & 1 & 1 & 1 & 1
\end{array}
\right\}
$$

Javidi and Horner [31] give a very good discussion of the effects of the nonlinearity in
the BJTC using a "Fourier" approach similar to that used in the previous section for the
nonlinear BPOF.

We mentioned earlier that the JTC and BJTC give rise to a number of
cross/autocorrelation and convolution signals in the "correlation plane." Grycewicz [32]
describes a clever technique that is, in principle, capable of eliminating all the unwanted
autocorrelation signals and leaving the desired cross-correlation signals. This is accom-

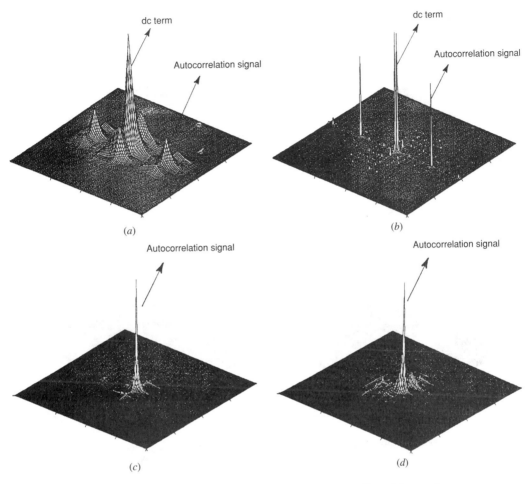

Figure 14.26 Autocorrelation results: (a) Classic correlator; (b) binary joint transform correlator; (c) continuous phase-only correlator; and (d) binary phase-only correlator. (After Javidi.)

plished by time-modulating the input images. (In *practice,* the technique serves primarily to "clean up" the correlation plane response of the JTC, while still leaving the autocorrelation signal intact.) The author proposes amplitude, phase, and polarization modulation of the images, though we'll concentrate on amplitude modulation. Figure 14.28 shows the conventional JTC, and Fig. 14.29 shows the RF modulated JTC setup.

Assuming amplitude modulation, with a 100% modulation depth, the inputs to the correlator are

$$\text{input}(x, y) = r\,(x - x_1, y)\,(1 + \cos\omega_r t) + s(x - x_2, y)\,(1 + \cos\omega_s t)$$

where the subscripts s and r refer to signal and reference fields, respectively. The field in the joint transform plane is then

$$E(\alpha, \beta) = R(\alpha, \beta)e^{j2\pi\alpha x_1}\,[1 + \cos\omega_r t] + S(\alpha, \beta)e^{j2\pi\alpha x_2}\,[1 + \cos\omega_s t]$$

The square of this field distribution is detected by the detector camera. Out of all the terms in the squared joint transform, only one will oscillate at the difference frequency

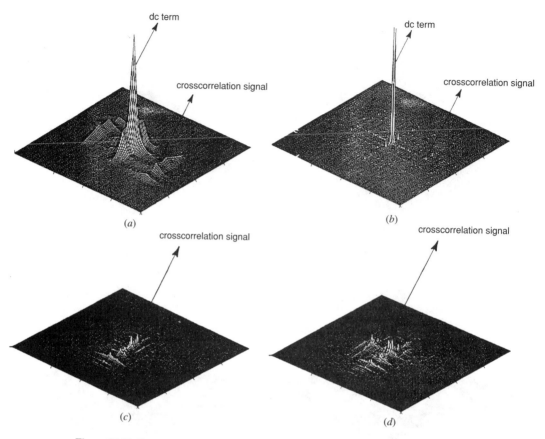

Figure 14.27 Cross-correlation results: (a) Classic correlator; (b) binary joint transform correlator; (c) continuous phase-only correlator; and (d) binary phase-only correlator. (After Javidi.)

$$\omega_r - \omega_s$$

and that is the term

$$E(\alpha, \beta) = 2\mathrm{Re}[R^*(\alpha, \beta)S(\alpha, \beta)e^{j2x\alpha(x_2-x_1)}] \cos(\omega_r - \omega_s)t$$

The joint power spectrum can be filtered to isolate this term. Unfortunately, the amplitude of this term contains a modulation term proportional to the difference $x_1 - x_2$. This must be compensated for somehow—perhaps electronically—before a clean cross-correlation signal can be obtained.

In the actual implementation of this scheme, the author used square-wave, rather than sinusoidal, modulation of the input scene (obtained by turning the inputs on and off), which naturally produces an infinite number of frequency components. However, the term of interest may still be obtained using the filtering scheme described above. He then removed the autocorrelation terms not by an actual electrical filtering operation, but by simply subtracting them from the total joint transform plane signal. Experimental results (and improvements over the standard BJTC) are shown in Fig. 14.30.

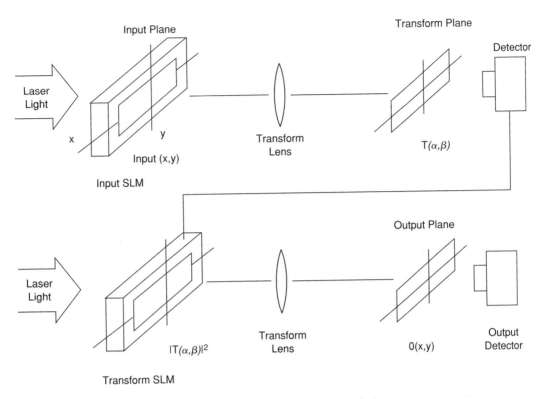

Figure 14.28 Optical JTC. (After Grycewicz.)

14.6 OPTICAL EVALUATION OF WAVELET TRANSFORMS USING OPTICAL CORRELATOR ARCHITECTURES
(with an introduction to wavelets)

The Fourier transform has long been—and still continues to be—used as a means of determining the "frequency content" of time-varying quantities and the "spatial-frequency content" of spatially varying quantities. For certain types of functions, however (e.g., musical pieces in which frequency content may vary widely from one passage to the next), the time-bandwidth product of the piece may be so large as to make practical calculation of the Fourier transform next to impossible. In addition, where wide changes in spectral content arise, it makes much more sense to analyze the primarily high-frequency passages separately from the primarily low-frequency passages, rather than trying to analyze all passages at once.

Loosely speaking, we might say that the Fourier spectrum of such a function tends to vary as a function of time. Now, in the strict sense of the Fourier transform, wherein we assume an infinite exponential kernel, it makes no sense to talk about a spectrum that "varies with time." The spectrum of the entire function is its spectrum—period. However, from a practical standpoint, it is sometimes useful to look at what may be termed a "moving spectrum" or an "instantaneous spectrum," that is, the frequency content of music as a function of time. It's pretty hard to calculate an instantaneous spectrum, but it is certainly

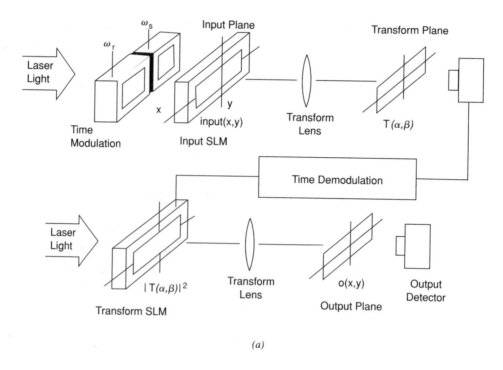

(a)

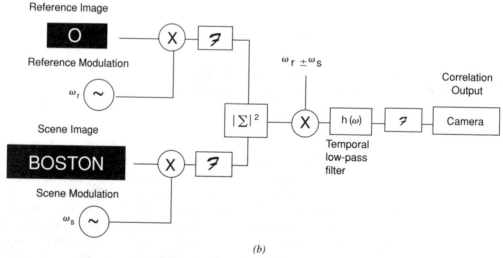

(b)

Figure 14.29 (a) Optical JTC, using superheterodyne image mixing and (b) system block
diagram of the superheterodyne image mixer. (After Grycewicz.)

possible to calculate the spectrum of a finite-length chunk of music. All that's done is to
imagine that the music suddenly begins (in a step function fashion from dead silence) at
the start of our measurement period and stops just as suddenly at the end of our measurement period. Mathematically, this is equivalent to multiplying the entire musical signal

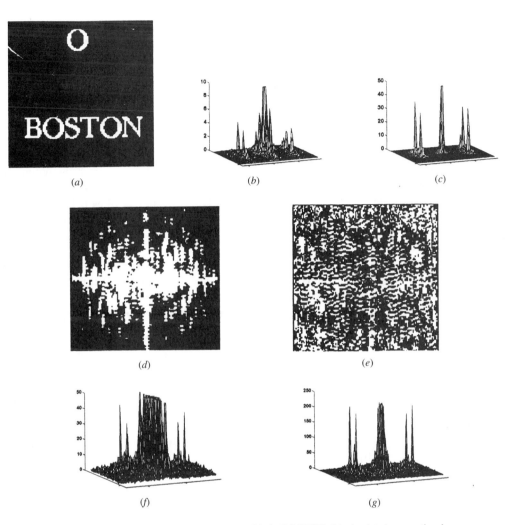

(a) (b) (c)

(d) (e)

(f) (g)

Figure 14.30 Results: (a) Input: find the O's in BOSTON; (b) simulated conventional
BJTC output; (c) simulated time-modulated JTC output; (d) experimental
Fourier plane input for conventional BJTC; (e) experimental Fourier plane
input for conventional BJTC; and (g) experimental output for BJTC using
time modulation. (After Grycewicz.)

by a finite-length rectangular pulse function known as a windowing function. This type
of Fourier transform operation is termed a windowed Fourier transform.

Now, Fourier-analyzing a function multiplied by a rectangular window may seem
innocent enough, but in reality, it leads to all sorts of problems. For as we know by the
convolution theorem, the spectrum of some signal (that is multiplied by a window function)
is equal to the true signal spectrum convolved with the spectrum of the windowing function.
So, the spectrum we calculate from such a windowed measurement is not the true spectrum
of the original underlying signal. It is corrupted (via convolution) by the spectrum of the
window. So, in order to obtain the true transform of the signal, it's necessary to deconvolve
the spectrum of the windowed function from the calculated spectrum.

When the rectangular window results in discontinuities in the signal at the ends of the window interval, the spectrum of the windowed function often exhibits a "ringing" phenomenon (known as the *Gibbs phenomenon* [4]) which is superimposed on the true spectrum and results from truncation in the time domain.

The Gibbs phenomenon can be mitigated somewhat by replacing the rectangular window function by various tapered windows [4,21]. The ultimate goal in selecting a window function is to get one whose transform is as close to a delta function as possible (which is the transform of the infinite exponential Fourier kernel). Convolution of the actual spectrum with the delta function will, of course, simply return the original spectrum. Convolution of the true spectrum with the near-delta-function transform of an optimized window function will cause a minimum amount of distortion to the true spectrum. The spectra of the rectangular and triangular windows are shown in Fig. 14.31 [21]. The spectra of some slightly more optimum windows are shown in Fig. 14.32 [21]. These latter spectra far more closely approximate a delta function than the sinc function spectrum of the rectangular pulse.

Of all the possible types of window functions, only one has the same shape in both the time and frequency domains (i.e., the window function is equal to its transform). This window is the Gaussian-magnitude window, which we discussed at length in connection with the space-bandwidth product. Dennis Gabor, the discoverer of holography, suggested

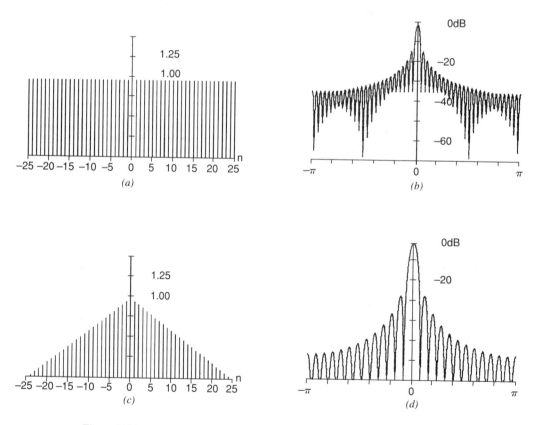

Figure 14.31 (a) Rectangular window; (b) log-magnitude of transform; (c) triangle window; and (d) log-magnitude of transform. (After Harris.)

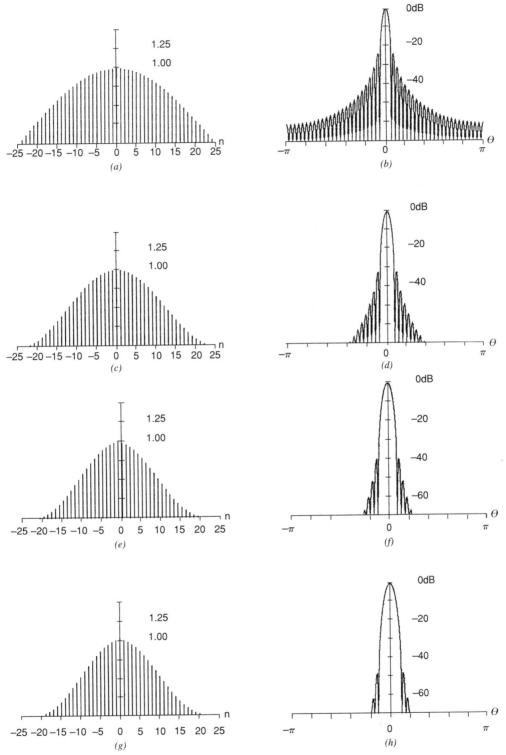

Figure 14.32 (a) Cos $(n\pi/N)$ window; (b) log-magnitude of transform; (c) Cos^2 $(n\pi/N)$ window; (d) log-magnitude of transform; (e) Cos^3 $(n\pi/N)$ window; (f) log-magnitude of transform; (g) Cos^4 $(n\pi/N)$ window; and (h) log-magnitude of transform. (After Harris.)

a type of "altered" Fourier transform (known as the *Gabor transform*), which employed an infinite exponential Fourier kernel multiplied by a Gaussian envelope. The Gabor transform kernel is defined [22] as

$$g_{m,n}(t) = g(t - nt_0)e^{j(2m\pi/T)t}$$

where

$$g(t) = \frac{1}{\sqrt{2\pi}\sigma} e^{-t^2/(2\sigma^2)}$$

is the Gaussian window.

Thus, the Gabor kernel is a moving Gaussian pulse window of constant width. Unlike the ordinary infinite sinusoid Fourier transform kernel, the Gabor kernel is defined by two integers *m, n*—one to determine the center of the window and one to determine the frequency. In the case of the Fourier transform, only the latter (frequency) property of the kernel is varied.

The wavelet transform kernel is a generalization of the Gabor kernel concept. As in the case of the Gabor kernel, the wavelet kernel is determined by two parameters—one representing the center of the window and one representing the frequency of oscillation within the window. Unlike the Gabor transform, however, the width of the wavelet window is inversely proportional to the spatial frequency, so that different kernels are simply expanded and contracted versions of one another and all have exactly the same space-bandwidth (or time-bandwidth) product. That is to say, all wavelets contain the same number of oscillations within a window, unlike the Gabor transform window wherein the window width is fixed, and different frequencies of oscillation produce different numbers of oscillation cycles within the fixed Gaussian envelope.

Mathematically, the wavelet kernel is

$$h_{mn}(t) = \frac{1}{\sqrt{a_m}} h[(t - b_n)/a_m],$$

where

$$b_n = nb_0a_m$$

where the spacing between successive envelopes, b_n, is proportional to the envelope width, given by a_m. The wavelet function $h(\cdot)$ is called the *mother wavelet* and $h_{mn}(\cdot)$ (the translated or expanded version) termed the *daughter wavelet*. The difference between different wavelet kernels and different Gabor kernels is shown in Fig. 14.33 [22]. The wavelet has an adaptive window of width inversely proportional to the oscillation frequency. A good description of a number of common wavelet kernel functions is given in [22], and the following text and figures are taken from that reference.

WAVELET FUNCTIONS

Optical implementation of the wavelet transform (WT) uses the explicit wavelet functions in the optical correlator. In the following, we review some popular wavelets:

Haar's Wavelet

The Haar's wavelet was first described by Haar in 1910. It is a bipolar step function:

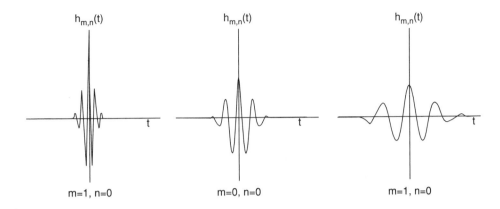

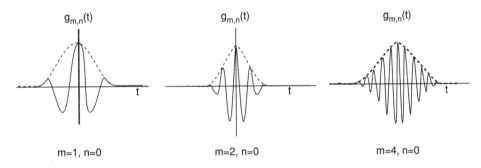

Figure 14.33 Wavelet bases (top) and fixed window Gabor bases (bottom). (After Sheng et al.)

$$h(t) = \text{rect}\left[2\left(t - \frac{1}{4}\right)\right] - \text{rect}\left[2\left(t - \frac{3}{4}\right)\right] \qquad (14.4)$$

which is real and antisymmetric about $t = 1/2$. The wavelet admissible condition is satisfied. The FT of the Haar's wavelet is

$$H(f) = 2i \exp(-i\pi f) \frac{1 - \cos \pi f}{\pi f} \qquad (14.5)$$

Its modulus is positive and even and symmetric to $f = 0$. The phase factor $\exp(-i\pi f)$ is related only to the fact that the $h(t)$ is antisymmetric about $t = 1/2$ as shown in Fig. 14.34.

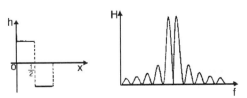

Figure 14.34 Haar's wavelet $h(x)$ and its fourier spectrum $H(f)$.

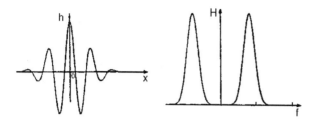

Figure 14.35 Morlet's wavelet (real part) $h(x)$ and its fourier spectrum $H(f)$.

The Haar's wavelet is orthogonal with discrete dilation and translation parameters. Its spectrum amplitude $|H(f)|$ converges to zero very slowly as $1/f$, because of the irregularity of $h(t)$. The first-order moment of the Haar's function is not zero.

Morlet's Wavelet

This wavelet was used by Martinet, Morlet, and Grossmann for the analysis of sound patterns:

$$h(t) = \exp(i2\pi f_0 t)\, \exp\left(-\frac{t^2}{2}\right) \qquad (14.6)$$

Its real part is an even cos-Gaussian function. The Fourier spectrum of the Morlet wavelet is the Gaussian functions shifted to f_0 and $-f_0$:

$$H(f) = 2\pi\{\exp[-2\pi^2(f - f_0)^2] + \exp[-2\pi^2(f + f_0)^2]\} \qquad (14.7)$$

which is even and real positive valued. Figure 14.35 shows the real part of the $h(t)$ and its Fourier spectrum. The wavelet admissible condition is not satisfied, because

$$H(0) \geq 0. \qquad (14.8)$$

But the value of the $H(0)$ is very close to zero for larger f_0 and can be approximately considered as zero in the numerical computation.

Mexican-hat Wavelet

The Mexican-hat-like wavelet was first introduced by Gabor and is well known as the Laplacian operator, which is widely used for zero-crossing multiresolution edge detection. This is in fact the second derivative of the Gaussian function:

$$h(t) = (1 - |t|^2)\, \exp\left(-\frac{|t|^2}{2}\right) \qquad (14.9)$$

It is even and real valued. The wavelet admissible condition is satisfied. The FT of the Mexican-hat wavelet is

$$H(f) = 4\pi^2 f^2 \exp(-2\pi f^2) \qquad (14.10)$$

It is also even and real valued, as shown in Fig. 14.36.

High-order derivatives of the Gaussian function can also be used as wavelets. The Fourier spectrum $H(f)$ of the nth-order derivative of the Gaussian will be a Gaussian function multi-

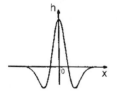
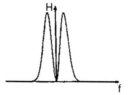

Figure 14.36 Mexican-hat wavelet $h(x)$ and its Fourier spectrum $H(f)$.

plied by $(i2\pi f)^n$ so that $H(0) = 0$, the wavelet admissible condition is satisfied, and $H^{(n-1)}(0) = 0$, the wavelet is of order $n - 1$. The $h(t)$ and $H(f)$ are indefinitely derivable $C\infty$ functions. They can be wavelets of very high order with the WT coefficient decaying very fast with increasing of the factor $1/a$.

Meyer's Wavelet

The FT of the Meyer's wavelet is shown in Fig. 14.37 and is defined as

$$H(f) = \exp(-i\pi f) \sin[\omega(f)] \tag{14.11}$$

with an even and symmetric function $\omega(f)$, in which the arc AB has a center of symmetry at $f = 1/2$ and $\omega(1/2) = \pi/4$:

$$\omega(1 - f) = \frac{\pi}{2} - \omega(f) \quad \text{for } 1/3 \le f \le 2/3 \tag{14.12}$$

and the arc BC has the same form but overturned and stretched:

$$\omega(2f) = \frac{\pi}{2} - \omega(f) \quad \text{for } 1/3 \le f \le 2/3 \tag{14.13}$$

The FT of the Meyer's wavelet is a compactly supported $C\infty$ function with rapid polynomial decay. If the phase factor $\exp(-i\pi f)$ in Eq. (14.11) corresponding to the shift in time of the $h(t)$ to $t = 1/2$ can be omitted, the $H(f)$ is real and even.

The Meyer's wavelet is expressed as

$$h(t) = 2 \int_0^\infty \sin[\omega(f)] \cos\left[2\pi\left(t - \frac{1}{2}\right)f\right] df \tag{14.14}$$

This is also a real function symmetric to $t = 1/2$, as shown in Fig. 14.37. Despite the rapid decay of $h(t)$, this wavelet has large support. Meyer has proven that the wavelets in Eq. (14.14) with discrete dilations and translations construct an orthonormal basis.

Lemarie-Battle's Wavelet

The Lemarie-Battle's wavelet is based on cadial splines with exponential decay. This is a C^k function with the first k moments of $h(t)$ vanishing. The wavelets and its FT are symmetric but still have large support, as shown in Fig. 14.38.

Daubechies's Wavelet

Daubechies built an orthogonal wavelet family with compact support by directly solving a discrete dilation equation with the approximation and orthogonality conditions. The Daubechies wavelet is a C^n function with small n and with compact support. The first-order moment is zero. As shown in Fig. 14.39, the Daubechies's wavelet is real and continuous but not very regular. They show no symmetry or antisymmetry.

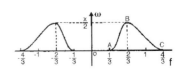

Figure 14.37 Meyer's wavelet $h(x)$ and $\omega(f)$ used to form its fourier spectrum $H(f) = \sin[\omega(f)]$.

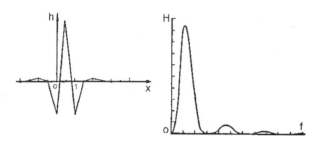

Figure 14.38 Lemarie-Battle's wavelet $h(x)$ and its Fourier spectrum $H(f)$.

Calculation of the wavelet transform involves evaluating convolution integrals; hence, optics appears to be a promising avenue for such calculations through the use of Fourier plane masks. The wavelet transform in one (temporal) dimension is given [22] as

$$W_s(a, b) = \frac{1}{\sqrt{a}} \int_{-\infty}^{\infty} h\left(\frac{t-b}{a}\right) s(t) \, dt$$

and in two (spatial) dimensions [23] as

$$W_s(a_x, a_y, b_x, b_y) = \frac{1}{\sqrt{a_x a_y}} \int \int_{-\infty}^{\infty} h\left(\frac{x-b_x}{a_x}, \frac{y-b_y}{a_y}\right) s(x, y) \, dx \, dy$$

which in both instances is a convolution. Since all wavelets are bipolar, real-valued functions, they can be formed using an amplitude mask cascaded with a binary phase mask. The FT's of most wavelets are real and positive functions, and so may be obtained with amplitude transmittance filters.

Using the convolution theorem, we may express the wavelet transform in terms of Fourier transform quantities [22] as

$$W_s(a, b) = \sqrt{a} \int_{-\infty}^{\infty} H(af) S(f) \, e^{-j2\pi fb} \, df$$

in 1-D (time) and [23]

$$W_s(a_x, a_y, b_x, b_y) = \sqrt{a_x a_y} \int \int_{-\infty}^{\infty} H(\alpha a_x, \beta a_y) S(\alpha, \beta) e^{-j2\pi(\alpha b_x + \beta b_y)} \, d\alpha \, d\beta$$

in 2-D (space), where the wavelet Fourier transforms are given as

$$H_{a,b}(f) = \int_{-\infty}^{\infty} h_{a,b}(t) e^{-j2\pi ft} \, dt = \sqrt{a} \, e^{-j2\pi fb} \, H(af)$$

in 1-D (time) and similarly in 2-D (space).

Sheng et al. [22] describe a 4f correlator implementation of the 1-D wavelet transform. In that implementation, the wavelet transform is taken for discrete dilation values, a, and continuous translation/shift values, b. A bank of Vander Lugt (offset Fourier transform hologram) filters is used for various discrete dilation values, a. These filters are arranged in a 2-D array in the FTP; hence, the area available to any one filter is significantly reduced.

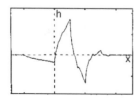 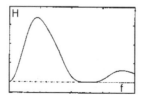

Figure 14.39 Daubechies's wavelet $h(x)$ and its Fourier spectrum.

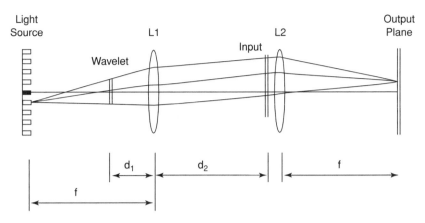

Figure 14.40 Shadow-casting system for 2-D wavelet transform. (After Yang et al.)

The coordinate in the correlation plane will be the continuous translation parameter, b. As mentioned in [22], the FTP filters for many typical wavelet transform kernels (including the Haar, Morlet, Mexican-hat, Meyer, and Lemarie-Battle wavelets) are positive real functions that may be implemented by an optically addressed liquid crystal light valve.

Yang et al. [23] show how different dilation factors may be obtained using the shadow casting system shown in Fig. 14.40. In this scheme, the dilation factors are obtained by moving the wavelet mask in the axial direction. By Yang's equation (4a), the dilation factor a is given as

$$a = \frac{f}{f - d_1}$$

Rather than using a bank of FTP filters in parallel in the FTP plane, Yang et al. propose an angular multiplexing scheme in which the different filters are recorded using reference waves of different offset angles. This is shown in Fig. 14.41. Since the various filters are recorded at different offset angles, the correlation outputs will be delivered to different portions of the correlation plane, wherein the local coordinates in each portion will represent the continuous translation variable.

Note that in all cases, whether the FTP filter is obtained by using N^2 FTP filters (where N^2 is the number of wavelet translational parameters) or whether the FTP filter is obtained by superimposing N^2 masks (with different reference field directions), the space-bandwidth requirement of the FTP filter is N^2 times what it would normally be for a single filter. In the case where the filters are placed in parallel in the FTP plane, the space side of the space-bandwidth product is increased. In the case of the angularly multiplexed filters, the bandwidth side of the space-bandwidth product is increased.

Sheng et al. [24] describe yet another implementation of the wavelet transform. In their scheme, they use N^2 filters (where N^2 is the number of dilations), all placed in parallel in the FTP plane. Their scheme for creating the FTP mask is shown in Fig. 14.42. This scheme imposes increased requirements on both the space and the bandwidth sides of the space-bandwidth product, since N^2 filters are placed side-by-side in the FTP plane *and* the filters all use different reference field directions. As in the case of the scheme of Yang et al., the different reference beams cause the different correlation plane outputs to be delivered to different regions of the correlation plane. The correlation system is shown in Fig. 14.43.

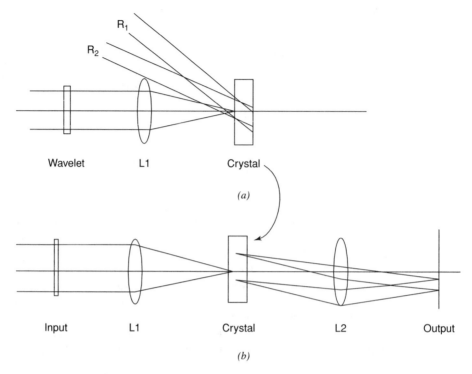

Figure 14.41 Optical WT processor based on frequency domain spatial filtering: (a) Recording of the angular multiplexed filter bank and (b) performing 2-D WT. (After Yang et al.)

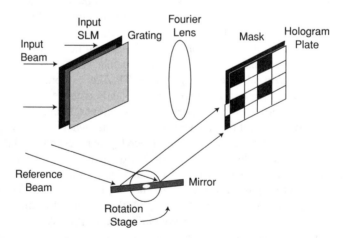

Figure 14.42 Recording architecture of a 2-D holographic wavelet transform filter. (After Sheng et al.)

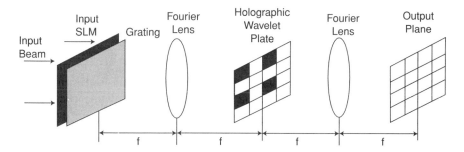

Figure 14.43 An N^4 correlator architecture for a 2-D holographic wavelet transform. (After Sheng et al.)

An optical implementation of the 2-D Haar wavelet transform is described by Burns et al. [25] in which the wavelet itself (not its Fourier transform) is implemented using a ternary phase and amplitude spatial light modulator. As described in Chapter 7, this device allows the three transmission coefficients of 0, ±1. The ±1 terms implemented the Haar transform, and the 0 terms implemented the background. This is shown in Fig. 14.44. In one system design, the various dilations were obtained by simply increasing the size of the square ±1 region of the transform. In the other case, the entire input plane was set to the ±1 transmission of the Haar transform, and the zero transmission region was obtained through the use of a variable-size square window. The two designs are shown in Fig. 14.45.

Note that the Haar wavelet is closely similar in form to the gradient operator discussed earlier in this chapter. Thus, the Haar transform is useful in edge enhancement applications.

14.7 MORPHOLOGICAL IMAGE PROCESSING USING OPTICAL FILTERING TECHNIQUES

Certain types of image processing operations are also possible via optical filtering [33]. Two common morphological operations are the *dilation* and *erosion* operations on binary images, shown in Fig. 14.46. These operations are realized by convolving an input image

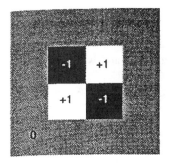

Figure 14.44 2-D Haar mother wavelet (*left*) and its Fourier transform (*right*). (After Burns et al.)

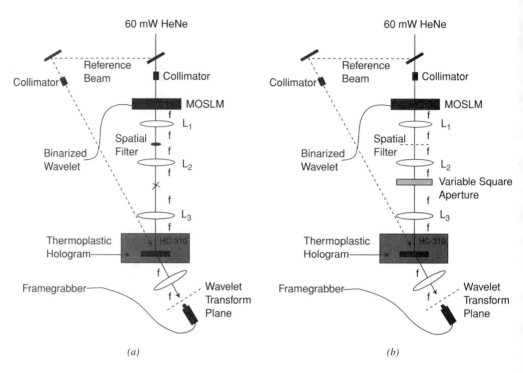

Figure 14.45 Vander Lugt optical correlation designs: (a) Wavelet dilations controlled
electronically using ternary-state MOSLM and spatial filter and (b) wavelet
dilations controlled by a variable square aperture. (After Burns et al.)

with the shape functions shown in the figure. (Note that the input image for the erosion
operation is presented in contrast-reversed mode.) The operations can be realized using
an ordinary 4f correlator architecture. In the language of Section 14.3, the shape element
is an (integrating) digital filter of the form

$$m(x, y) = \begin{pmatrix} 0 & 1 & 0 \\ 1 & 1 & 1 \\ 0 & 1 & 0 \end{pmatrix}$$

As stated in [33], two other common morphological operations are the *opening*
operation (consisting of an erosion followed by a dilation) and the *closing* operation
(consisting of a dilation followed by an erosion). An opening followed by a closing operation is effective in removing isolated black and white pixels ("salt and pepper" noise)
from a binary image.

To obtain Figs. 14.46 (d), (e) from Figs. 14.46 (a), (b) via (c), we place the *center
pixel* of the structuring element (c) over each pixel of (a), (b). Then we multiply (c) by
(a), (b) on a pixel-by-pixel basis, wherein a white pixel is regarded as a "1" and a shaded
pixel is regarded as a "0." Then we sum all of the products of pixels in (c) with pixels
in (a), (b). If the result is nonzero, the pixel in (d) or (e) corresponding to the location of
the center pixel of (c) is painted white, otherwise it is shaded.

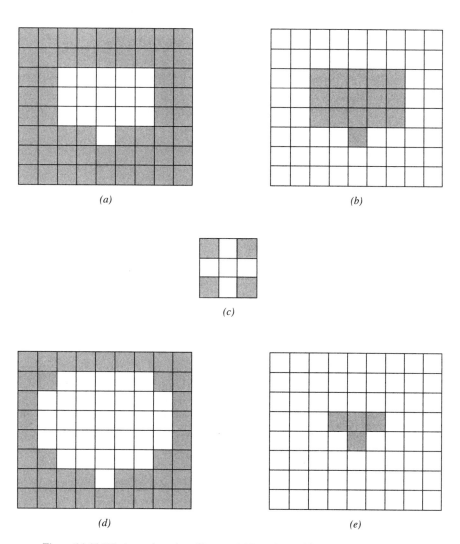

Figure 14.46 Dilation and erosion of images: (a) Input image for dilation; (b) input image for erosion; (c) structuring element; (d) dilated image; and (e) eroded image.

REFERENCES

[1] Goodman, J. W., *Introduction to Fourier Optics,* New York: McGraw-Hill, 1968.

[2] O'Neill, E. L., "Spatial Filtering in Optics," *IRE Trans. Inform. Theory,* Vol. IT-2, 1956, p. 56.

[3] Vander Lugt, A. B., "Signal Detection by Complex Spatial Filtering," *IEEE Trans. Inform. Theory,* IT-10, 1964, p. 2.

[4] Hamming, R. W., *Digital Filters,* Prentice-Hall, 1977.

[5] Hall, J. E., "Real Time Image Enhancement Using 3×3 Pixel Neighborhood Operator Functions," *Opt. Eng.,* Vol. 19, No. 3, May/June 1980, pp. 421–424.

[6] Strickland, R. N., and Aly, Maged Youssef, "Image Sharpness Enhancement Using Adaptive 3×3 Convolution Masks," *Opt. Eng.,* Vol. 24, No. 4, July/August 1985, pp. 683–686.

[7] Lee, Sing H., "Optical Implementations of Digital Algorithms for Pattern Recognition," *Opt. Eng.,* Vol. 25, No. 1, January 1986, pp. 69–75.

[8] Petrosky, K. J., and Lee, S. H., "New Method of Producing Gradient Correlation Filters for Signal Detection," *Appl. Opt.,* Vol. 10, No. 8, August 1971, pp. 1968–1969.

[9] Horner, J. L., "Light Utilization in Optical Correlatore," *Appl. Opt.,* Vol. 21, No. 24, December 15, 1982, pp. 4511–4514.

[10] Caulfield, H. J., "Role of the Horner Efficiency in the Optimization of Spatial Filters for Optical Pattern Recognition," *Appl. Opt.,* Vol. 21, No. 24, December 15, 1982, pp. 4391, 4392.

[11] Oppenheim, A. V., and Lim, J. S., "The Importance of Phase in Signals," *Proc. IEEE,* Vol. 69, No. 5, May 1981, pp. 529–541.

[12] Horner, J. L., and Gianino, P. D., "Phase-Only Matched Filtering," *Appl. Opt.,* Vol. 23, No. 6, March 15, 1984, pp. 812–816.

[13] Gianino, P. G., and Horner, J. L., "Additional Properties of the Phase-Only Correlation Filter," *Opt. Eng.,* Vol. 23, No. 6, November/December 1984, pp. 695–697.

[14] Horner, J. L., and Leger, J. R., "Pattern Recognition with Binary Phase-Only Filters," *Appl. Opt.,* Vol. 24, No. 5, March 1, 1985, pp. 609–611.

[15] Horner J. L., and Bartelt, H. O., "Two-Bit Correlation," *Appl. Opt.,* Vol. 24, No. 18, September 15, 1985, pp. 2889–2893.

[16] Horner, J. L., Javidi, B., and Zhang, Guanshen, "Analysis of Method to Eliminate Undesired Responses in a Binary Phase-Only Filter," *Opt. Eng.,* Vol. 33, No. 6, June 1994, pp. 1774–1776.

[17] Horner, J. L., and Gianino, P. D., "Applying the Phase-Only Filter Concept to the Synthetic Discriminant Function Correlation Filter," *Appl. Opt.,* Vol. 24, No. 6, March 15, 1985, pp. 851–855.

[18] Barnes, T. H., et al, "Optoelectronic Determination of Binary Phase-Only Filters for Optical Correlation," *Opt. Eng.,* Vol. 31, No. 9, September 1992, pp. 1936–1945.

[19] Scholl, Marija S., "Digital-Electronic/Optical Apparatus Would Recognize Targets," *NASA Tech Briefs,* December 1994, p. 40.

[20] Weaver, C. S., and Goodman, J. W., "A Technique for Optically Convolving Two Functions," *Appl. Opt.,* Vol. 5, No. 7, July 1966, pp. 1248–1249.

[21] Harris, F. J., "On the Use of Windows for Harmonic Analysis with the Discrete Fourier Transform," *Proc. IEEE,* Vol. 46, No. 1, January 1978, pp. 51–83.

[22] Sheng, Y., Roberge, D., and Szu, H. H., "Optical Wavelet Transform," *Opt. Eng.,* Vol. 31, No. 9, September 1992, pp. 1840–1845.

[23] Yang, X., Szu, H. H., Sheng, Y., and Caulfield, H. J., "Optical Haar Wavelet Transforms of Binary Images," *Opt. Eng.,* Vol. 31, No. 9, September 1992, pp. 1846–1851.

[24] Sheng, Y., Lu, T., Roberge, D., and Caulfield, H. J., "Optical N4 Implementation of a Two-Dimensional Wavelet Transform," *Opt. Eng.,* Vol. 31, No. 9, September 1992, pp. 1859–1864.

[25] Burns, T. J., et al., "Optical Haar Wavelet Transform," *Opt. Eng.,* Vol. 31, No. 9, September 1992, pp. 1852–1858.

[26] Javidi, B., and Odeh, S., "Multiple Object Identification by Bipolar Joint Transform Correlation," *Opt. Eng.,* Vol. 27, No. 4, April 1988, pp. 295–300.

[27] Javidi, B., "Comparison of Binary Joint Transform Correlators and Phase-Only Matched Filter Correlators," *Opt. Eng.,* Vol. 28, No. 3, March 1989, pp. 267–272.

[28] Rogers, S. K., et al., "New Binarizarion Techniques for Joint Transform Correlators," *Opt. Eng.,* Vol. 29, No. 9, September 1990, pp. 1088–1093.

[29] Hahn, W. B., and Flannery, D. L., "Design Elements of Binary Joint Transform Correlation and Selected Optimization Techniques," *Opt. Eng.,* Vol. 31, No. 5, May 1992, pp. 896–905.

[30] Javidi, B., "Comparison of the Nonlinear Joint Transform Correlator and the Nonlinearly Transformed Matched Filter Based Correlator for Noisy Input Scenes," *Opt. Eng.,* Vol. 29, No. 9, September 1990, pp. 1013–1020.

[31] Javidi, B., and Horner, J. L., "Multifunction Nonlinear Signal Processor: Deconvolution and Correlation," *Opt. Eng.,* Vol. 28, No. 8, August 1989, pp. 837–843.

[32] Grycewicz, T. J., "Applying Time Modulation to the Joint Transform Correlator," *Opt. Eng.,* Vol. 33, No. 6, June 1994, pp. 1813–1819.

[33] Chao, T-H, "Dynamically Reconfigurable Optical Morphological Processor," *NASA Tech Briefs,* Vol. 20, No. 2, February 1996.

PART VI
APPENDIXES

Elements of Vector Analysis

A.1 VECTOR ALGEBRA

For vector functions **A** and **B,** the following identities hold:

$$|\mathbf{A}|^2 = \mathbf{A} \cdot \mathbf{A}^* \tag{A.1}$$

$$\mathbf{A} + \mathbf{B} = \mathbf{B} + \mathbf{A} \tag{A.2}$$

$$\mathbf{A} \cdot \mathbf{B} = \mathbf{B} \cdot \mathbf{A} \tag{A.3}$$

$$\mathbf{A} \times \mathbf{B} = -\mathbf{B} \times \mathbf{A} \tag{A.4}$$

$$(\mathbf{A} + \mathbf{B}) \cdot \mathbf{C} = \mathbf{A} \cdot \mathbf{C} + \mathbf{B} \cdot \mathbf{C} \tag{A.5}$$

$$(\mathbf{A} + \mathbf{B}) \times \mathbf{C} = \mathbf{A} \times \mathbf{C} + \mathbf{B} \times \mathbf{C} \tag{A.6}$$

$$\mathbf{A} \cdot (\mathbf{B} \times \mathbf{C}) = \mathbf{B} \cdot (\mathbf{C} \times \mathbf{A}) = \mathbf{C} \cdot (\mathbf{A} \times \mathbf{B}) \tag{A.7}$$

$$\mathbf{A} \times (\mathbf{B} \times \mathbf{C}) = (\mathbf{A} \cdot \mathbf{C})\mathbf{B} - (\mathbf{A} \cdot \mathbf{B})\mathbf{C} \tag{A.8}$$

A.2 VECTOR DIFFERENTIAL CALCULUS

For scalar functions ϕ and ψ and vector functions **A** and **B,** the following identities hold:

$$\nabla(\phi + \psi) = \nabla\phi + \nabla\psi \tag{A.9}$$

$$\nabla \cdot (\mathbf{A} + \mathbf{B}) = \nabla \cdot \mathbf{A} + \nabla \cdot \mathbf{B} \tag{A.10}$$

Source: C.R. Scott, ''Field Theory of Acousto-Optic Signal Processing Devices,'' Norwood MA: Artech House, 1992.

$$\nabla \times (\mathbf{A} + \mathbf{B}) = \nabla \times \mathbf{A} + \nabla \times \mathbf{B} \tag{A.11}$$

$$\nabla(\phi\psi) = \phi\nabla\psi + \psi\nabla\phi \tag{A.12}$$

$$\nabla \cdot (\phi\mathbf{A}) = \phi\nabla \cdot \mathbf{A} + \mathbf{A} \cdot \nabla\phi \tag{A.13}$$

$$\nabla \times (\phi\mathbf{A}) = \phi\nabla \times \mathbf{A} - \mathbf{A} \times \nabla\phi \tag{A.14}$$

$$\nabla \cdot (\mathbf{A} \times \mathbf{B}) = \mathbf{B} \cdot (\nabla \times \mathbf{A}) - \mathbf{A} \cdot (\nabla \times \mathbf{B}) \tag{A.15}$$

$$\nabla^2\mathbf{A} = \nabla(\nabla \cdot \mathbf{A}) - \nabla \times \nabla \times \mathbf{A} \tag{A.16}$$

$$\nabla \times (\phi\nabla\psi) = \nabla\phi \times \nabla\psi \tag{A.17}$$

$$\nabla \times \nabla\phi = 0 \tag{A.18}$$

$$\nabla \cdot \nabla \times \mathbf{A} = 0 \tag{A.19}$$

A.3 VECTOR INTEGRAL CALCULUS

$$\iiint_v \nabla \cdot \mathbf{A}\, dv = \oiint_s \mathbf{A} \cdot \mathbf{ds}$$

$$\iint_s (\nabla \times \mathbf{A}) \cdot \mathbf{ds} = \oint_c \mathbf{A} \cdot \mathbf{dl}$$

$$\iiint_v (\nabla \times \mathbf{A})\, dv = - \oiint_s \mathbf{A} \times \mathbf{ds}$$

$$\iint_s (\hat{\mathbf{n}} \times \nabla\phi)\, \mathbf{ds} = \oint_c \phi\mathbf{dl}$$

$$\iiint_v \nabla\phi\, dv = \oiint_s \phi\mathbf{ds}$$

where the vectors $\hat{\mathbf{n}}$, $\mathbf{ds}$ point *out* of the volume v.

A.4 VECTOR DIFFERENTIAL OPERATORS IN RECTANGULAR, CYLINDRICAL, AND SPHERICAL COORDINATES

Rectangular Coordinates

$$\nabla\Phi = \hat{x}\frac{\partial\Phi}{\partial x} + \hat{y}\frac{\partial\Phi}{\partial y} + \hat{z}\frac{\partial\Phi}{\partial z}$$

$$\nabla \cdot \mathbf{A} = \frac{\partial A_x}{\partial x} + \frac{\partial A_y}{\partial y} + \frac{\partial A_z}{\partial z}$$

$$\nabla \times \mathbf{A} = \hat{x}\left(\frac{\partial A_z}{\partial y} - \frac{\partial A_y}{\partial z}\right) + \hat{y}\left(\frac{\partial A_x}{\partial z} - \frac{\partial A_z}{\partial x}\right) + \hat{z}\left(\frac{\partial A_y}{\partial x} - \frac{\partial A_x}{\partial y}\right)$$

$$\nabla^2 \Phi = \frac{\partial^2 \Phi}{\partial x^2} + \frac{\partial^2 \Phi}{\partial y^2} + \frac{\partial^2 \Phi}{\partial z^2}$$

$$\nabla^2 \mathbf{A} = \hat{x}\nabla^2 A_x + \hat{y}\nabla^2 A_y + \hat{z}\nabla^2 A_z$$

Cylindrical Coordinates

$$\nabla \Phi = \hat{\rho}\frac{\partial \Phi}{\partial \rho} + \hat{\phi}\frac{1}{\rho}\frac{\partial \Phi}{\partial \phi} + \hat{z}\frac{\partial \Phi}{\partial z}$$

$$\nabla \cdot \mathbf{A} = \frac{1}{\rho}\frac{\partial}{\partial \rho}(\rho A_\rho) + \frac{1}{\rho}\frac{\partial A_\phi}{\partial \phi} + \frac{\partial A_z}{\partial z}$$

$$\nabla \times \mathbf{A} = \hat{\rho}\left(\frac{1}{\rho}\frac{\partial A_z}{\partial \phi} - \frac{\partial A_\phi}{\partial z}\right) + \hat{\phi}\left(\frac{\partial A_\rho}{\partial z} - \frac{\partial A_z}{\partial \rho}\right) + \hat{z}\left(\frac{1}{\rho}\frac{\partial}{\partial \rho}(\rho A_\phi) - \frac{1}{\rho}\frac{\partial A_\rho}{\partial \phi}\right)$$

$$\nabla^2 \Phi = \frac{1}{\rho}\frac{\partial}{\partial \rho}\left(\rho \frac{\partial \Phi}{\partial \rho}\right) + \frac{1}{\rho^2}\frac{\partial^2 \Phi}{\partial \phi^2} + \frac{\partial^2 \Phi}{\partial z^2}$$

$$\nabla^2 \mathbf{A} = \nabla(\nabla \cdot \mathbf{A}) - \nabla \times \nabla \times \mathbf{A}$$

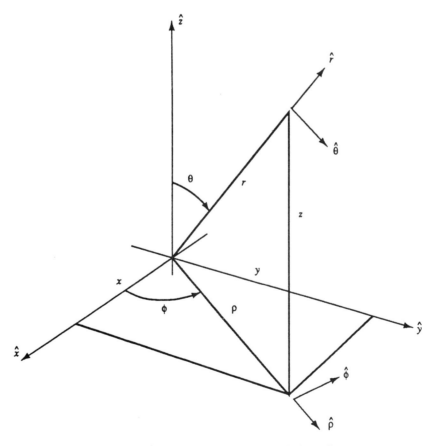

Figure A.1 Rectangular, cylindrical, and spherical coordinates.

Spherical Coordinates

$$\nabla \Phi = \hat{r} \frac{\partial \Phi}{\partial r} + \hat{\theta} \frac{1}{r} \frac{\partial \Phi}{\partial \theta} + \hat{\phi} \frac{1}{r \sin \theta} \frac{\partial \Phi}{\partial \phi}$$

$$\nabla \cdot \mathbf{A} = \frac{1}{r^2} \frac{\partial}{\partial r} (r^2 A_r) + \frac{1}{r \sin \theta} \frac{\partial}{\partial \theta} (\sin \theta A_\theta) + \frac{1}{r \sin \theta} \frac{\partial A_\phi}{\partial \phi}$$

$$\nabla \times \mathbf{A} = \frac{\hat{r}}{r \sin \theta} \left[\frac{\partial}{\partial \theta} (A_\phi \sin \theta) - \frac{\partial A_\theta}{\partial \phi} \right] + \hat{\theta} \frac{1}{r} \left[\frac{1}{\sin \theta} \frac{\partial A_r}{\partial \phi} - \frac{\partial}{\partial r} (r A_\phi) \right]$$

$$+ \hat{\phi} \frac{1}{r} \left[\frac{\partial}{\partial r} (r A_\theta) - \frac{\partial A_r}{\partial \theta} \right]$$

$$\nabla^2 \Phi = \frac{1}{r^2} \frac{\partial}{\partial r} \left(r^2 \frac{\partial \Phi}{\partial r} \right) + \frac{1}{r^2 \sin \theta} \frac{\partial}{\partial \theta} \left(\sin \theta \frac{\partial \Phi}{\partial \theta} \right) + \frac{1}{r^2 \sin^2 \theta} \frac{\partial^2 \Phi}{\partial \phi^2}$$

$$\nabla^2 \mathbf{A} = \nabla (\nabla \cdot \mathbf{A}) - \nabla \times \nabla \times \mathbf{A}$$

The variables for the three coordinate systems are shown in Fig. A.1.

B

Theorems and Relations from Fourier Analysis

TABLE B.1 Fourier Transform Theorems

$f(t) = \dfrac{1}{2\pi}\displaystyle\int_{-\infty}^{\infty} F(\omega)e^{j\omega t}\,d\omega$	$F(\omega) = \displaystyle\int_{-\infty}^{\infty} f(t)e^{-j\omega t}\,dt$
$f(at)$	$\dfrac{1}{\lvert a \rvert} F\!\left(\dfrac{\omega}{a}\right)$
$f^*(t)$	$F^*(-\omega)$
$F(t)$	$2\pi f(-\omega)$
$f(t - t_0)$	$F(\omega)e^{-jt_0\omega}$
$f(t)e^{j\omega_0 t}$	$F(\omega - \omega_0)$
$f(t)\cos\omega_0 t$	$\dfrac{1}{2}\,[F(\omega + \omega_0) + F(\omega - \omega_0)]$
$f(t)\sin\omega_0 t$	$\dfrac{j}{2}\,[F(\omega + \omega_0) - F(\omega - \omega_0)]$
$\dfrac{d^n f(t)}{dt^n}$	$(j\omega)^n F(\omega)$
$(-jt)^n f(t)$	$\dfrac{d^n F(\omega)}{d\omega^n}$
$m_n = \displaystyle\int_{-\infty}^{\infty} t^n f(t)\,dt$	$F(\omega) = \displaystyle\sum_{n=0}^{\infty} \dfrac{m_n}{n!}\,(-j\omega)^n$
$\displaystyle\int_{-\infty}^{\infty} f_1(\tau)f_2(t - \tau)\,d\tau$	$F_1(\omega)F_2(\omega)$
$\displaystyle\int_{-\infty}^{\infty} f(t + \tau)f^*(\tau)\,d\tau$	$\lvert F(\omega)\rvert^2$
$\displaystyle\int_{-\infty}^{\infty} \lvert f(t)\rvert^2\,dt = \dfrac{1}{2\pi}\displaystyle\int_{-\infty}^{\infty} \lvert F(\omega)\rvert^2\,d\omega$	
$f(t) + jf(t)$	$2F(\omega)U(\omega)$
$f(t)$	$-j\,\mathrm{sgn}\,\omega F(\omega)$
$\displaystyle\sum_{n=-\infty}^{\infty} f(t + nT) = \dfrac{1}{T}\displaystyle\sum_{n=-\infty}^{\infty} F\!\left(\dfrac{2\pi n}{T}\right)e^{j2\pi nt/T}$	

Source: A Papoulis, ''Systems and Transforms with Applications in Optics,'' Malabar, FL: Krieger, 1981.

TABLE B.2 Examples of Fourier Transforms

$f(t) = \frac{1}{2\pi} \int_{-\infty}^{\infty} F(\omega)e^{j\omega t}d\omega$	$F(\omega) = \int_{-\infty}^{\infty} f(t)e^{-j\omega t}dt$

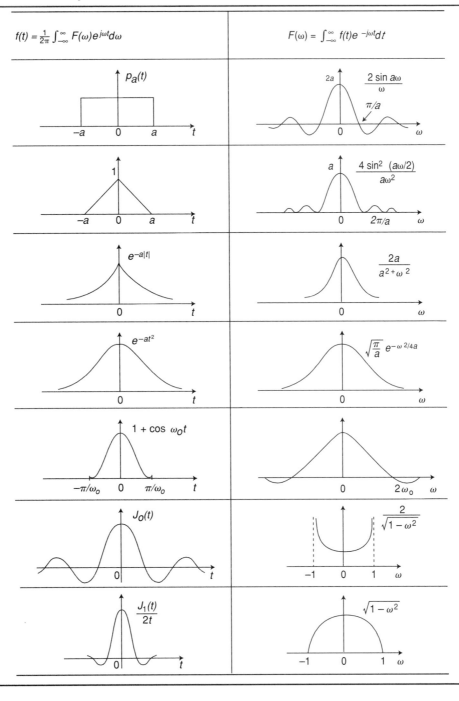

TABLE B.2 Examples of Fourier Transforms *(Continued)*

$f(t) = \frac{1}{2\pi}\int_{-\infty}^{\infty} F(\omega)e^{j\omega t}d\omega$	$F(\omega) = \int_{-\infty}^{\infty} f(t)e^{-j\omega t}dt$
$e^{-\alpha t}U(t)$	$\dfrac{1}{\alpha + j\omega}$
$\dfrac{j}{\pi t}$	$\text{sgn }\omega$
$t^{\alpha}U(t)\quad \alpha > -1$	$\dfrac{\Gamma(\alpha+1)}{|\omega|^{\alpha+1}} e^{\pm\frac{j\pi(\alpha+1)}{2}}\quad \begin{array}{l}-\text{ if }\omega > 0\\ +\text{ if }\omega < 0\end{array}$
$t^{n}e^{-\alpha t}U(t)\quad \alpha > 0$	$\dfrac{n!}{(\alpha + j\omega)^{n+1}}$
$J_n(t)$	$\begin{cases}\dfrac{2\cos(n\arcsin\omega)}{\sqrt{1-\omega^2}} & n\text{ even }|\omega| < 1\\[2mm] \dfrac{-2j\sin(n\arcsin\omega)}{\sqrt{1-\omega^2}} & n\text{ odd }|\omega| < 1\\[2mm] 0 & |\omega| > 1\end{cases}$
$\dfrac{J_n(t)}{t^n}$	$\begin{array}{ll}\dfrac{2(1-\omega^2)^{n-1/2}}{1\cdot 3\cdot 5\cdots(2n-1)} & |\omega| < 1\\[2mm] 0 & |\omega| > 1\end{array}$
$e^{j\alpha t^2}$	$\sqrt{\dfrac{\pi}{\alpha}}\, e^{j\pi/4}e^{-j\omega^2/4\alpha}$
$\cos \alpha t^2$	$\sqrt{\dfrac{\pi}{\alpha}}\cos\left(\dfrac{\omega^2}{4\alpha} - \dfrac{\pi}{4}\right)$
$\sin \alpha t^2$	$\sqrt{\dfrac{\pi}{\alpha}}\sin\left(\dfrac{\omega^2}{4\alpha} - \dfrac{\pi}{4}\right)$
$e^{j\alpha t^2}\quad 0 < t < T$ $0\qquad$ otherwise	$\sqrt{\dfrac{\pi}{2\alpha}}\, e^{-j\omega^2/4\alpha}\left[F\left(\sqrt{\alpha}\,T - \dfrac{\omega}{2\sqrt{\alpha}}\right)\right.$ $\left. + F\left(\dfrac{\omega}{2\sqrt{\alpha}}\right)\right]$ $F(x) = \sqrt{\dfrac{2}{\pi}}\int_0^x e^{jy^2}dy$

Note that the definition of the Fourier transform given in this appendix differs from that given in the main text. In the text, the factor of $1/2\pi$ precedes the integral for the transform, whereas in this appendix, it precedes the integral for the inverse transform.

To use the transform pairs in this appendix with the transform pairs as defined in the main text, multiply the transforms in this appendix by $1/2\pi$.

C

Vector Calculations in Source and Field Coordinates

Transferring the vector differential operators from the source region coordinates to the observation region coordinates is a relatively straightforward matter. Denoting the source region points with primed variables and the observation region points with unprimed variables, we find that the equations for the electric and magnetic fields are

$$\mathbf{E}(\mathbf{r}) = \iiint\limits_{V'} \left[-j\omega\mu\psi(\mathbf{r}|\mathbf{r}')\mathbf{J}(\mathbf{r}') - \mathbf{J}_m(\mathbf{r}') \times \nabla'\psi(\mathbf{r}|\mathbf{r}') + \frac{1}{\epsilon}\rho(\mathbf{r}')\nabla'\psi(\mathbf{r}|\mathbf{r}') \right] dv'$$

(C.1)

$$\mathbf{H}(\mathbf{r}) = \iiint\limits_{V'} \left[-j\omega\epsilon\psi(\mathbf{r}|\mathbf{r}')\mathbf{J}_m(\mathbf{r}') + \mathbf{J}(\mathbf{r}') \times \nabla'\psi(\mathbf{r}|\mathbf{r}') + \frac{1}{\mu}\rho_m(\mathbf{r}')\nabla'\psi(\mathbf{r}|\mathbf{r}') \right] dv'$$

(C.2)

where

$$\psi(\mathbf{r}|\mathbf{r}') = \frac{e^{-jk|\mathbf{r}-\mathbf{r}'|}}{4\pi|\mathbf{r}-\mathbf{r}'|}$$

(C.3)

and

$$\mathbf{J}_m = \mathbf{E} \times \hat{\mathbf{n}}$$

(C.4)

$$\rho_m = \mu\hat{\mathbf{n}}\cdot\mathbf{H}$$

(C.5)

are magnetic surface current and charge, respectively. These are defined on the surface S which bounds the volume V defining the observation region.

Source: C. R. Scott, ''Field Theory of Acousto-Optic Signal Processing Devices,'' Norwoad MA: Artech House, 1992.

It can be readily shown that

$$\nabla'\psi(\mathbf{r}|\mathbf{r}') = -\nabla\psi(\mathbf{r}|\mathbf{r}') \tag{C.6}$$

That is, the gradient of ψ with respect to the source coordinates is the negative of that taken with respect to the field coordinates. With this, (C.1) becomes

$$\mathbf{E}(\mathbf{r}) = \iiint\limits_{V'} \left[-j\omega\mu\psi(\mathbf{r}|\mathbf{r}')\mathbf{J}(\mathbf{r}') + \mathbf{J}_m(\mathbf{r}') \times \nabla'\psi(\mathbf{r}|\mathbf{r}') - \frac{1}{\epsilon}\rho(\mathbf{r}')\nabla\psi(\mathbf{r}|\mathbf{r}') \right] dv' \tag{C.7}$$

The second term under the integral is equal to

$$-\nabla \times [\mathbf{J}_m(\mathbf{r}')\psi(\mathbf{r}|\mathbf{r}')]$$

because the current $\mathbf{J}_m(\mathbf{r}')$ is constant with respect to the differential operator acting on the field coordinates. Thus, (C.7) becomes

$$\mathbf{E}(\mathbf{r}) = -j\omega\mu \iiint\limits_{V'} \mathbf{J}(\mathbf{r}')\psi(\mathbf{r}|\mathbf{r}')dv'$$

$$- \nabla \times \iiint\limits_{V'} \mathbf{J}_m(\mathbf{r}')\psi(\mathbf{r}|\mathbf{r}')dv' \tag{C.8}$$

$$- \nabla \iiint\limits_{V'} \frac{1}{\epsilon}\rho(\mathbf{r}')\psi(\mathbf{r}|\mathbf{r}')dv'$$

Similarly, (C.2) becomes

$$\mathbf{H}(\mathbf{r}) = -j\omega\epsilon \iiint\limits_{V'} \mathbf{J}_m(\mathbf{r}')\psi(\mathbf{r}|\mathbf{r}')dv'$$

$$+ \nabla \times \iiint\limits_{V'} \mathbf{J}(\mathbf{r}')\psi(\mathbf{r}|\mathbf{r}')dv' \tag{C.9}$$

$$- \nabla \iiint\limits_{V,} \frac{1}{\mu}\rho_m(\mathbf{r}')\psi(\mathbf{r}|\mathbf{r}')dv'$$

where, as always, any applicable surface integrals (i.e., those involving magnetic currents and changes) are implied. The last two equations may be recast into the form

$$\mathbf{E}(\mathbf{r}) = -j\omega\mu\mathbf{A}(\mathbf{r}) - \nabla \times \mathbf{F}(\mathbf{r}) - \frac{1}{\epsilon}\nabla\Phi(\mathbf{r}) \tag{C.10}$$

$$\mathbf{H}(\mathbf{r}) = -j\omega\epsilon\mathbf{F}(\mathbf{r}) + \nabla \times \mathbf{A}(\mathbf{r}) - \frac{1}{\mu}\nabla\Phi_m(\mathbf{r}) \tag{C.11}$$

where

$$\mathbf{A}(\mathbf{r}) = \iiint\limits_{V'} \mathbf{J}(\mathbf{r}') \frac{e^{-jk|\mathbf{r}-\mathbf{r}'|}}{4\pi|\mathbf{r}-\mathbf{r}'|} \, dv' \tag{C.12}$$

$$\mathbf{F}(\mathbf{r}) = \iiint\limits_{V'} \mathbf{J}_m(\mathbf{r}') \frac{e^{-jk|\mathbf{r}-\mathbf{r}'|}}{4\pi|\mathbf{r}-\mathbf{r}'|} \, dv' \tag{C.13}$$

$$\Phi(\mathbf{r}) = \iiint\limits_{V'} \rho(\mathbf{r}') \frac{e^{-jk|\mathbf{r}-\mathbf{r}'|}}{4\pi|\mathbf{r}-\mathbf{r}'|} \, dv' \tag{C.14}$$

$$\Phi_m(\mathbf{r}) = \iiint\limits_{V'} \rho_m(\mathbf{r}') \frac{e^{-jk|\mathbf{r}-\mathbf{r}'|}}{4\pi|\mathbf{r}-\mathbf{r}'|} \, dv' \tag{C.15}$$

Because the "magnetic currents" and "magnetic charges" are tangential electric fields and normal magnetic fields over some boundary surface, the integrals in (C.13) and (C.15) are understood to be evaluated over a surface rather than a volume distribution. If we momentarily ignore the terms in (C.10) and (C.11) due to $\mathbf{F}$ and Φ_m, we may note the striking similarity between (C.10) and (C.11) and the electric and magnetic field wave equations. In light of the extra terms due to $\mathbf{F}$ and Φ_m in (C.10) and (C.11), Faraday's law is sometimes written in the generalized form

$$\nabla \times \mathbf{E} = -j\omega\mu\mathbf{H} - \mathbf{J}_m$$

Singularity Functions

The one-dimensional Dirac delta function, widely used in electric circuit analysis, can be defined as the limit of a sequence of pulses of decreasing width, increasing height, and unit area. Of course, a multitude of different pulse shapes can be used in the definition; three equally acceptable definitions are

$$\delta(t) = \lim_{N \to \infty} N \exp(-N^2 \pi t^2) \tag{D.1a}$$

$$\delta(t) = \lim_{N \to \infty} N \, \text{rect}(Nt) \tag{D.1b}$$

$$\delta(t) = \lim_{N \to \infty} N \, \text{sinc}(Nt) \tag{D.1c}$$

While the δ function is used in circuit analysis to represent a sharp, intense pulse of current or voltage, the analogous concept in optics is a point source of light, or a *spatial* pulse of unit area. The definition of a δ function on a two-dimensional space is a simple extension of the one-dimensional case, although there is even greater latitude in the possible choice for the functional form of the pulses. Possible definitions of the spatial δ function include

$$\delta(x, y) = \lim_{N \to \infty} N^2 \exp[-N^2 \pi (x^2 + y^2)] \tag{D.2a}$$

$$\delta(x, y) = \lim_{N \to \infty} N^2 \, \text{rect}(Nx)\text{rect}(Ny) \tag{D.2b}$$

$$\delta(x, y) = \lim_{N \to \infty} N^2 \, \text{sinc}(Nx)\text{sinc}(Ny) \tag{D.2c}$$

$$\delta(x, y) = \lim_{N \to \infty} \frac{N^2}{\pi} \, \text{circ}(N \sqrt{x^2 + y^2}) \tag{D.2d}$$

$$\delta(x, y) = \lim_{N \to \infty} N \frac{J_1(2\pi N \sqrt{x^2 + y^2})}{\sqrt{x^2 + y^2}} \tag{D.2e}$$

Definitions (D.2a) to (D.2c) are separable in rectangular coordinates, while definitions

Source: J. W. Goodman, "Introduction to Fourier Optics," New York: McGraw-Hill, 1968.

(D.2d) and (D.2e) are circularly symmetric. In some applications one definition may be more convenient than others, and the definition best suited for the problem can be chosen.

Each of the above definitions of the spatial δ function has the following fundamental properties:

$$\delta(x, y) = \begin{cases} \infty & x = y = 0 \\ 0 & \text{otherwise} \end{cases} \tag{D.3}$$

$$\iint_{-\epsilon}^{\epsilon} \delta(x, y)\,dx\,dy = 1 \qquad \text{any } \epsilon > 0 \tag{D.4}$$

$$\iint_{-\infty}^{\infty} g(\xi, \eta)\delta(x - \xi, y - \eta)\,d\xi\,d\eta = g(x, y) \tag{D.5}$$

at each point of continuity of **g**

Property (D.5) is often referred to as the *sifting* property of the δ function. An additional property of considerable importance can be proved from any of the definitions, namely,

$$\delta(ax, by) = \frac{1}{|ab|}\,\delta(x, y) \tag{D.6}$$

There is no reason why the δ function cannot be defined on a space of higher dimensionality than two, but the properties of such functions are exactly analogous to their counterparts on spaces of lower dimensionality.
In the above,

$$\text{rect}(x) = \begin{cases} 1 & |x| \le \dfrac{1}{2} \\ 0 & \text{otherwise} \end{cases}$$

$$\text{circ}(\sqrt{x^2 + y^2}) = \begin{cases} 1 & \sqrt{x^2 + y^2} \le 1 \\ 0 & \text{otherwise} \end{cases}$$

Fresnel Integrals

The Fresnel cosine and sine integrals are defined by

$$C(\tau) = \int_0^\tau \cos \frac{\pi}{2} y^2 dy \qquad S(\tau) = \int_0^\tau \sin \frac{\pi}{2} y^2 dy \tag{E.1}$$

These functions and their derivatives

$$\frac{dC(\tau)}{d\tau} = \cos \frac{\pi}{2} \tau^2 \qquad \frac{dS(\tau)}{d\tau} = \sin \frac{\pi}{2} \tau^2$$

are shown in Fig. E.1. Their extrema, τ_m^c and τ_m^s, coincide with the zero crossings of $\cos (\pi \tau^2/2)$ and $\sin (\pi \tau^2/2)$. Thus

$$\tau_{max}^c = \sqrt{4n + 1} \qquad \tau_{min}^c = \sqrt{4n + 3} \qquad n = 0, 1, 2, \ldots$$

$$\tau_{max}^s = \sqrt{4n + 2} \qquad \tau_{min}^s = \sqrt{4n}$$

We find it convenient also to introduce the complex function

$$F(x) = \sqrt{\frac{2}{\pi}} \int_0^x e^{jy^2} dy \tag{E.2}$$

It is easy to see that

$$F(x) = C\left(x \sqrt{\frac{2}{\pi}}\right) + jS\left(x \sqrt{\frac{2}{\pi}}\right) \tag{E.3}$$

From an integral identity it follows that

$$F(\infty) = \frac{1}{\sqrt{2}} e^{j\pi/4} \qquad \text{hence } C(\infty) = S(\infty) = 1/2 \tag{E.4}$$

Cornu spiral If $F(x)$ is plotted in the complex plane as the real parameter x takes values from 0 to ∞, the curve of Fig. E.2, known as the Cornu spiral, results. From

Source: A. Papoulis, ''Systems and Transforms with Applications in Optics,'' Malabar, FL: Krieger, 1981.

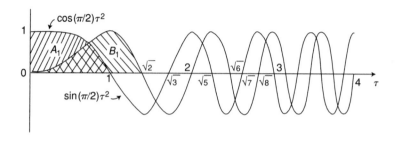

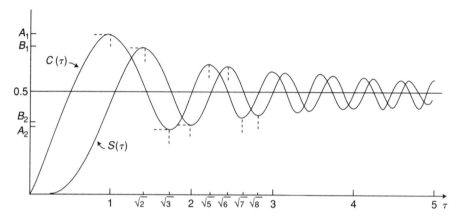

Figure E.1 Fresnel sine and cosine integrals.

$$\frac{d\mathrm{F}(x)}{dx} = \sqrt{\frac{2}{\pi}}\, e^{jx^2}$$

it follows that for $dx > 0$, the amplitude and phase of the increment $d\mathrm{F}$ are given by

$$|d\mathrm{F}| = \sqrt{\frac{2}{\pi}}\, dx \qquad \theta_{d\mathrm{F}} = x^2$$

Hence,

$$\int_0^x |d\mathrm{F}(x)|\, dx = \sqrt{\frac{2}{\pi}}\, x$$

Thus, the length of the spiral from the origin to the point $\mathrm{F}(x)$ equals $x\sqrt{2/\pi}$.

Series expansions   Expanding the exponential in (E.2) and integrating the resulting series termwise, we obtain

$$\mathrm{F}(x) = \sqrt{\frac{2}{\pi}}\left(x + j\frac{x^3}{3} + \frac{j^2}{2!}\frac{x^5}{5} + \cdots + \frac{j^n}{n!}\frac{x^{2n+1}}{2n+1} + \cdots\right) \qquad (\mathrm{E.5})$$

From the above and (E.3) it follows that

$$\sqrt{\frac{\pi}{2}}\, C\left(x\sqrt{\frac{2}{\pi}}\right) = x - \frac{x^5}{2!5} + \cdots + \frac{(-1)^n}{(2n)!}\frac{x^{4n+1}}{4n+1} + \cdots \qquad (\mathrm{E.6})$$

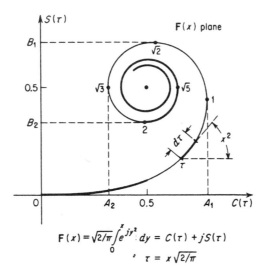

$$F(x) = \sqrt{2/\pi} \int_0^x e^{jy^2}\, dy = C(\tau) + jS(\tau)$$

$$\tau = x\sqrt{2/\pi}$$

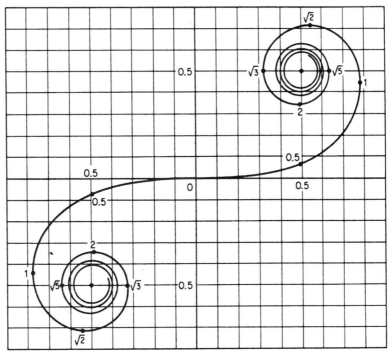

Figure E.2 Cornu spiral.

$$\sqrt{\frac{\pi}{2}} S\left(x \sqrt{\frac{2}{\pi}}\right) = \frac{x^3}{3} - \frac{x^7}{3!7} + \cdots + \frac{(-1)^n}{(2n+1)!} \frac{x^{4n+3}}{4n+3} + \cdots \tag{E.7}$$

LARGE X. $F(x)$ can be expanded into a series in powers of $1/x$:

$$F(x) = \frac{1}{\sqrt{2}} e^{j\pi/4} + \frac{1}{jx\sqrt{2\pi}} e^{jx^2} + \cdots \tag{E.8}$$

This can also be derived from (E.2) by integration by parts. From the real and imaginary parts of (E.8), the expansions

$$C(\tau) = 1/2 - \frac{1}{\pi\tau} \sin\left(\frac{\pi}{2} \tau^2\right) + \cdots \tag{E.9}$$

$$S(\tau) = 1/2 - \frac{1}{\pi\tau} \cos\left(\frac{\pi}{2} \tau^2\right) + \cdots \tag{E.10}$$

easily result.

If $F(x)$ is approximated by only two terms of the series (E.5) or (E.8), the resulting error is less than 0.01 for x in the heavy parts of the Cornu spiral of Fig. E.2.

F

Bessel Functions of Integer Order

Bessel's equation of order n is

$$x \frac{d}{dx}\left(x \frac{df}{dx}\right) + (x^2 - n^2)f(x) = 0 \tag{F.1}$$

or,

$$x^2 \frac{d^2 f}{dx^2} + x \frac{df}{dx} + (x^2 - n^2)f(x) = 0 \tag{F.2}$$

This is a second-order equation with a regular singular point at $x = 0$.

The last equation may be rewritten in the following form in order to emphasize the dependence on the integer parameter, n:

$$x^2 \frac{d^2 f_n}{dx^2} + x \frac{df_n}{dx} + (x^2 - n^2)f_n(x) = 0 \tag{F.3}$$

Each different value of n produces a different function f_n.

Because (F.3) is a second-order ordinary differential equation, it has two independent solutions, one that is analytic at $z = 0$ and one that is singular there. The function analytic at $z = 0$ is called the Bessel function, $J_n(x)$, whereas the singular solution is called the Neumann function, $N_n(x)$. Other solutions of interest in wave propagation problems are the Hankel functions of the first and second kinds defined by

$$H_n^{(1)}(x) = J_n(x) + jN_n(x) \tag{F.4a}$$

$$H_n^{(2)}(x) = J_n(x) - jN_n(x) \tag{F.4b}$$

These functions possess the following properties:

Negative Order

$$f_{-n}(x) = (-1)^n f_n(x) \tag{F.5}$$

where

Source: C. R. Scott, "Field Theory of Acousto-Optic Signal Processing Devices." Norwood MA: Artech House, 1992.

$$f_n = J_n, N_n, H_n^{(1)}, H_n^{(2)}$$

Recurrence Relations

$$f_{n-1}(x) + f_{n+1}(x) = \frac{2n}{x} f_n(x) \tag{F.6a}$$

$$f_{n-1}(x) - f_{n+1}(x) = 2f_n'(x) \tag{F.6b}$$

$$f_n'(x) = f_{n-1}(x) - \frac{n}{x} f_n(x) \tag{F.6c}$$

$$f_n'(x) = -f_{n+1}(x) + \frac{n}{x} f_n(x) \tag{F.6d}$$

$$f_0'(x) = -f_1(x) \tag{F.6e}$$

Wronskians

$$J_n(x)N_n'(x) - N_n(x)J_n'(x) = \frac{2}{\pi x} \tag{F.7}$$

$$J_{n+1}(x)J_{-n}(x) + J_n(x)J_{-(n+1)}(x) = 0 \tag{F.8}$$

$$J_{n+1}(x)N_n(x) - J_n(x)N_{n+1}(x) = \frac{2}{\pi x} \tag{F.9}$$

Asymptotic Expansions for $x \gg n$

$$J_n(x) \sim \sqrt{\frac{2}{\pi x}} \cos\left(x - \frac{n\pi}{2} - \frac{\pi}{4}\right) \tag{F.10}$$

$$N_n(x) \sim \sqrt{\frac{2}{\pi x}} \sin\left(x - \frac{n\pi}{2} - \frac{\pi}{4}\right) \tag{F.11}$$

$$H_n^{(1)}(x) \sim \sqrt{\frac{2}{\pi x}} \, e^{j[x-(n\pi/2)-(\pi/4)]} \tag{F.12}$$

$$H_n^{(2)}(x) \sim \sqrt{\frac{2}{\pi x}} \, e^{-j[x-(n\pi/2)-(\pi/4)]} \tag{F.13}$$

Asymptotic Expansions for $n \gg x$

$$J_n(x) \sim \frac{1}{\sqrt{2n\pi}} \left(\frac{ex}{2n}\right)^n \tag{F.14}$$

$$N_n(x) \sim -\sqrt{\frac{2}{n\pi}} \left(\frac{ex}{2n}\right)^{-n} \tag{F.15}$$

Evaluation of Bessel functions

A. If $n \gg x$ or $x \gg n$, use either of the asymptotic forms.

B. Otherwise, choose $N \gg x$ (say $N = 3x$) and calculate $J_{N+1}(x)$ and $J_N(x)$ using (B.14). Then use (B.6a) to calculate all $J_n(x)$ for $n < N$. Then normalize the Bessel functions so that $J_0(x) + 2J_2(x) + 2J_4(x) + \ldots = 1$.

Integral Relations

$$\int_0^z J_n(t)dt = 2 \sum_{k=0}^{\infty} J_{n+2k+1}(z) \qquad (n \geq 0) \tag{F.16}$$

$$\int_0^z J_{2n}(t)dt = \int_0^z J_0(t)dt - 2 \sum_{k=0}^{n-1} J_{2k+1}(z) \tag{F.17}$$

$$\int_0^z J_{2n+1}(t)dt = 1 - J_0(z) - 2 \sum_{k=1}^{n} J_{2k}(z) \tag{F.18}$$

$$\int_0^z J_{n+1}(t)dt = \int_0^z J_{n-1}(t)dt - 2J_n(z) \qquad (n > 0) \tag{F.19}$$

$$\int_0^z J_1(t)dt = 1 - J_0(z) \tag{F.20}$$

$$\int_0^z t^n J_{n-1}(t)dt = z^n J_n(z) \qquad (n > 0) \tag{F.21}$$

$$\int_0^z t J_{n-1}^2(t)dt = 2 \sum_{k=0}^{\infty} (n + 2k)J_{n+2k}^2(z) \tag{F.22}$$

$$\int_0^1 t J_n(\alpha_p t)J_n(\alpha_q t)dt = 0 \qquad (p \neq q, n > -1)$$

$$= \frac{1}{2} [J_n'(\alpha_p)]^2 \qquad (p = q, b = 0, n > -1)$$

$$= \frac{1}{2\alpha_p^2} \left[\left(\frac{a}{b}\right)^2 + \alpha_p^2 - n^2 \right] [J_n(\alpha_p)]^2 \tag{F.23}$$

$$(p = q, b \neq 0, n \geq -1)$$

where α_p, α_q are the pth, qth positive zeroes of the equation $aJ_n(x) + bxJ_n'(x) = 0$, and a and b are real constants.

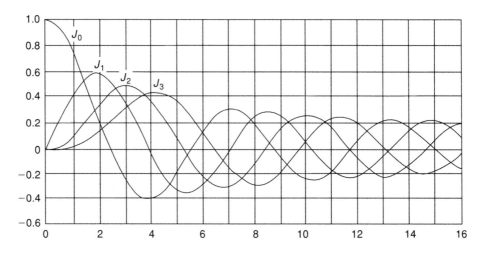

Figure F.1 Bessel functions of the first kind. (From Harrington, R. F., *Time-Harmonic Electromagnetic Fields,* New York: McGraw-Hill, 1960.)

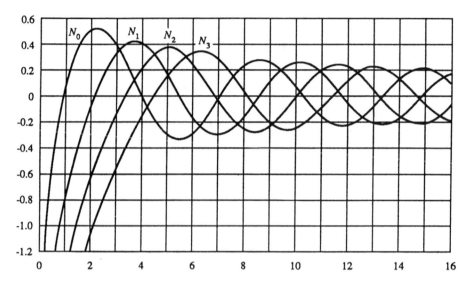

Figure F.2 Bessel functions of the second kind. (From Harrington, R. F., *Time-Harmonic Electromagnetic Fields*. New York: McGraw-Hill, 1960.)

The Lommel integral formula is

$$\int_0^a J_n(k_m\rho)J_n(k\rho)\rho d\rho =$$

$$\frac{1}{k^2 - k_m^2}\left[k_m aJ_{n-1}(k_m a)J_n(ka) - kaJ_{n-1}(ka)J_n(k_m a)\right] \quad \text{(F.24)}$$

where $k_m a$ is the mth zero of $J_n'(x)$ and $k^2 \neq k_m^2$.

Plots of the Bessel and Neumann functions are shown in Figs. F.1 and F.2.

Index

About the Author

Craig Scott holds a B.S. and an M.S. degree in electrical engineering from UCLA. He is an electrical engineer at Northrop Grumman Corporation, Military Aircraft Systems Division. He is currently working primarily in the areas of conformal antenna analysis and design, photonic bandgap (PBG) structure analysis, design, and testing, and multilayer frequency selective surface (FSS) radome analysis, design, and testing. He has developed spectral domain method-of-moments (MoM) theory and software tools for analyzing and designing cavity-backed airborne antennas having reduced radar cross-section (RCS) levels. These tools have enabled the design of low RCS airborne antennas without the use of lossy magram materials that can degrade antenna gain performance. He has also developed spectral domain MoM theory and software tools for analyzing conductive and dielectric, infinite- and finite-thickness, and doped and undoped PBG structures. In addition, he has led an R&D effort to experimentally characterize PBG lattices with and without lattice site doping. Preliminary analytical and test data on these structures have demonstrated their significant potential for future use in passive and active integrated structural antennas and radomes.

From 1988 to 1995, Mr. Scott was an electrical engineer at Rockwell International, North American Aircraft Division, working in the areas of frequency selective surface and printed phased array analysis and design. He has developed a number of low RCS FSS structures and technologies for ultra-wideband performance.

From 1983 to 1988, Mr. Scott was an electrical engineer at TRW, working in the areas of reflector antenna analysis and design, FSS analysis and design, and body-of-revolution (BOR) MoM software development. He has developed extensive software for calculating the radiation pattern of symmetrical and offset parabolic reflector antennas using a variety of integration algorithms, including the Jacobi-Bessel method, the Fourier-Bessel method, and the Ludwig integration algorithm. He has also developed software for calculating the radiation pattern from symmetric and offset hyperbolic subreflectors using geometrical optics combined with the geometrical theory of diffraction (GTD). In addition, he has produced software for calculating the reflection and transmission properties of a wide variety of FSS structures using the spectral domain method. FSSs designed using this software have been successfully built, tested, and flown. He has also done extensive software development in the the area of applying MoM techniques to BOR scattering objects using the electric field, magnetic field, and combined field integral equation formulations. This software involved the calculation of plane and spherical wave scattering from BORs as well as radiation from slot antennas mounted on BORs.

Mr. Scott has authored three other books in electromagnetics, antenna theory, and optics. These are *The Spectral Domain Method in Electromagnetics, Modern Methods of Reflector Antenna Analysis and Design,* and *Field Theory of Acousto-Optic Signal Processing Devices.*